OXFORD MATHEMATICAL MONOGRAPHS

Series Editors

E. M. FRIEDLANDER   I. G. MACDONALD
L. NIRENBERG   R. PENROSE   J. T. STUART

OXFORD MATHEMATICAL MONOGRAPHS

A. Belleni-Morante: *Applied semigroups and evolution equations*
I. G. Macdonald: *Symmetric functions and Hall polynomials*
J. W. P. Hirschfeld: *Projective geometrics over finite fields*
N. Woodhouse: *Geometric quantization*
A. M. Arthurs: *Complementary variational principles* Second edition
P. L. Bhatnagar: *Nonlinear waves in one-dimensional dispersive systems*
N. Aronszain, T. M. Creese, and L. J. Lipkin: *Polyharmonic functions*
J. A. Goldstein: *Semigroups of linear operators*
M. Rosenblum and J. Rovnyak: *Hardy classes and operator theory*
J. W. P. Hirschfeld: *Finite projective spaces of three dimensions*
K. Iwasawa: *Local class field theory*
A. Pressley and G. Segal: *Loop groups*
J. C. Lennox and S. E. Stonehewer: *Subnormal subgroups of groups*
D. E. Edmunds and W. D. Evans: *Spectral theory and differential operators*
Wang Jianhua: *The theory of games*
S. Omatu and J. H. Seinfeld: *Distributed parameter systems: theory and applications*
D. Holt and W. Plesken: *Perfect groups*
J. Hilgert, K. H. Hofmann and J. D. Lawson: *Lie groups, convex cones, and semigroups*
S. Dineen: *The Schwarz lemma*
B. Dwork: *Generalized hypergeometric functions*
R. J. Baston and M. G. Eastwood: *The Penrose transform: its interaction with representation theory*
S. K. Donaldson and P. B. Kronheimer: *The geometry of four-manifolds*
T. Petrie and J. Randall: *Connections, definite forms, and four-manifolds*
R. Henstock: *The general theory of integration*

The General Theory of Integration

RALPH HENSTOCK
Department of Mathematics
University of Ulster

CLARENDON PRESS · OXFORD
1991

Oxford University Press, Walton Street, Oxford OX2 6DP

*Oxford New York Toronto
Delhi Bombay Calcutta Madras Karachi
Petaling Jaya Singapore Hong Kong Tokyo
Nairobi Dar es Salaam Cape Town
Melbourne Auckland
and associated companies in
Berlin Ibadan*

Oxford is a trade mark of Oxford University Press

*Published in the United States
by Oxford University Press, New York*

© *Ralph Henstock 1991*

*All rights reserved. No part of this publication may be reproduced,
stored in a retrieval system, or transmitted, in any form or by any means,
electronic, mechanical, photocopying, recording, or otherwise, without
the prior permission of Oxford University Press*

*British Library Cataloguing in Publication Data
Henstock, Ralph
The general theory of integration.
Calculus. Integration. Functional
I. Title
515.42
ISBN 0-19-853566-X*

*Library of Congress Cataloging in Publication Data
Henstock, Ralph.
The general theory of integration/Ralph Henstock.
(Oxford mathematical monographs)
Includes bibliographical references.
1. Integrals. I. Series.
QA311.H54   1991   515' 4–dc20   90-20901
ISBN 0-19-853566-X*

*Typeset by Macmillan India Ltd
Printed in Great Britain by
Bookcraft (Bath) Ltd
Midsomer Norton, Avon*

PREFACE

Every good mathematical book stands like a tree with its roots in the past and its branches stretching out towards the future. Whether the fruits of this tree are desirable and whether the branches will be quarried for mathematical wood to build further edifices, I will leave to the judgement of history. The roots of this book take nourishment from the concept of definite integration of continuous functions, where Riemann's method is the high-water mark of the simpler theory.

The essence of his method of integrating a bounded function f of a real variable x over a finite real interval $[a, b]$, is this. He cuts up $[a, b]$ into a finite number of smaller intervals such as $[u, v]$ and takes a value x in $[u, v]$ at which to evaluate f, finding the sum s of all products $f(x)(v - u)$. If s tends to a limit as the greatest of the $v - u$ tends to 0, wherever the x lie, he says that f is (Riemann) integrable over $[a, b]$.

We generalize this simple process, altering the limit process on s and obtaining generalized Riemann integrals, some of which have Lebesgue properties for limits under the integral sign. The two simplest Lebesgue properties that the ordinary Riemann integral does not have, are as follows.

(a) If the convergent sequence $(f_n(x))$ of functions integrable over $[a, b]$ with $b - a$ finite, is bounded by M independent of n and x, then the limit function is integrable and

(b) $$\int_a^b \lim_{n \to \infty} f_n(x)\,dx = \lim_{n \to \infty} \int_a^b f_n(x)\,dx.$$

(c) If, instead of being bounded, the $f_n(x)$ are monotone increasing in n and integrable to F_n over $[a, b]$, with (F_n) bounded above, then $f_n(x)$ tends to a finite limit $f(x)$ for almost all x as $n \to \infty$, $f(x)$ is integrable over $[a, b]$, and its integral is the limit of (F_n).

Three generalized Riemann integrals satisfy (a), (b), and (c), they are more powerful than Lebesgue's integral, they are equivalent respectively to Denjoy's special and general integrals and Burkill's approximate Perron integral, and their definitions use Riemann sums s. We can even give conditions that are necessary as well as sufficient for (b). We can generalize the $[u, v]$ on the real line. For example, Lebesgue's way of defining his integral of a function f satisfying $t \leq f \leq w$ for constants t, w, is to take the limit of s on replacing the $[u, v]$ by the measurable sets for which f takes values in $[u, v)$, where the $[u, v)$ form a division of the range $[t, w)$ of values of f. The limit is taken as the greatest $v - u$ tends to 0. This approach needs the measure of every measurable set used in the limit process. But there is no need to go to all this trouble, we only need intervals to define Lebesgue integrals, as will be shown in Chapter 0, Section 1.

More general spaces naturally need more than ordinary intervals. For example, n-dimensional Euclidean spaces need n-dimensional rectangles (*bricks*) while sequence and function spaces need *cylinder sets* based on bricks in some sense, and usually topological spaces need even more general sets. The values of functions are often taken as general as seems possible at this time, but if necessary the theory could be restricted to real or complex values, or values in some Banach space.

A distinguishing feature of modern mathematics is the establishment of patterns of behaviour in mathematical systems, leading to a unified set of proofs of results instead of a heterogeneous collection of proofs, one set for each system. In integration theory, J. C. Burkill (1924a,b,c) unified the variety of ordinary Riemann-type and Stieltjes-type integrals by using functions of intervals, while Jeffery and Miller (1945) unified the proofs for many convergence-factor integrals. The unification produced by division systems and division space integrals began with a most surprising result. For years it was thought that Riemann-type integration had been superseded by the integration first introduced by Lebesgue. But independently, Kurzweil (1957) and Henstock (1961b, 1963c) found that Perron integration could be given by a generalized Riemann integral. Perron integration, being equivalent to special Denjoy integration, includes Lebesgue integration, so that it was clear that the latter is included in generalized Riemann integration (see Henstock (1968a)), and Davies and Schuss (1970) proved this directly, for a quite general system. Then McShane (1969, 1973) gave the generalized Riemann integral exactly equivalent to Lebesgue's. The division system and division space integrals of this book were shown to include many other integrals, see Henstock (1968b, pp. 219–25), and now the list is still longer.

An earlier unification in other areas of analysis was carried out using the methods of functional analysis. Here, three principles have come to the fore. The *Hahn–Banach theorem* concerns the existence of extensions of a linear functional to regions over which it was not originally defined, and in a sense belongs to the algebraic side of functional analysis. The principle of *uniform boundedness* or of *uniform convergence*, developed by Banach, Steinhaus, and Sargent, often leads to a proof that the limit of a sequence of continuous linear operators is continuous. The *interior mapping principle* is that continuous linear mappings between certain types of spaces map open sets onto open sets. This book uses the first two principles and division systems and spaces in the study of integration theory.

Some integrals have not yet been put as generalized Riemann integrals, one large group being those integrals defined by using convergence (or smoothing) factors, such as J. C. Burkill's (1932, 1935, 1936a,b, 1951a,b) Çesaro–Perron integrals, the James (1946, 1950, 1954, 1955, 1956) integrals, the Marcinkiewicz–Zygmund (1936) integral, and Taylor's (1955) Abel–Perron integral, together with the N- and N-variational integrals, a unified system that includes most of the integrals outside the generalized

Riemann system. Cross (1985/1986, 1986, 1987/1988) has recently shown how to define some convergence-factor integrals using generalized Riemann integrals, so that the demarcation line between the two types could be getting blurred.

Wiener and Feynman integration need their own chapter, they are usually based on a continuum number of copies of the real line so that, for the second, the measure produced is not of generalized bounded variation. One of the simplest measures is based on $\exp(iax^2)$ for various real constants a, which causes the trouble. Wiener integration, based on $\exp(-ax^2)$, is far easier to work with.

In other chapters we study the differentiation of integrals, and repeated integrals, in part using general limits under the integral sign, extending (b).

This book was partly written during the tenure of a Leverhulme Trust Emeritus Fellowship, and my thanks go to the Trust. My thanks also go to the staff of the Oxford University Press for their efficient handling of the printing and publishing work.

Ulster   R. H.
1990

CONTENTS

0 Introduction and prerequisites — 1

- 0.1 Introduction: the calculus, Lebesgue, and gauge (Kurzweil–Henstock) integrals — 1
- 0.2 Basic definitions for simple set theory — 7
- 0.3 Basic definitions for algebra — 11
- 0.4 Basic definitions and theory for topology — 13
- 0.5 Basic definitions and theory for order — 26
- 0.6 Short history of integration — 32

1 Division systems and division spaces — 40

- 1.1 Definitions — 40
- 1.2 The norm integral — 47
- 1.3 The refinement integral — 49
- 1.4 The gauge (Kurzweil–Henstock) integral — 50
- 1.5 The gauge integral, associated points at vertices, and infinite intervals — 52
- 1.6 McShane's modification, giving absolute integrals — 54
- 1.7 Symmetric intervals — 54
- 1.8 The divergence theorem — 55
- 1.9 The general Denjoy integral — 56
- 1.10 Burkill's approximate Perron integral — 57
- 1.11 Division systems and spaces in a topology — 58

2 Generalized Riemann and variational integration in division systems and division spaces — 59

- 2.1 Free division systems, the free p-variation, free norm variation, and corresponding variational integral, and the free variation set — 59
- 2.2 Freely decomposable division system — 71
- 2.3 Division systems, the generalized Riemann and variational integrals, the variation, and the variation set — 74
- 2.4 Decomposable division systems — 81
- 2.5 Division spaces — 83
- 2.6 Decomposable division spaces and integration by substitution — 91
- 2.7 Additive division spaces — 92
- 2.8 Decomposable additive division spaces — 101

	2.9	Star-sets and the intrinsic topology	103
	2.10	Special results	112

3 Limits under the integral sign, functions depending on a parameter 120

3.1	Introduction and necessary and sufficient conditions	120
3.2	Monotone sequences and functions	126
3.3	The bounded Riemann sums test and the majorized (dominated) convergence test of Arzelà and Lebesgue	134

4 Differentiation 142

4.1	The differentiation of strong variational integrals	142
4.2	Further results on differentiation	147

5 Cartesian products of a finite number of division systems (spaces) 149

5.1	Fubini-type results	149
5.2	Tonelli-type results on the reversal of order of double integrals	154

6 Integration in infinite-dimensional spaces 157

6.1	Introduction	157
6.2	Division space integration	158

7 Perron-type, Ward-type, and convergence-factor integrals 167

7.1	Perron-type integrals in fully decomposable division spaces	167
7.2	Ward integrals in decomposable division systems and spaces	171
7.3	Convergence-factor integrals in fully decomposable division spaces	173

8 Functional analysis and integration theory 184

8.1	Introduction	184
8.2	Young's inequality and Orlicz spaces	185

8.3	The functional analysis of continuous generalized Riemann integrals	199
8.4	Density integration	202

References 208

Name index 257

Subject index 260

CHAPTER 0

INTRODUCTION AND PREREQUISITES

0.1 INTRODUCTION: THE CALCULUS, LEBESGUE, AND GAUGE (KURZWEIL–HENSTOCK) INTEGRALS

The idea behind this book's theory can be approached from the calculus definition of the integral inverse of the derivative, thus stepping directly from the mathematics of Newton and Leibnitz into the post-Lebesgue modern mathematics. The indefinite Newton integral $F(x)$ of the function $f:[a,b] \to R$, is any solution of the differential equation $dy/dx = f(x)$ in $[a, b]$. Thus, given $\varepsilon > 0$, a $\delta = \delta(x) > 0$ exists (now called a *gauge*) such that

(0.1.1) $\quad |\{F(x+h) - F(x)\}/h - f(x)| < \varepsilon, \quad |F(x+h) - F(x) - f(x)h| < \varepsilon|h|$
$$(a \leq x \leq b, 0 < |h| \leq \delta(x)).$$

For $b - a$ finite, a partition of $[a, b]$ into a finite number n of smaller intervals can be written

(0.1.2) $\qquad a = u_0 < u_1 < \cdots < u_n = b.$

We choose a point x_j in each smaller interval $[u_{j-1}, u_j]$, so obtaining a collection of interval–point pairs $([u_{j-1}, u_j]; x_j)$ $(j = 1, 2, \ldots, n)$ called a *division* $\mathscr{E}$ of $[a, b]$. We say that $\mathscr{E}$ is compatible with $\delta(x)$, or is δ-*fine* (following McShane 1969, 1973) if $|u_j - u_{j-1}| < \delta(x_j)$ $(j = 1, \ldots, n)$. To fit the later additive division space we sometimes take x_j to be either u_{j-1} or u_j $(j = 1, \ldots, n)$, calling such a division *restricted*. For the $\delta(x)$ of (0.1.1) and a δ-fine restricted division we put $h = u_j - u_{j-1}$ when $x = x_j = u_{j-1}$, and $h = u_{j-1} - u_j$ when $x = x_j = u_j$. Then for

$$s = \sum_{j=1}^n f(x_j)(u_j - u_{j-1}), \ |F(b) - F(a) - s|$$

$$= \left| \sum_{j=1}^n \{F(u_j) - F(u_{j-1}) - f(x_j)(u_j - u_{j-1})\} \right|$$

$$\leq \sum_{j=1}^n \varepsilon(u_j - u_{j-1}) = \varepsilon(b - a).$$

Thus the sum s tends to $F(b) - F(a)$ as $\delta(x) \to 0$, so that we can take $\varepsilon \to 0$, for all restricted δ-fine divisions of $[a, b]$.

This result leads to the *gauge* (or *Kurzweil–Henstock*) integral H over $[a, b]$ of a function $h(u_{j-1}, u_j; x_j)$, as a number with the following property. Given $\varepsilon > 0$, there is a positive function δ on $[a, b]$ such that for all δ-fine restricted divisions of $[a, b]$,

$$(0.1.3) \qquad \left| \sum_{j=1}^{n} h(u_{j-1}, u_j; x_j) - H \right| < \varepsilon.$$

We write $H = \int_a^b dh$, or $\int_a^b f\,dx$ if $h(u, v; x) = f(x)(v - u)$. This integral is the special case in Sections 1.4 and 1.5 of the more general integral of the rest of the book. One essential requirement is that for each gauge $\delta(x) > 0$ in $[a, b]$, there is a δ-fine restricted division of $[a, b]$. See Section 1.4, Theorem 1.4.1.

The integral is uniquely defined. For if H_1 and H_2 have the property of H for the same h and $[a, b]$, then given $\varepsilon > 0$, there are gauges $\delta_k(x)$ on $[a, b]$ such that for δ_k-fine restricted divisions of $[a, b]$,

$$(0.1.4) \qquad \left| \sum_{j=1}^{n} h(u_{j-1}, u_j; x_j) - H_k \right| < \varepsilon \qquad (k = 1, 2).$$

If $\delta(x) \equiv \min(\delta_1(x), \delta_2(x))$, $\delta > 0$ on $[a, b]$ and there is a δ-fine, and so δ_k-fine ($k = 1, 2$) restricted division of $[a, b]$. Hence (0.1.4) holds for $k = 1, 2$ and

$$|H_1 - H_2| = \left| \left\{ \sum_{j=1}^{n} h - H_2 \right\} - \left\{ \sum_{j=1}^{n} h - H_1 \right\} \right| < 2\varepsilon.$$

Thus $H_1 - H_2 = 0$, being independent of $\varepsilon > 0$.

Earlier, the Newton (calculus) integral was proved to be a gauge integral. The Riemann integral is, too, using constant gauges. By Borel's covering theorem, if δ is continuous there is a constant $c > 0$ with $\delta(x) \geq c$ in $[a, b]$, and we have the ordinary Riemann arrangement such as occurs for Burkill integration. However, $\delta(x) > 0$ normally varies from point to point, and a typical construction is as follows. Let

$$(0.1.5) \qquad 0 < \delta(x) < |x - w| \qquad (x \neq w, a < w < b).$$

Then δ tends to 0 as $x \to w$, even though $\delta(w) > 0$, and δ is discontinuous at w. There is a δ-fine restricted division of $[a, b]$. If $x = u_j \neq w$, $\delta(x) < |u_j - w|$, and neither $[u_{j-1}, u_j]$ nor $[u_j, u_{j+1}]$ can contain w when w is not the point associated with the interval. Thus w is only contained in an interval if it is the point where f is evaluated, and so has to be one of the points of division, say u_k, and $[u_{k-1}, w]$ and $[w, u_{k+1}]$ are intervals of the δ-fine division. If $w = a$ or b then one interval is missing. More generally, for a sequence (w_j) of such w and a sequence (δ_j) of gauges (0.1.5), we can use

$$\delta^{(n)}(x) = \inf_{1 \leq j \leq n} \delta_j(x) > 0.$$

Then $w_1, \ldots, w_n$ are included in the division-points of every $\delta^{(n)}$-fine division of $[a, b]$. Note that (w_j) could be dense in $[a, b]$ in the sense of Section 0.4 on topology.

For another construction let (X_k) by a sequence of mutually disjoint sets in $(-\infty, \infty)$ of measure zero, i.e. for each k, $\varepsilon > 0$, there is a countable union G_k of disjoint open intervals I with $G_k \supseteq X_k$ and $mG_k < \varepsilon 4^{-k}$ (the 4^{-k} is needed later), where mG_k is the sum of lengths of the I of G_k. If X is the union of sets X_k and if

$$|f(x)| \leq 2^k \quad (x \in X_k, k = 1, 2, \ldots), \qquad f(x) = 0 \quad (x \notin X),$$

we call f a *null function*. Its gauge integral over $[a, b]$ is zero. For, given $\varepsilon > 0$, choose $\delta(x) > 0$ so that $\delta(x) \leq 1$ $(x \notin X)$, and

$$(x - \delta(x), x + \delta(x)) \subseteq G_k \quad (x \in X_k, k = 1, 2, \ldots).$$

For the usual δ-fine division of $[a, b]$, the non-zero $f(x_j)$ have $x_j \in X$ so that

$$\left| \sum_{j=1}^n f(x_j)(u_j - u_{j-1}) - 0 \right| \leq \sum_{k=1}^\infty 2^k mG_k < \sum_{k=1}^\infty \varepsilon 2^{-k} = \varepsilon.$$

A typical null function is the indicator of the rationals, 1 at each rational and 0 at each irrational. For, writing the rationals u/v $(v > 0)$ as a sequence in which $|u| + v$ takes the successive values 1 once, 2 twice, 3 four times, 4 six times, ..., while in each group with constant $|u| + v$, $|u|$ goes from its maximum down to 1, we have

$$0/1, 1/1, -1/1, 2/1, -2/1, 1/2, -1/2, 3/1, -3/1, 2/2, -2/2, 1/3, -1/3,$$
$$4/1, \ldots$$

Removing the second and later appearances of each rational, we have a disjoint sequence (r_k) of all rationals and X_k can be the singleton of r_k alone. Such null functions are typical Lebesgue-integrable functions. We go further.

Theorem 0.1.1. *For B a Banach space with norm $\|\cdot\|$, M a measurable set of finite measure, and $f: M \to B$, Lebesgue–Bochner integrable to $H(M_1)$ over measurable subsets M_1 of M, then, given $\varepsilon > 0$, $x \in M$, there is an open set $G(x)$ containing x such that if (I_j) is a disjoint sequence of measurable subsets of M with $m(M \setminus \bigcup_{j=1}^\infty I_j) = 0$, and (x_j) a sequence of points with $I_j \subseteq G(x_j)$ $(j = 1, 2, \ldots)$, then*

(0.1.6) $$\sum_{j=1}^\infty \| f(x_j) m(I_j) - H(I_j) \| < \varepsilon.$$

Naturally the proof needs properties of the Bochner integral that are found in books on the subject. Of course m here denotes Lebesgue measure.

THE GENERAL THEORY OF INTEGRATION

Proof. First, a Bochner integrable f is Pettis integrable and so measurable. Thus there are a separable set $B_1 \subseteq B$ and a set $X \subseteq M$ of measure zero, such that $f \in B_1$ on $M \setminus X$. Let $(s_n) \subseteq B_1$ be a sequence dense in B_1. Then if $S(s, \varepsilon)$ denotes the sphere $\{b \in B_1 : \|b - s\| < \varepsilon\}$ with centre s and radius ε, B_1 is the union of the $S(s_n, \varepsilon)$ ($n = 1, 2, \ldots$). By the absolute continuity of the Bochner integral there is an $a > 0$ such that for measurable sets $Y \subseteq M$, $m(Y) < a$ implies

$$(L) \int_Y \|f\| \, dm < \varepsilon/3,$$

(B) and (L) denoting the Bochner and Lebesgue integrals. For $\varepsilon' = \varepsilon/(6m(M))$ let

$$X_1 = \{x : f(x) \in S(s_1, \varepsilon')\}, \quad X_n = \left\{x : f(x) \in S(s_n, \varepsilon') \setminus \bigcup_{j=1}^{n-1} S(s_j, \varepsilon')\right\} \quad (n = 2, 3, \ldots).$$

Choose for each n an open set $G_n \supseteq X_n$ such that

(0.1.7) $\qquad m(G_n \setminus X_n) < \min\left[\varepsilon/\{3 \cdot 2^n(\|s_n\| + \varepsilon')\}, a/2^n\right],$

and take $G(x) = G_n (x \in X_n)$. If (I_j), (x_j) are as given, with $x_j \in X_{n(j)}$, then

$$I_j \subseteq G_{n(j)}, \quad I_j \setminus X_{n(j)} \subseteq G_{n(j)} \setminus X_{n(j)},$$

$$\sum_{j=1}^{\infty} \|f(x_j)m(I_j) - H(I_j)\| = \sum_{j=1}^{\infty} \left\|(B)\int_{I_j} \{f(x_j) - f(x)\} \, dm\right\|$$

$$\leq \sum_{j=1}^{\infty} (L) \int_{I_j} \|f(x_j) - f(x)\| \, dm$$

$$\leq \sum_{j=1}^{\infty} (L) \int_{I_j \cap X_{n(j)}} \|f(x_j) - f(x)\| \, dm$$

$$+ \sum_{j=1}^{\infty} (L) \int_{I_j \setminus X_{n(j)}} \|f(x_j)\| \, dm$$

$$+ \sum_{j=1}^{\infty} (L) \int_{I_j \setminus X_{n(j)}} \|f(x)\| \, dm = R + S + T,$$

say. If $x \in I_j \cap X_{n(j)}$, then $f(x)$ and $f(x_j)$ lie in $S(s_{n(j)}, \varepsilon')$, so differing by less than $2\varepsilon'$, and

$$R \leq \sum_{j=1}^{\infty} (L) \int_{I_j \cap X_{n(j)}} 2\varepsilon' \, dm \leq 2\varepsilon' \sum_{j=1}^{\infty} m(I_j) = 2\varepsilon' m(M) = \varepsilon/3.$$

In S we collect those terms (if any) for which $n(j)$ has a given value n. By (0.1.7),

$$S = \sum_{n=1}^{\infty} \sum_{n(j)=n} (L) \int_{I_j \setminus X_n} \|f(x_j)\| \, dm \leq \sum_{n=1}^{\infty} \sum_{n(j)=n} (\|s_n\| + \varepsilon') m(I_j \setminus X_n)$$
$$= \sum_{n=1}^{\infty} (\|s_n\| + \varepsilon') m(G_n \setminus X_n) \leq \sum_{n=1}^{\infty} \varepsilon/(3 \cdot 2^n) = \varepsilon/3.$$

Finally, for Y the disjoint union of sets $I_j \setminus I_{n(j)}$,

$$m(Y) = \sum_{n=1}^{\infty} \sum_{n(j)=n} m(I_j \setminus X_{n(j)}) \leq \sum_{n=1}^{\infty} m(G_n \setminus X_n) < a.$$

By definition of a, the proof finishes with

$$T = (L) \int_Y \|f(x)\| \, dm < \varepsilon/3, \ R + S + T < \varepsilon. \quad \square$$

In n dimensions, taking the I_j as n-dimensional intervals forming a partition of a larger interval M, we see that the gauge integral includes the Lebesgue integral for real and complex values, and the Bochner integral for Banach space values. It is wider, it includes the Newton integral, which is not always a Lebesgue integral since the modulus of a derivative need not be finitely integrable. For example,

(0.1.8) $$G(x) = x^2 \sin(x^{-2}) (x \neq 0), \ G(0) = 0,$$
$$G'(x) = 2x \sin(x^{-2}) - 2x^{-1} \cos(x^{-2}) (x \neq 0), \ G'(0) = 0,$$

unbounded near $x = 0$ due to the $2x^{-1}$ even though $G'(0)$ exists. Now $|\cos x| \geq 2^{-1/2}$ in $[n\pi - \pi/4, n\pi + \pi/4]$ $(n = 1, 2, \ldots)$. For x^{-2} in that range and

$$\int |f(x)| \, dx \geq 2^{1/2} \log_e \sqrt{\{(n\pi + \pi/4)/(n\pi - \pi/4)\}}$$
$$= 2^{-1/2} \log_e \{(1 + 1/(4n))/(1 - 1/(4n))\}$$
$$= 2^{-3/2} n^{-1} \{1 - o(1)\}.$$

where $f(x) = 2x^{-1} \cos(x^{-2}) \ (x \neq 0)$,

Summing for $n = 1, 2, \ldots$ we have $+\infty$. The continuous first term in $G'(x)$ does not affect the result that $|G'(x)|$ is not finitely integrable over $[0, 1]$. Away from 0, $G'(x)$ is continuous and takes large positive and large negative values because of its first term. Hence there is a $b > 0$ with $G'(b) = 0$ (so $b^2 \tan(b^{-2}) = 1$). Let G_1 be the integral of the function equal to G' in $[-b, b]$ and 0 elsewhere, so that G_1' exists everywhere, let $(s_n), (t_n), (u_n)$ be sequences in

R, and

$$G_2 \equiv \sum_{n=1}^{\infty} G_1(t_n x - s_n)/u_n.$$

If the intervals $[(s_n - b)/t_n, (s_n + b)/t_n]$ $(n = 1, 2, \ldots)$ are mutually disjoint, then G_2 is differentiable in those intervals. By choice of (u_n) it may be possible to have G_2 differentiable everywhere.

In Sections 1.4 and 1.5 we show that the gauge integral is a generalized Riemann integral from a fully decomposable division space, equivalent to the Denjoy–Perron integral. For the Cauchy limit and the Harnack extension see Theorems 2.10.6, 2.10.7. For the first part of this section, with Levi's weak monotone convergence theorem proved for the gauge integral, see Henstock (1968a). For Theorem 0.1.1, based on the work of R. O. Davies, see Schuss (1969), Davies and Schuss (1970), and Henstock (1980b), pp. 226–7, Lemma 4. Much of the theory of the gauge integral can be found in McLeod (1980), Kurzweil (1980), and Henstock (1988a). The last book was designed to supersede most of the book Henstock (1963c) which is now out of print. McShane (1983) gives the theory of a generalized Riemann integral equivalent to the Lebesgue integral and very near to the gauge integral, and it covers many areas of use of integrals.

Example 0.1.1 There are bounded functions not integrable by Newton's nor Riemann's integrals: for example, f, the indicator of the rationals. Then every upper Darboux sum over $[a, b]$ is $b - a$ and every lower Darboux sum is 0, and f is not Riemann integrable. Darboux's theorem shows that if g is finitely differentiable everywhere in $[a, b]$ and $g'(a) \neq g'(b)$, with k a number between $g'(a)$ and $g'(b)$, then for a c in $a < c < b$, $g'(c) = k$. (The easy proof considers $g(x) - kx$.) Thus f is not a derivative and so is not Newton integrable. However, f is a null function with Lebesgue integral 0. More simply,

$$h(x) = 0 \quad (x < \tfrac{1}{2}(a + b)), \qquad h(x) = 1 \quad (x \geq \tfrac{1}{2}(a + b))$$

is Riemann but not Newton integrable over $[a, b]$.

Exercise 0.1.2 Using (0.1.5) for $w = 0$, prove that if

$$f(x) = 1 - g(x) = \begin{cases} 1 & (x \leq 0) \\ 0 & (x > 0) \end{cases}, \quad \text{then} \quad \int_{-1}^{1} f \, dg = 1, \quad \int_{-1}^{1} g \, df = 0.$$

Exercise 0.1.3 Discuss for various numbers p, q the continuity and differentiability of

$$F(x) = x^p \sin(x^{-q}) \quad (x \neq 0), \qquad F(0) = 0,$$

and the boundedness and integrability of $F'(x)$.

Exercise 0.1.4 Let a set X be dense in $[a, b]$. In the second proof of Section 1.4, Theorem 1.4.1, case $n = 1$, using sums over divisions of

$$h(u_{j-1}, u_j; x_j) \equiv f(x_j)(g(u_j) - g(u_{j-1})),$$

with

$$y_j \in (u_j, u_{j+1}) \cap I(u_j) \cap I(u_{j+1}) \cap X,$$

$$f(u_j)(g(u_j) - g(y_{j-1})) + f(u_j)(g(y_j) - g(u_j)) = f(u_j)(g(y_j) - g(y_{j-1})).$$

Hence prove that if the two gauge integrals exist with $g = g_1$ on X, then

$$\int_a^b f\,dg = \int_a^b f\,dg_1 + f(b)\{g(b) - g_1(b)\} - f(a)\{g(a) - g_1(a)\}.$$

Given that $G \neq G_1$ are constants with $g(x) = G$ in a set dense in $[a, b]$, $g(x) = G_1$ in another set dense in $[a, b]$, prove that $f(b) = f(a)$. Use Theorem 2.5.2 to prove that the only functions f that are integrable with respect to this last g are constant, even though the set where $g \neq G$ can be of measure zero. Thus a useful g can vary little. But f can vary much more, and if f is constant almost everywhere, with g absolutely continuous, the Lebesgue integral of f with respect to g exists and f is gauge integrable with respect to g.

Example 0.1.5 For F differentiable at all points of $[-1, 1]$, we construct a function G equal to F in $|x| \leq 1$, equal to 0 in $|x| \geq 2$, and differentiable everywhere, by defining G as a polynomial in $[-2, -1]$, with the same value and derivative at -1 as F, and that is 0 with derivative 0, at $x = -2$, and similarly in $[1, 2]$, e.g.

$$G(x) = (x - 2)^2 \{(2F(1) + F'(1))x - (F(1) + F'(1))\} \qquad (1 \leq x \leq 2)$$

with a similar formula in $[-2, -1]$. If

$$((a_n - 4.3^{-n}, a_n + 4.3^{-n})) \qquad (n = 1, 2, \ldots)$$

is a sequence of disjoint open intervals, H is differentiable everywhere, where

$$H(x) \equiv \sum_{n=1}^{\infty} 4^{-n} G((x - a_n) \cdot 3^n).$$

For if the nth interval is to the right of an x outside the given intervals and $x \leq a_n - 4.3^{-n} < a_n - 2.3^{-n} < x + h$, then $2.3^{-n} < h$, $4^{-n}/h < \frac{1}{2}(\frac{3}{4})^n \to 0$. Similarly for $-h$, giving $H'(x) = 0$ outside the intervals. This example could be used with (0.1.8).

0.2 BASIC DEFINITIONS FOR SIMPLE SET THEORY

Most readers can avoid this section, it gives the ordinary language of simple set theory, avoiding a precise logical approach.

If an object X is a collection of other objects, it is called a *set* or *family*, two interchangeable terms being used in the interests of readability. To define X we need only know which objects are in X (writing $x \in X$, or x *is in* X, or x *is of* X, or x is a *member* or *element of* X), and which objects x are not in X (writing $x \notin X$ or $x \bar{\in} X$). An empty set $\emptyset$ is a set such that $x \in \emptyset$ is false for all objects x. Logically, for a set X we cannot handle with safety the collection of all $x \notin X$. However, we suppose given a definable universal set U that contains all objects considered in a given theory, and then can define the complement $U \setminus X$ (or $\setminus X$ if U is understood) as the set of all $x \in U$ with $x \notin X$. For example, U could be the set $\mathbb{R}$ of all real numbers (the real line) or the set $\mathbb{C}$ of all complex numbers (the complex plane).

If $x \in Y$ when $x \in X$, we write $X \subseteq Y$, or $Y \supseteq X$, or X *is a subset of* Y, or X *is contained in* Y, or Y *contains* X. If $X \subseteq Y$ and $Y \subseteq X$ we write $X = Y$. For example, the set X of a, a, b is the set of a, b. It does not matter how many times we prove that $a \in X$, once is enough. If X, Y are empty sets then $X = Y$ and we suppose that there is just one empty set. If $X = Y$ is false, we write $X \neq Y$, or, X and Y *are distinct*. If $X \subseteq Y$, $X \neq Y$, we write $X \subset Y$, or $Y \supset X$, or X is a *proper* subset of Y, or X is contained in Y *properly*, or Y contains X *properly*.

If x, y are objects let $sing(x)$, $pair(x, y)$ be respectively the *singleton of* x (i.e. x alone) and the *pair of* x *and* y (i.e. x, y alone). Logically, $x \in sing(x)$, and if $z \in sing(x)$ then $z = x$. $x \in pair(x, y)$, $y \in pair(x, y)$, and if $z \in pair(x, y)$ then $z = x$ or $z = y$. The *ordered pair* (x, y) of x and y is $pair(sing(x), pair(x, y))$, so introducing the required asymmetry. Logically, $pair(x, x) = sing(x)$ and $(x, x) = sing(sing(x))$.

If $\mathscr{P}(x)$ is a collection of symbols involving an object x, such that the substitution of a particular object y for x gives a statement that is either true or false, we call $\mathscr{P}(x)$ a propositional function of objects x, often writing $\mathscr{P}(x)$ for '$\mathscr{P}(x)$ is true'. We write $\{x : \mathscr{P}(x)\}$ for the set of all x for which $\mathscr{P}(x)$ is true.

A *relation* R is a set of ordered pairs (x, y), writing xRy if $(x, y) \in R$. The relation R is an *equivalence* relation if *reflexive* (xRx if xRy), *symmetric* (yRx if xRy), and *transitive* (if xRy and yRz, then xRz), the R-*coset* (or R-*equivalence class*) $R(x)$ being the set of all y satisfying xRy, written $\{y : xRy\}$.

The *Cartesian product* $X \otimes Y$ of sets X, Y, is the family of ordered pairs (x, y) for all $x \in X$, all $y \in Y$, i.e. if $x \in X$, $y \in Y$, then $(x, y) \in X \times Y$, while if $z \in X \times Y$ then $z = (x, y)$ for some $x \in X$, some $y \in Y$.

A *function* (mapping, operator) f on a set X to a set Y, written $f : X \to Y$ or $y = f(x)$, is a subset Z of $X \otimes Y$ such that if $x \in X$, there is a $y \in Y$ with $(x, y) \in Z$, and that if, for the same x, $(x, z) \in Z$ for some $z \in Y$, then $z = y$. If $X_1 \subseteq X$, $Y_1 \subseteq Y$, we write

$$f(X_1) = \{y : y = f(x), x \in X_1\}, \qquad f^{-1}(Y_1) = \{x : y = f(x), y \in Y_1, x \in X\},$$

respectively the *f-image of* X_1 and the *inverse f-image of* Y_1.

0.2 BASIC DEFINITIONS FOR SIMPLE SET THEORY

We can now give a more general definition of a Cartesian product. For each $a \in A$ let $Y_a \subseteq Y$. Then the *Cartesian product* $\bigotimes_A Y_a$ is the set of all $f: A \to Y$ for which $f(a) \in Y_a$ when $a \in A$. If $Y_a = Y (a \in A)$ we write the Cartesian product as Y^A, so that $f \in Y^A$ if and only if $f: A \to Y$. When A is the set $1, 2, 3, \ldots, n$ of positive integers, the Cartesian product is often written

$$\bigotimes_{j=1}^{n} Y_j = Y_1 \otimes Y_2 \otimes \cdots \otimes Y_n.$$

The *union* (*join, sum*) of those Y for which $\mathscr{P}(Y)$, is the set of all objects, each of which is in some set Y with $\mathscr{P}(Y)$, and the *intersection* (*meet, product*) of those Y, is the set of all objects, each of which lies in all Y with $\mathscr{P}(Y)$, and we write, respectively,

$$\cup \{Y: \mathscr{P}(Y)\}, \quad \cap \{Y: \mathscr{P}(Y)\}.$$

Variations of this notation occur. The union and intersection of two sets X, Y are written, respectively, $X \cup Y$ and $X \cap Y$. For n sets or a sequence of sets we use

$$\bigcup_{j=1}^{n} X_j, \quad \bigcap_{j=1}^{n} X_j, \quad \bigcup_{j=1}^{\infty} X_j, \quad \bigcap_{j=1}^{\infty} X_j$$

Note that no limit process is used to define the last two. If $\mathscr{X}$ is a family of sets X, or if A is a family of suffixes a, then

$$\bigcup_{\mathscr{X}} X = \bigcup \{X: X \in \mathscr{X}\}, \quad \bigcap_{\mathscr{X}} X = \bigcap \{X: X \in \mathscr{X}\},$$

$$\bigcup_{A} X_a = \bigcup \{X_a: a \in A\}, \quad \bigcap_{A} X_a = \bigcap \{X_a: a \in A\}.$$

Two sets with empty intersection are called *disjoint*. If each pair of distinct sets from a family $\mathscr{X}$ of sets is disjoint, we say that the sets of $\mathscr{X}$ are *mutually disjoint*.

The de Morgan rules for operating with complements are as follows:

$$\setminus \cup \{Y: \mathscr{P}(Y)\} = \cap \{\setminus Y: \mathscr{P}(Y)\}$$
$$\setminus \cap \{Y: \mathscr{P}(Y)\} = \cup \{\setminus Y: \mathscr{P}(Y)\}$$

Example 0.2.1 If $Y \supseteq X$ for all X satisfying $\mathscr{P}(X)$, then $Y \supseteq \cup \{X: \mathscr{P}(X)\}$.
If $Y \subseteq X$ for all X satisfying $\mathscr{P}(X)$, then $Y \subseteq \cap \{X: \mathscr{P}(X)\}$.

Example 0.2.2 For the universal set U, and, of course, $X \subseteq U$, then $X \cup \setminus X = U$, $X \cap \setminus X = \varnothing$.

Example 0.2.3 If $X \subseteq U$, $Y \subseteq U$, we write the complement of Y relative to X as $X \setminus Y$, the set of points in X that are not in Y, so that

$X \setminus Y = X \cap (U \setminus Y)$. Then $X \setminus (X \setminus Y) = X \cap Y$,

$$\cup \{X : \mathscr{P}(X)\} \setminus Y = \cup \{X \setminus Y : \mathscr{P}(X)\},$$
$$\cap \{X : \mathscr{P}(X)\} \setminus Y = \cap \{X \setminus Y : \mathscr{P}(X)\}.$$

Example 0.2.4

$$\cup \{X : \mathscr{P}(X)\} \cap \cup \{Y : \mathscr{P}_1(Y)\} = \cup \{X \cap Y : \mathscr{P}(X), \mathscr{P}_1(Y)\}.$$
$$\cap \{X : \mathscr{P}(X)\} \cup \cap \{Y : \mathscr{P}_1(Y)\} = \cap \{X \cup Y : \mathscr{P}(X), \mathscr{P}_1(Y)\}.$$

Example 0.2.5 The *indicator* $\chi(X; x)$ of a set X is the function that is 1 when $x \in X$, and 0 when $x \notin X$. Denoting pair $(0, 1)$ by 2, the indicators of all subsets of X form 2^X. Thus it is usual to write 2^X for the family of all subsets of X. Or one can interpret 2^X in terms of cardinal numbers.

Exercise 0.2.6 Let A, B be two index sets, and let $X : A \otimes B \to 2^U$. Given that $\mathscr{P}(a)$, $\mathscr{Q}(b)$ are propositional functions of $a \in A$, $b \in B$, respectively, show that

$$\bigcap_B \left\{ \bigcap_A \{X(a, b) : \mathscr{P}(a)\} : \mathscr{Q}(b) \right\} = \bigcap_{A^B} \left\{ \bigcap_B \{X(a(b), b) : \mathscr{Q}(b)\} : \mathscr{P}(a(b)), \mathscr{Q}(b) \right\}$$

which generalizes *Ex. 0.2.4*.

Example 0.2.7 $\displaystyle\bigcup_A \{X_a : \mathscr{P}(a)\} \setminus \bigcup_A \{Y_b : \mathscr{P}(b)\} \subseteq \bigcap_A \{X_a \setminus Y_a : \mathscr{P}(a)\}$

where X_a, Y_a are functions of $a \in A$, their values being sets. (Use *Ex. 0.2.3*.)

Example 0.2.8 The *symmetric difference* of two sets X and Y is

$$d(X, Y) = (X \setminus Y) \cup (Y \setminus X).$$

If $X \subseteq U$, $Y \subseteq U$, $W \subseteq U$, then

$$d(\setminus X, \setminus Y) = d(X, Y), \quad d(W, X) \cup d(X, Y) \supseteq d(W, Y).$$

Example 0.2.9 $\displaystyle d\left(\mathbf{M}_A \{X_a : \mathscr{P}(a)\}, \mathbf{M}_A \{Y_b : \mathscr{P}(b)\} \right) \subseteq \bigcup_A \{d(X_a, Y_a) : \mathscr{P}(a)\}$

when M is the symbol $\cup$ and the symbol $\cap$.

Example 0.2.10 Let $f : X \to Y$. Then

(0.2.1) $\quad f(\cup \{Z : Z \subseteq X, \mathscr{P}(Z)\}) = \cup \{f(Z) : Z \subseteq X, \mathscr{P}(Z)\},$

(0.2.2) $\quad f(\cap \{Z : Z \subseteq X, \mathscr{P}(Z)\}) \subseteq \cap \{f(Z) : Z \subseteq X, \mathscr{P}(Z)\},$

Example 0.2.11 Let $f(x) = 0$ $(x > 0)$, $X = (0, 1)$, and let $\mathscr{P}(Z)$ denote that $Z = (0, \varepsilon)$ for some $\varepsilon > 0$. Then in (0.2.2) the left side is empty, while the right side is sing(0).

Exercise 0.2.12 Let proj: $X \otimes Y \to X$ with $\text{proj}(x, y) = x$. For A a set of suffixes with $X_a \subseteq X$, $Y_a \subseteq Y$, and for some $y \in Y$ let $y \in Y_a$ (all $a \in A$). Prove that

$$\left(\bigcap_A X_a\right) \otimes \text{sing}(y) \subseteq X_b \otimes Y_b \text{ (each } b \in A\text{)}, \text{proj}\left(\bigcap_A X_a \otimes Y_a\right)$$

$$= \bigcap_A X_a = \bigcap_A \text{proj}(X_a \otimes Y_a).$$

Example 0.2.13 There is equality in Ex. 0.2.10(0.2.2) when $f(x) = y = f(z)$ always implies $x = z$, which is when an inverse function can be defined. Equality also occurs in Ex. 0.2.12 when $f = \text{proj}$, the intersection of the Y_a not being empty, even though proj^{-1} cannot be defined.

Example 0.2.14 If $f: X \to Y$, then for various $Z \subseteq Y$,

$$f^{-1}(\cup Z) = \cup f^{-1}(Z), \quad f^{-1}(\cap Z) = \cap f^{-1}(Z).$$

0.3 BASIC DEFINITIONS FOR ALGEBRA

A structure involved in generalized Riemann integration is the algebra of the space K of values of the functions to be integrated, sometimes more general than $\mathbb{R}$ (real line) or $\mathbb{C}$ (complex plane), and so we turn to definitions, beginning with a function $m: K \times K \to K$, called *multiplication* and written $m(x, y) = x \cdot y$, the *product of x and y*. If $x, y \in K$ and $X, Y \subseteq K$ we write

$$x \cdot X = \{x \cdot z : z \in X\}, \quad X \cdot y = \{z \cdot y : z \in X\}, \quad X \cdot Y = \{z \cdot w : z \in X, w \in Y\},$$

$$x \cdot (X \cdot y) = \{x \cdot (z \cdot y) : z \in X\},$$

etc. K has an *identity* (or *unit*) u, relative to m, if there is a $u \in K$ with

(0.3.1) $\qquad x \cdot u = u \cdot x = x \qquad$ (all $x \in K$).

For two units z, u, $z = z \cdot u = u$, and so the unit is unique if it exists. K is a *semigroup* if the multiplication is *associative*, i.e.

(0.3.2) $\qquad x \cdot (y \cdot z) = (x \cdot y) \cdot z \qquad$ (all $x, y, z \in K$),

so that we can omit brackets and write $x \cdot y \cdot z$, and can write $x \cdot (X \cdot y)$ as $x \cdot X \cdot y$, etc.

The multiplication is *commutative* if

(0.3.3) $\qquad x \cdot y = y \cdot x \qquad$ (all $x, y \in K$).

Then, and only then, we usually write the multiplication as *addition*, $m(x, y) = x + y$ with various sets $x + X$ etc. (unless a second operation of addition is already used). If the unit exists it is then called a *zero*. If also K is a semigroup, we say that K is an *additive* semigroup.

K is a *group*, if it is a semigroup with unit u, such that each $x \in K$ has an inverse

(0.3.4) $\quad x^{-1} \in K$ with $x \cdot x^{-1} = u = x^{-1} \cdot x; \; X^{-1} \equiv \{z^{-1} : z \in X\}$.

When K is also additive, it is an *additive* group and we write x^{-1} as $-x$, $x \cdot y^{-1}$ as $x + (-y)$ or $x - y$, and X^{-1} as $-X$.

For x in a group K, x^{-1} is uniquely defined. For if $x \cdot y = u$ then

$$x^{-1} = x^{-1} \cdot u = x^{-1} \cdot (x \cdot y) = (x^{-1} \cdot x) \cdot y = u \cdot y = y,$$

and similarly, if $y \cdot x = u$ then $y = x^{-1}$. Further, for all $x, y \in K$,

$$(y^{-1} \cdot x^{-1}) \cdot (x \cdot y) = y^{-1} \cdot \{x^{-1} \cdot (x \cdot y)\} = y^{-1} \cdot \{(x^{-1} \cdot x) \cdot y\} = y^{-1} \cdot (u \cdot y)$$
$$= y^{-1} \cdot y = u,$$

and the uniqueness of the inverse gives (0.3.5), (0.3.6).

(0.3.5) $\quad\quad\quad\quad\quad\quad (x \cdot y)^{-1} = y^{-1} \cdot x^{-1},$

(0.3.6) $\quad\quad\quad\quad\quad\quad (x^{-1})^{-1} = x.$

A *ring* is an additive group K with group operation $x + y$, together with a multiplication for which K is a semigroup (thus $(ab)c = a(bc)$) *distributive on the left and right*, i.e.

$$a(b + c) = ab + ac, \quad (b + c)a = ba + ca.$$

Then $a + b$ is the *sum of a, b*, and ab their *product*, and aa is written a^2.

A ring K is *commutative* if, for all $a, b \in K$, $ab = ba$. The unit of the addition operation of K is called the *zero*, 0. By its uniqueness,

$$aa = (a + 0)a = aa + 0a, \quad aa = a(a + 0) = aa + a0, \quad 0a = 0 = a0$$

(all $a \in K$).

If the non-zero elements of a ring K are a commutative group under multiplication, K is called a *field*. Then the unit of the commutative group under multiplication is written 1 and called the *unit*. The *inverse* of $x \neq 0$ in the multiplication group keeps the same name in the field and is written x^{-1}. There is no 0^{-1} since $a0 = 0a = 0$, never 1.

A *linear space* (*vector space*, *linear vector space*) K over a field F, is an additive group K with unit θ, inverse $-x$ of $x \in K$, and a function $f : F \otimes K \to K$, written $f(a, x) = ax$, that is distributive in a and x,

$$a(x + y) = ax + ay, \quad (a + b)x = ax + bx, \quad a(bx) = (ab)x,$$
$$1x = x (a, b, 1 \in F, x, y \in K).$$

0.4 BASIC DEFINITIONS AND THEORY FOR TOPOLOGY

'Nearness' is the idea behind that of the *limit* x of a sequence (x_n); x_n is 'near' to x for all sufficiently large integers n. To clarify this idea we put another structure on the space K of values, a *topology* (here) or an *order* (next section).

A *topology in K* is a collection $\mathscr{G}$ of some subsets G of K, called $\mathscr{G}$-*sets*, with the properties:

(a) K and the empty set are in $\mathscr{G}$;
(b) if $\mathscr{H} \subseteq \mathscr{G}$, the union of the $G \in \mathscr{H}$ lies in $\mathscr{G}$;
(c) the intersection of any finite number of $G \in \mathscr{G}$, is in $\mathscr{G}$.

Then $(K, \mathscr{G})$ is called a *topological space*. Subsequent definitions often mention $\mathscr{G}$. If in a long stretch of theory $\mathscr{G}$ is unaltered, it is safe later to omit $\mathscr{G}$, calling $\mathscr{G}$-sets *open sets*. The $\mathscr{G}$-*interior* X° of a set $X \subseteq K$, is the union of all $\mathscr{G}$-sets $G \subseteq X$, which union by (b) lies in $\mathscr{G}$. A $\mathscr{G}$-*neighbourhood* of a point $x \in K$ is any $G \in \mathscr{G}$ with $x \in G$. A $\mathscr{G}$-*neighbourhood* of a set $X \subseteq K$, is any $G \in \mathscr{G}$ with $X \subseteq G$. A family $\mathscr{H} \subseteq \mathscr{G}$ is a *local base for* $\mathscr{G}$ *at a point* $x \in K$ if, for each $G \in \mathscr{G}$ with $x \in G$, there is an $H \in \mathscr{H}$ with $x \in H \subseteq G$, i.e. every $\mathscr{G}$-neighbourhood of x contains an $\mathscr{H}$-neighbourhood of x. $\mathscr{H}$ *is a base for* $\mathscr{G}$ if it is a local base at each point of K. $\mathscr{H}$ *is a subbase for* $\mathscr{G}$ if $\mathscr{H} \subseteq \mathscr{G}$ and if all intersections of a finite number of members of $\mathscr{H}$ form a base for $\mathscr{G}$.

A $\mathscr{G}$-*closed set* F is the complement (with respect to K) of a $\mathscr{G}$-set. By repeated use of the de Morgan rules on complements, the family $\mathscr{F}$ of all $\mathscr{G}$-closed sets satisfies:

(a)* The empty set and K are in $\mathscr{F}$,
(b)* if $\mathscr{H} \subseteq \mathscr{F}$, the intersection of the $F \in \mathscr{H}$ lies in $\mathscr{F}$,
(c)* the union of any finite number of $F \in \mathscr{F}$, is in $\mathscr{F}$.

Conversely, if a family $\mathscr{F}$ of subsets of K has these properties, the family of complements is a topology. The $\mathscr{G}$-*closure* $\bar{X} = \mathscr{G}X$ of $X \subseteq K$ is the intersection of all $\mathscr{G}$-closed sets $F \supseteq X$, $\mathscr{G}$-closed by (b)*, and contains X as each such $F \supseteq X$. An $x \in K$ is a $\mathscr{G}$-*limit-point* (or $\mathscr{G}$-*cluster point*) of a set $X \subseteq K$ if each $G \in \mathscr{G}$ with $x \in G$ contains a $y \in X$ with $y \neq x$. The set X' of all $\mathscr{G}$-*limit-points of* X is called the $\mathscr{G}$-*derived set of* X. An $x \in K$ is a $\mathscr{G}$-*contact point of* X if each $G \in \mathscr{G}$ with $x \in G$ contains a point of X, which in this case can be x. A set $X \subseteq K$ is $\mathscr{G}$-*dense* in a set $Y \subseteq K$, if $Y \subseteq \bar{X}$. A set $X \subseteq K$ is *nowhere* $\mathscr{G}$-*dense* (or $\mathscr{G}$-*rare*) in $Y \subseteq K$, if the $\mathscr{G}$-set $\setminus \bar{X}$ is $\mathscr{G}$-dense in Y. If $Y = K$ we omit the Y in the definition. The set $X \subseteq K$ is of the *first* $\mathscr{G}$-*category*, or $\mathscr{G}$-*meagre*, in K, if X is the union of an at most countable family of nowhere $\mathscr{G}$-dense sets in K. The complement of a set of the first $\mathscr{G}$-category is $\mathscr{G}$-residual or $\mathscr{G}$-comeagre. A set $X \subseteq K$ is of the *second $\mathscr{G}$-category*, or *non-$\mathscr{G}$-meagre*, in K, if not of the first $\mathscr{G}$-category in K. A set X is of the *second $\mathscr{G}$-category at a point* $x \in K$, if X intersects each $\mathscr{G}$-neighbourhood of x in a set of the second $\mathscr{G}$-category. For $X \subseteq K$, $G \in \mathscr{G}$, and $G \cap X$ not empty, $G \cap X$ is called a *$\mathscr{G}$-portion of X*. An $x \in X \setminus X'$ is called a *G-isolated point of X*. Then some $\mathscr{G}$-portion of X contains only x. The set X is *$\mathscr{G}$-dense-in-itself* if $X' \supseteq X$, while X is *$\mathscr{G}$-perfect* if $\mathscr{G}$-closed and $\mathscr{G}$-dense-in-itself, so $X' = X$. The set X is *nowhere $\mathscr{G}$-dense-in-itself* if each $\mathscr{G}$-portion of X has a $\mathscr{G}$-isolated point of X.

Theorem 0.4.1.

(0.4.1) *G is a $\mathscr{G}$-set if and only if G contains a $\mathscr{G}$-neighbourhood of each $x \in G$.*

(0.4.2) *A set is $\mathscr{G}$-closed if and only if it contains all its $\mathscr{G}$-limit-points.*

(0.4.3) $\setminus \bar{X} = (\setminus X)^\circ$.

(0.4.4) *If $X \subseteq Y \subseteq K$ then $X' \subseteq Y'$ and $\bar{X} \subseteq \bar{Y}$.*

(0.4.5) *If $X \subseteq K$ then $\bar{X} = X \cup X'$, the set of all $\mathscr{G}$-contact points of X.*

(0.4.6) *If $X, Y \subseteq K$ then $(X \cup Y)' = X' \cup Y'$, $\overline{X \cup Y} = \bar{X} \cup \bar{Y}$.*

Proof. For (0.4.1), if $G \in \mathscr{G}$ then G is a $\mathscr{G}$-neighbourhood of all its points. Conversely, if $G(x) \subseteq G$ for a $\mathscr{G}$-neighbourhood $G(x)$ of each $x \in G$, the union of the $G(x)$ is G, and by (b), the union is in $\mathscr{G}$. For (0.4.2) let $x \notin F$, F $\mathscr{G}$-closed. Then $x \in \setminus F$, a $\mathscr{G}$-neighbourhood of x free from points of F. Thus x cannot be a $\mathscr{G}$-limit-point of F and all such points lie in F. Conversely, if F contains all its $\mathscr{G}$-limit points and $x \notin F$, x is not a $\mathscr{G}$-limit point. So a $\mathscr{G}$-neighbourhood of x is free from points of F and lies in $\setminus F$. By (0.4.1), $\setminus F$ is a $\mathscr{G}$-set and F is $\mathscr{G}$-closed. For (0.4.3), for all $\mathscr{G}$-closed sets $F \supseteq X$, and so all $\mathscr{G}$-sets $\setminus F \subseteq \setminus X$,

$$\setminus \bar{X} = \setminus \cap F = \cup \setminus F = (\setminus X)^\circ.$$

(0.4.4) is trivial. For (0.4.5), if $X \subseteq F \subseteq K$, F $\mathscr{G}$-closed, (0.4.4) gives

$$X \cup X' \subseteq F \cup F' = F, \quad X \cup X' \subseteq \bar{X}.$$

If $x \notin X \cup X'$, a $\mathscr{G}$-neighbourhood of x has no points of X nor of X', and (0.4.1) gives

$$\setminus (X \cup X') \in \mathscr{G}, \quad X \subseteq X \cup X', \quad \bar{X} \subseteq X \cup X', \quad \bar{X} = X \cup X'.$$

For (0.4.6), if $x \in (X \cup Y)'$ and $x \notin X'$, then a $\mathscr{G}$-neighbourhood G of x has no points of X. For G_1 a $\mathscr{G}$-neighbourhood of x, $G \cap G_1$ is a $\mathscr{G}$-set containing x and so a $g \in X \cup Y$ with $g \neq x$. As $g \in G$, $g \notin X$ and so $g \in Y$ and $x \in Y'$ as G_1 is arbitrary. Hence

$$(X \cup Y)' \subseteq X' \cup Y', \quad X' \subseteq (X \cup Y)', \quad Y' \subseteq (X \cup Y)', \quad (X \cup Y)' = X' \cup Y'.$$

0.4 BASIC DEFINITIONS AND THEORY FOR TOPOLOGY

Using (0.4.5),

$$\overline{X \cup Y} = (X \cup Y) \cup (X \cup Y)' = X \cup Y \cup X' \cup Y' = (X \cup X') \cup (Y \cup Y')$$
$$= \bar{X} \cup \bar{Y}. \quad \square$$

Theorem 0.4.2.

(0.4.7) X is nowhere $\mathscr{G}$-dense in Y if, and only if, each $\mathscr{G}$-portion of Y contains a $\mathscr{G}$-portion $G \cap Y$ ($G \in \mathscr{G}$) with $G \cap \bar{X}$ empty.

(0.4.8) For $X, Y, Z \subseteq K$, if X, Y, are nowhere $\mathscr{G}$-dense in Z, so is $X \cup Y$.

(0.4.9) A set X is not nowhere $\mathscr{G}$-dense in a set Y, if and only if there is a $\mathscr{G}$-portion of Y contained in $\bar{X}$.

(0.4.10) If $G \in \mathscr{G}$ then $\bar{G} \setminus G$ is nowhere $\mathscr{G}$-dense.

(0.4.11) If F is $\mathscr{G}$-closed then $F \setminus F^\circ$ is nowhere $\mathscr{G}$-dense.

(0.4.12) If the union of a sequence (Y_n) of sets is of the second $\mathscr{G}$-category in X, then for an integer n, Y_n is $\mathscr{G}$-dense in a $\mathscr{G}$-portion of X.

Proof. For (0.4.7) let $G \in \mathscr{G}$ with $G \cap Y$ not empty. As $\setminus \bar{X} \in \mathscr{G}$, $G_1 = G \setminus \bar{X} \in \mathscr{G}$ by (c). If X is nowhere $\mathscr{G}$-dense in Y then $G_1 \cap Y$ is not empty. The converse is clear. For (0.4.8), if $G \in \mathscr{G}$ with $G \cap Z$ not empty, then by (0.4.7) G contains a $G_2 \in \mathscr{G}$ with $G_2 \cap Z$ not empty but $G_2 \cap \bar{X}$ empty. In turn, G_2 contains a $G_3 \in \mathscr{G}$ with $G_3 \cap Z$ not empty but $G_3 \cap \bar{Y}$ empty. Hence by (0.4.6), $G_3 \cap \overline{X \cap Y}$ is empty, proving (0.4.8). For (0.4.9), if X is nowhere $\mathscr{G}$-dense in Y, then

$$Y \subseteq (\setminus \bar{X})',$$

so that every $\mathscr{G}$-portion of Y contains a point of $\setminus \bar{X}$. If this is false, there is a $\mathscr{G}$-portion of Y lying in $\bar{X}$, and conversely. For (0.4.10), $\bar{G} \setminus G$ is $\mathscr{G}$-closed. If not nowhere $\mathscr{G}$-dense, then by (0.4.9) there is a non-empty $G_4 \in \mathscr{G}$ lying in $\bar{G} \setminus G$, so that $G_4 \subseteq \bar{G}$. As G_4 is a $\mathscr{G}$-neighbourhood of each of its points, G_4 contains a point of G, contradicting $G_4 \subseteq \bar{G} \setminus G$. Hence (0.4.10). For (0.4.11) we replace G by F°, $\bar{G}$ by F, and the same argument gives a non-empty $G_5 \subseteq F$, so that $G_5 \subseteq F^\circ$, a contradiction. For (0.4.12), by definition at least one Y_n is not nowhere $\mathscr{G}$-dense in X, so a $\mathscr{G}$-portion of X is contained in $\bar{Y}_n$, giving the result. $\square$

We now turn to metric spaces, in which these ideas are used. A function $d: K \times K \to \mathbb{R}$ (real line) is a *pseudometric* in K if, for all $x, y, z \in K$,

(0.4.13) $\qquad d(x, y) = d(y, x) \geq 0;$

(0.4.14) $\qquad d(x, y) = 0 \quad \text{if } x = y;$

(0.4.15) $\qquad d(x, y) + d(y, z) \geq d(x, z) \quad \text{(triangle inequality)}.$

Then we can consider cosets X such that x and y are in the same X if $d(x, y) = 0$, and x and y are in different cosets if $d(x, y) \neq 0$. If for points we substitute these cosets we obtain a metric which with a reasonable definition of a suitable function d, satisfies (0.4.13) to (0.4.15) together with

(0.4.16) If $d(x, y) = 0$ then $x = y$.

A *sphere* $S(x, r)$ centre x and radius $r > 0$, is

$$\{y : d(x, y) < r\}.$$

The family of all spheres in K is a sub-base for a topology in K called the *metric topology*, say $\mathscr{G}$, and (K, d) is then called a *metric space*. A *closed sphere* $CS(x, r)$ centre x, radius $r > 0$, is

$$\{y : d(x, y) \leq r\}.$$

A sequence (x_n) K is *d-convergent to a d-limit* x if, given $\varepsilon > 0$, there is an integer N, depending on ε, such that $d(x_n, x) < \varepsilon$ (all $n \geq N$). (x_n) is *d-fundamental* if, given $\varepsilon > 0$, there is an integer M depending on ε such that $d(x_m, x_n) < \varepsilon$ (all $m, n \geq M$). If every d-fundamental sequence in K is d-convergent to some d-limit we say that (K, d) is *d-complete*, and K (or (K, d)) is *a complete metric space*.

Theorem 0.4.3. $\mathscr{G} S(x, r) \subseteq CS(x, r)$. Let $(x_n) \subseteq S(x, r)$ with d-limit y. Then $y \in CS(x, r)$.

Proof. Given $\varepsilon > 0$ and $y \in \mathscr{G} S(x, r)$, there is a point $z \in S(x, r)$ satisfying $d(y, z) < \varepsilon$. By (0.4.15),

$$d(y, x) \leq d(y, z) + d(z, x) < \varepsilon + r, \quad d(y, x) \leq r,$$

since y and x are independent of $\varepsilon > 0$. Hence $y \in CS(x, r)$. A similar proof gives the second result. □

Theorem 0.4.4. (Baire's category theorem) *Every complete metric space* (K, d) *is of the second $\mathscr{G}$-category in the metric topology $\mathscr{G}$.*

Proof. For (X_n) a sequence of nowhere $\mathscr{G}$-dense sets in K let $x_1 \in K$ and $S_1 = S(x_1, 1)$. By mathematical induction, given $S_n \equiv S(x_n, r_n)$, there are x_{n+1} and $r_{n+1} > 0$ such that

$$S_{n+1} \equiv S(x_{n+1}, r_{n+1}) \subseteq S_n, \, r_{n+1} < \min\{1/(n + 1), (r_n - d(x_n, x_{n+1}))/2\},$$

with S_{n+1} free from all points of X_n ($n = 1, 2, \ldots$). Then

$$d(x_m, x_n) < \max(1/(m + 1), 1/(n + 1)),$$

so that (x_n) is d-fundamental and so d-convergent to some point x. As

0.4 BASIC DEFINITIONS AND THEORY FOR TOPOLOGY 17

$x_{n+1} \in S_{n+1}$, by Theorem 0.4.3,

$$d(x, x_{n+1}) \leqslant r_{n+1} < (r_n - d(x_n, x_{n+1}))/2,$$

$$d(x, x_n) \leqslant d(x, x_{n+1}) + d(x_{n+1}, x_n) < (r_n + d(x_{n+1}, x_n))/2 < r_n,$$

and $x \in S_n$, so that $x \notin X_n$ ($n = 1, 2, \ldots$), and K is not contained in the union of the X_n. Hence the result. □

Returning to general topological spaces, a $\mathscr{G}$-cover C of a set $C \subseteq K$, is a family of $\mathscr{G}$-sets whose union contains C. Then C is $\mathscr{G}$-compact if, for each $\mathscr{G}$-cover $\mathscr{C}$ of C, there is a $\mathscr{G}$-cover $\mathscr{C}_0 \subseteq \mathscr{C}$ of C containing a finite number of sets (i.e. a *finite cover*). Borel's and Goursat's covering theorems show that finite closed real intervals and finite-radius closed circles on the complex plane are compact in their respective metric topologies.

Theorem 0.4.5.
(0.4.17) *If $G \in \mathscr{G}$, C $\mathscr{G}$-compact, then $C \backslash G$ is $\mathscr{G}$-compact.*
(0.4.18) *If F is $\mathscr{G}$-closed, C $\mathscr{G}$-compact, and $F \subseteq C$, then F is $\mathscr{G}$-compact.*

Proof. For $\mathscr{C}$ a $\mathscr{G}$-cover of $C \backslash G$ in (0.4.17), G with $\mathscr{C}$ is a $\mathscr{G}$-cover $\mathscr{C}_1$ of C. Thus there is a finite $\mathscr{G}$-cover $\mathscr{C}_0$ of C in $\mathscr{C}_1$. Taking G from $\mathscr{C}_0$ if $G \in \mathscr{C}_0$, we have in $\mathscr{C}$ a finite $\mathscr{G}$-cover of $C \backslash G$, which is therefore $\mathscr{G}$-compact.
 In (0.4.18) $\backslash F \in \mathscr{G}$ and $F = C \cap F = C \backslash (\backslash F)$. Use (0.4.17). □

A family $\mathscr{F}$ of sets in K has *the finite intersection property relative to a set* $X \subseteq K$ if each finite collection from $\mathscr{F}$ has a non-empty intersection with X.

Theorem 0.4.6. *A set C is $\mathscr{G}$-compact if and only if every family $\mathscr{F}$ of $\mathscr{G}$-closed sets with the finite intersection property relative to C, has a non-empty intersection with C.*

Proof. For C satisfying the second condition let $\mathscr{G}_1$ be a family of $\mathscr{G}$-sets such that no finite union covers C. The complements of sets of $\mathscr{G}_1$ are a family $\mathscr{F}_1$ of $\mathscr{G}$-closed sets with the finite intersection property relative to C, so that a point $c \in C$ lies in the intersection of all sets of $\mathscr{F}_1$, c cannot be in any set of $\mathscr{G}_1$, and $\mathscr{G}_1$ cannot be a $\mathscr{G}$-cover of C. Hence each $\mathscr{G}$-cover of C contains a finite sub-$\mathscr{G}$-cover and C is $\mathscr{G}$-compact. Conversely, if C is $\mathscr{G}$-compact we reverse the argument to finish the proof. □

For $(K_j, \mathscr{G}_j)$ ($j = 1, 2$) two topological spaces and $f: K_1 \to K_2$, f is $\mathscr{G}_1$-$\mathscr{G}_2$-*continuous at a point* $v \in K_1$ when, given a $\mathscr{G}_2$-neighbourhood G_2 of $f(v)$, there is a $\mathscr{G}_1$-neighbourhood G_1 of v with $f(G_1) \subseteq G_2$, and f is $\mathscr{G}_1$-$\mathscr{G}_2$-*continuous in* K_1 if $\mathscr{G}_1$-$\mathscr{G}_2$-continuous at every point of K_1.

Theorem 0.4.7.

(0.4.19) $f: K_1 \to K_2$ is $\mathcal{G}_1$-$\mathcal{G}_2$-continuous in K_1 if and only if $f^{-1}(G_2) \in \mathcal{G}_1$ $(G_2 \in \mathcal{G}_2)$.

(0.4.20) If in K_1, C is $\mathcal{G}_1$-compact and f $\mathcal{G}_1$-$\mathcal{G}_2$-continuous, then $f(C)$ is $\mathcal{G}_2$-compact.

Proof. For (0.4.19) let f be $\mathcal{G}_1$-$\mathcal{G}_2$-continuous and $f(v) \in G_2 \in \mathcal{G}_2$. Then a $\mathcal{G}_1$-neighbourhood G_1 of v has $f(G_1) \subseteq G_2$ and so $G_1 \subseteq f^{-1}(G_2)$. By Theorem 0.4.1(0.4.1), $f^{-1}(G_2) \in \mathcal{G}_1$. For the converse we reverse the argument. For (0.4.20) let $\mathcal{G}_3$ be a $\mathcal{G}_2$-cover of $f(C)$. By (0.4.19), if $G_3 \in \mathcal{G}_3 \subseteq \mathcal{G}_2$, $f^{-1}(G_3) \in \mathcal{G}_1$. Also, as the $G_3 \in \mathcal{G}_3$ are a $\mathcal{G}_2$-cover of $f(C)$, the $f^{-1}(G_3)$ $\mathcal{G}_1$-cover C, which is $\mathcal{G}_1$-compact. So a finite number, say $f^{-1}(G_j)$ $(4 \leq j \leq n)$, $\mathcal{G}_1$-cover C, and each point of $f(C)$ lies in a G_j $(4 \leq j \leq n)$, giving a finite $\mathcal{G}_2$-cover of $f(C)$, which is therefore $\mathcal{G}_2$-compact. $\square$

Theorem 0.4.8. For $f: K_1 \to K_2$, $\mathcal{G}_1$-$\mathcal{G}_2$-continuous in K_1, $f(\mathcal{G}_1 X) \subseteq \mathcal{G}_2 f(X)$ $(X \subseteq K_1)$, or $f(\bar{X}) \subseteq \overline{f(X)}$ if the two topologies are assumed.

Proof. Let $x \in \mathcal{G}_1 X$ with G_2 a $\mathcal{G}_2$-neighbourhood of $f(x)$. There are G_1, v with

$$x \in G_1 \in \mathcal{G}_1, \quad f(G_1) \subseteq G_2, \quad v \in X \cap G_1, \quad f(v) \in f(X) \cap G_2.$$

As true for all such G_2, $f(x) \in \mathcal{G}_2 f(X)$, proving the result. $\square$

The topology $\mathcal{G}$ is a *Hausdorff topology* if every two distinct points of K have disjoint $\mathcal{G}$-neighbourhoods. Then $(K, \mathcal{G})$ is called *Hausdorffian*, or *Hausdorff*, or a *Hausdorff space*.

Theorem 0.4.9. If $(K, \mathcal{G})$ is Hausdorff, a $\mathcal{G}$-compact set $C \subseteq K$ is $\mathcal{G}$-closed, there are disjoint $\mathcal{G}$-neighbourhoods of C and each $x \notin C$, and if $X \subseteq C$ then $\mathcal{G}X \subseteq C$.

Proof. If fixed $x \notin C$, $y \in C$, there are disjoint $\mathcal{G}$-neighbourhoods $G(x, y)$ of x, $G^*(x, y)$ of y, the $G^*(x, y)$ being a $\mathcal{G}$-cover of the $\mathcal{G}$-compact C. So, for a finite number $y_1, \ldots, y_n$ of y, the union of $G^*(x, y_j)$ $(j = 1, \ldots, n)$ is a $\mathcal{G}$-neighbourhood N^* of C. The intersection N of $G(x, y_j)$ $(j = 1, \ldots, n)$ is a $\mathcal{G}$-neighbourhood of x, and

$$N \cap N^* = \bigcup_{j=1}^{n} N \cap G^*(x, y_j) \subseteq \bigcup_{j=1}^{n} G(x, y_j) \cap G^*(x, y_j) = \emptyset.$$

Hence $x \notin \bar{C}$, $\bar{C} = C$. If $X \subseteq C$ then $\mathcal{G}X \subseteq \mathcal{G}C = C$. Hence the theorem. $\square$

0.4 BASIC DEFINITIONS AND THEORY FOR TOPOLOGY

Theorem 0.4.10. *Let $(K_j, \mathscr{G}_j)$ ($j = 1, 2$) be Hausdorff with $f: K_1 \to K_2$, $\mathscr{G}_1$-$\mathscr{G}_2$-continuous in K_1. If $X \subseteq C \subseteq K_1$, C $\mathscr{G}_1$-compact, then $f(\mathscr{G}_1 X) = \mathscr{G}_2 f(X)$; loosely, $f(\bar{X}) = \overline{f(X)}$.*

Proof. Let $y \in \mathscr{G}_2 f(X)$, $y \in G \in \mathscr{G}_2$. Then $G \cap f(X)$ has an $f(x)$ with $x \in X$. For $\mathscr{G}_1$ Hausdorff, if

$$X(G) \equiv f^{-1}(G) \cap X, \quad X(G) \subseteq X \subseteq C, \quad \mathscr{G}_1 X(G) \subseteq \mathscr{G}_1 C = C$$

by Theorem 0.4.9. The family of sets $\mathscr{G}_1 X(G)$ with $y \in G \in \mathscr{G}_2$, has the finite intersection property and so has a common point z by Theorem 0.4.6. If G_2 is a $\mathscr{G}_2$-neighbourhood of $f(z)$, there is a G_1 with $z \in G_1 \in \mathscr{G}_1$ and $f(G_1) \subseteq G_2$, and there is an $x \in X(G) \cap G_1$, so that $f(x) \in G$. As $x \in G_1$, $f(x) \in G_2$ and $G \cap G_2$ is not empty, G and G_2 being arbitrary $\mathscr{G}_2$-neighbourhoods of y and $f(z)$, respectively. As $\mathscr{G}_2$ is Hausdorff,

$$f(z) = y, \quad z \in \mathscr{G}_1 X(G) \subseteq \mathscr{G}_1 X, \quad \mathscr{G}_2 f(X) \subseteq f(\mathscr{G}_1 X).$$

By Theorem 0.4.8 the last two sets are the same. □

Using product topologies, Theorems 0.4.8, 0.4.10 extend to functions of two or more variables. Let A be a set of suffixes and, for each $a \in A$, let $(K_a, \mathscr{G}_a)$ be a topological space. For L the union of the K_a so that $K_a \subseteq L$ ($a \in A$), we construct

$$K = \bigotimes_A K_a, \quad G = \bigotimes_A G_a, \quad G_a \in \mathscr{G}_a \quad (a \in A).$$

The family $\mathscr{H}$ of all G with $G_a = K_a$, except possibly for just one $a \in A$, is a sub-base for the required product topology, a base is the family $\mathscr{J}$ of all G with $G_a = K_a$ except possibly for a finite number of $a \in A$, and the product topology $\mathscr{G}$ is the family of all unions of subfamilies of $\mathscr{J}$, with the empty set.

We now bring in the algebra with the topology. First we say that *K is a topological group with topology $\mathscr{G}$*, if $(K, \mathscr{G})$ is Hausdorff, a group using $x \cdot y$, and with $x^{-1} \cdot y$ $\mathscr{G} \otimes \mathscr{G}$-$\mathscr{G}$-continuous. Normally the mention of $\mathscr{G}$ is suppressed. There are weaker conditions that lead to the same conditions, see the examples.

Theorem 0.4.11. *If K is a topological group with topology $\mathscr{G}$, and G a $\mathscr{G}$-neighbourhood of a point $x \in K$, there is a $\mathscr{G}$-neighbourhood of x whose $\mathscr{G}$-closure is in G. Such a topological space is called regular.*

Proof. For u the unit, $u^{-1} \cdot x = x$, so that by continuity there are $\mathscr{G}$-neighbourhoods G_1 of u, G_2 of x, with $G_1^{-1} \cdot G_2 \subseteq G$. If $y \in \bar{G}_2$, $G_1 \cdot y$ is a $\mathscr{G}$-neighbourhood of y, and so has a point $z \in G_2$. Thus, for some

$$g \in G_1, \quad g \cdot y = z, \quad y = g^{-1} \cdot z \in G_1^{-1} \cdot G_2 \subseteq G, \quad \bar{G}_2 \subseteq G. \quad \square$$

Theorem 0.4.12. *If K is a topological group with topology $\mathscr{G}$, and $X \subseteq K$, $G \in \mathscr{G}$, then $X.G^{-1} = \bar{X}.G^{-1}$.*

Proof. If $y \in \bar{X}.G^{-1}$ then $y = x.g^{-1}$ $(x \in \bar{X}, g \in G)$, $x \in x.g^{-1}.G$, a $\mathscr{G}$-neighbourhood of x. Hence for $v \in X \cap (x.g^{-1}.G)$, $h \equiv y^{-1}.v = g.x^{-1}.v \in G$, $y = v.h^{-1} \in X.G^{-1}$, $\bar{X}.G^{-1} \subseteq X.G^{-1}$. The opposite inclusion is trivially true, so that we have the result. $\square$

Theorem 0.4.13. *In a topological group K with unit $u \in G \in \mathscr{G}$ (the topology), there is a sequence $(G_j) \subseteq \mathscr{G}$ with*

$$G_1 = G, \quad u \in G_j, \quad G_j.G_j^{-1} \subseteq G_{j-1} \quad (j = 2, 3, \ldots),$$

$$G_{N+1}.G_{N+2}.\cdots.G_q \subseteq G_N \quad (q > N \geq 1).$$

Proof. By continuity of $x.y$ and $x.y^{-1}$, given $u \in G_j \in \mathscr{G}$, there is a G_{j+1} with

$$u \in G_{j+1} \in \mathscr{G}, \quad G_{j+1}.G_{j+1} \subseteq G_j, \quad G_{j+1}.G_{j+1}^{-1} \subseteq G_j \quad (j = 1, 2, \ldots).$$

$$G_{N+1}.\cdots.G_{q-1}.G_q \subseteq G_{N+1}.\cdots.G_{q-1}.G_q.G_q \subseteq G_{N+1}.\cdots.G_{q-1}.G_{q-1}$$

$$\subseteq \ldots \subseteq G_{N+1}.G_{N+1} \subseteq G_N. \quad \square$$

Theorem 0.4.14. *In a topological group K with topology $\mathscr{G}$, if $y \in X.G$ for each $G \in \mathscr{G}$ with $u \in G$, then $y \in \bar{X}$.*

Proof. If $y = x.g$ for $x \in X$, $g \in G$, then $x = y.g^{-1}$ and $X \cap (y.G^{-1})$ is not empty. Given an arbitrary $\mathscr{G}$-neighbourhood G_0 of y, a $\mathscr{G}$-neighbourhood G of u has $y.G^{-1} \subseteq G_0$, $X \cap G_0$ is not empty, and $y \in \bar{X}$. $\square$

Beyond additive groups we have linear spaces in which we impose topologies. First, a linear space K is *a normed linear space*, or a *normed space*, if to each $x \in K$ corresponds a real number $|x|$ called the *norm of x*, which satisfies

(0.4.21) $\qquad |0| = 0; \; |x| > 0 \; (x \neq 0),$

(0.4.22) $\qquad |x + y| \leq |x| + |y| \; (x, y \in K),$

(0.4.23) $\qquad |ax| = |a||x| \, (x \in K, a \in F \text{ (scalar field)}).$

Thus $d(x, y) \equiv |x - y|$ is a metric in K, giving the *norm* (or *strong*) *topology*.

A linear space K is a *linear topological space* (*topological vector space*) if K is a topological group under addition, if the scalar field F has a norm, and if the mapping $(a, x) \to ax$ of $F \times K \to K$ is continuous in the product topology of F and K, and the topology of K.

For F the real line $\mathbb{R}$, a set $X \subseteq K$ is *convex* if $x, y \in X$ and $0 < a < 1$ imply

$$ax + (1 - a)y \in X.$$

0.4 BASIC DEFINITIONS AND THEORY FOR TOPOLOGY

A set X is *balanced* (*équilibré*) if $x \in X$ and $|a| \leq 1$ imply $ax \in X$. A set X is *absorbing* if for each $x \in K$ there is an $a > 0$ with $a^{-1}x \in X$. Then

$$p_X(x) \equiv \inf\{a : a > 0, a^{-1}x \in X\}$$

is the *Minkowski functional* of the convex balanced and absorbing set $X \subseteq K$.

A linear topological space is called a *locally convex linear topological space* (or a *locally convex space*) if each $\mathscr{G}$-neighbourhood of 0 contains a convex, balanced and absorbing $\mathscr{G}$-neighbourhood of 0. In such a space a convex, balanced and absorbing closed set is called a *barrel* (*tonneau*). Then also K is called a *barrel space* if each of its barrels contains a $\mathscr{G}$-neighbourhood of 0.

Theorem 0.4.15. *A locally convex space K of the second category is a barrel space.*

Proof. Let X be a barrel in K. As X is balanced and absorbing, K is the union of closed sets nX ($n = 1, 2, \ldots$). As K is of the second category with nx continuous, one nX, and so X, have interior points, say nx_0 and x_0. If $x_0 = 0$, X contains a $\mathscr{G}$-neighbourhood of 0. If $x_0 \neq 0$ then $-x_0 \in X$ as X is balanced. By convexity X contains $\frac{1}{2}(x_0 + (-x_0)) = 0$ as a $\mathscr{G}$-interior point. □

A linear space is an *F-space* if

(0.4.24) *it is a metric space with an invariant metric d, i.e.*

$$d(x, y) = d(x - y, 0);$$

(0.4.25) *the mapping $(a, x) \to ax$ of $F \otimes K \to K$ is continuous in a for each x, and continuous in x for each a;*

(0.4.26) *the metric space K is complete.*

Here, $|x| \equiv d(x, 0)$ is a norm.

A *complete normed linear space* (*LNC-space*, or *Banach space*) is a normed linear space that is complete in its norm topology. Or, a Banach space is an F-space with $|ax| = |a| \cdot |x|$ for the norm of the general vector in the space.

Theorem 0.4.16. *An F-space K is a topological group of the second category.*

Proof. By (0.4.26) and Theorem 0.4.4 (Baire's category theorem) the space is of the second category. Given $\varepsilon > 0$ and an integer k, the set $Q(a, k, \varepsilon)$ where

$$|2^{-k}ax| + |-2^{-k}ax| \leq \varepsilon$$

is closed since $2^{-k}ax$ and $-2^{-k}ax$ are continuous in x for each fixed a, k. Then

$$Q_1(k, \varepsilon) \equiv \bigcap_{|a|<1} Q(a, k, \varepsilon), \quad P_m(\varepsilon) \equiv \bigcap_{k=m}^{\infty} Q_1(k, \varepsilon)$$

are also closed. These are the respective sets where

$$N_k(x) \equiv \sup_{|a|<1} \{|2^{-k}ax| + |-2^{-k}ax|\} \leqslant \varepsilon, \quad N_k(x) \leqslant \varepsilon \quad (\text{all } k \geqslant m).$$

Since ax is continuous in a, given $x \in K$, there is an integer m depending on x, ε, such that $N_k(x) \leqslant \varepsilon$ (all $k \geqslant m$) and so $x \in P_m(\varepsilon)$, and K is the union of the $P_m(\varepsilon)$ for each $\varepsilon > 0$. Since K is of the second category, there is an m for which $P_m(\varepsilon)$ is dense in some sphere S. As $P_m(\varepsilon)$ is closed, it contains S. Now

$$|2^{-k}a(x-y)| + |-2^{-k}a(x-y)| \leqslant |2^{-k}ax| + |-2^{-k}ay| + |-2^{-k}ax|$$
$$+ |2^{-k}ay| \leqslant N_k(x) + N_k(y),$$
$$N_k(x-y) \leqslant N_k(x) + N_k(y),$$

taking the supremum for $|a| < 1$. Hence in the sphere S_1 with centre the origin and with the same radius r as S, we have

$$|2^{-k}ax| + |-2^{-k}ax| \leqslant 2\varepsilon(|a|<1), \quad |ax| + |-ax| \leqslant 2\varepsilon(|a|<2^{-k}, |x|<\delta)$$

and ax is continuous in (a, x) at the origin, and so everywhere in K, and K is a topological group. □

Note that the use of 2^{-k} is to ensure a countable number of $P_m(\varepsilon)$ for each $\varepsilon > 0$, for the use of Baire's category theorem.

Theorem 0.4.17. *All locally convex F-spaces and all Banach spaces are barrel spaces.*

Proof. By Theorem 0.4.16 an F-space is a topological group of the second category. By Theorem 0.4.4 the same is true of Banach spaces. We now use Theorem 0.4.15. □

Some barrel spaces are locally convex spaces of the first category. A linear topological space K is said to be an α-*space* if there is a sequence (X_n) of subsets of K with union K, such that

(0.4.27) $\quad 0 \in X_1, \ X_n - X_n \subseteq X_{n+1} \quad (n = 1, 2, 3, \ldots),$

(0.4.28) *and every set X_n is nowhere dense in K.*

A linear topological space K that is not an α-space, is called a β-*space*. Then every sequence (X_n) of subsets of K satisfying (0.4.27) does not satisfy (0.4.28), and at least one X_n is dense in some neighbourhood.

Theorem 0.4.18. *Every locally convex β-space K is a barrel space.*

Proof. For X a barrel in K and so closed, and $X_n = 2^n X$ $(n = 1, 2, \ldots), (X_n)$ satisfies (0.4.27), so that an X_n is dense in some neighbourhood, and the same

0.4 BASIC DEFINITIONS AND THEORY FOR TOPOLOGY

is true of X. Being closed, X contains the neighbourhood. Now apply the end of the proof of Theorem 0.4.15. □

Sargent (1953) gives examples of spaces of integrable functions that are locally convex β-spaces of the first category, Thomson (1970b) brings in barrel spaces, and Henstock (1963c), pp. 142–9, gives a theory for topological groups. See Section 8.3 later.

A linear function or functional $F: K \to H$ with K, H normed linear spaces each with the scalar field $\mathbb{R}$ (real line), is an F with the properties

(0.4.29) $\qquad F(f+g) = F(f) + F(g) \quad$ (all $f, g \in K$),

(0.4.30) $\qquad F(af) = aF(f) \qquad$ (all $a \in \mathbb{R}, f \in K$).

The norm of F is defined to be $\sup |F(f)|/|f| (f \in K, |f| \neq 0)$, written $|F|$ as for the norm of f. The first and second norm signs refer to the norms in K, H, respectively.

Theorem 0.4.19. *A linear function F on K that is continuous at a point of K has a finite norm, and conversely, a linear function F on K that has a finite norm is continuous everywhere in K.*

Proof. For linear F, if continuous at $f \in K$, then given $\varepsilon > 0$, there is a $\delta > 0$ such that for all $g \in K$ with $|g - f| < \delta$, we have

(0.4.31) $\qquad |F(g) - F(f)| < \varepsilon, \quad |F(g-f)| < \varepsilon$

on using (0.4.29). As $\varepsilon > 0$ is arbitrary, F is continuous at the zero 0 of K, and then everywhere in K. Noting the scalar multiplication in (0.4.30), (0.4.31) for just one $\varepsilon > 0$ gives

$$\left|\tfrac{1}{2}\delta g/|g|\right| = \tfrac{1}{2}\delta |g|/|g| = \tfrac{1}{2}\delta < \delta (|g| \neq 0, g \in K),$$

$$\left|\tfrac{1}{2}\delta F(g)/|g|\right| = \left|F(\tfrac{1}{2}\delta g/|g|)\right| < \varepsilon,$$

(0.4.32) $\qquad |F(g)| / |g| < 2\varepsilon/\delta \quad (g \neq 0, g \in K),$

and $|F|$ is finite. Conversely, (0.4.31) follows for each $f \in K$, from (0.4.32). □

Theorem 0.4.20. *Let K be a normed barrel space, H a normed linear space, both with scalar field $\mathbb{R}$, and let $\mathcal{F}$ be a collection of continuous linear functionals $F: K \to H$. If for each $g \in K$ the corresponding collection of $F(g)$ is bounded, then the corresponding collection of norms $|F|$ is also bounded.*

This is (an extension of) the celebrated theorem of Hahn, Banach and Steinhaus, see Banach and Steinhaus (1927). H can be just a locally convex topological vector space, until at the last we need a norm.

Proof. For V a closed convex balanced set in H containing a neighbourhood of 0, put

$$B = \bigcap_{F \in \mathscr{F}} F^{-1}(V), \quad \mathscr{F}(g) \equiv \{F(g): F \in \mathscr{F}\}.$$

Then B is closed, convex and balanced. It is also absorbing, for if $g \in K$, $F(g)$ is given to be bounded and so lies in aV for some $a > 0$. Hence $g \in aB$ and B is absorbing. So B is a barrel in the barrel space K and so contains a neighbourhood of the zero. As $F(B) \subseteq V$ for all $F \in \mathscr{F}$, and as V is arbitrary with K, H normed spaces, the argument of Theorem 0.4.19 shows that the collection of norms $|F|$, for the $F \in \mathscr{F}$, is bounded. □

A subset X of a locally convex topological vector space K is called *precompact* if, for every convex balanced neighbourhood U of the zero, X is covered by a finite number of sets of the form $a + U$ $(a \in K)$.

Theorem 0.4.21. *Let K be a barrel space and H a Hausdorff locally convex topological vector space. If (F_n) is a sequence of continuous linear functionals of K into H that is pointwise convergent to a functional F_0, then F_0 is a continuous linear functional and the convergence is uniform on every precompact subset of K.*

Proof. For each $f \in K$, $(F_n(f))$ is bounded and so by the proof of Theorem 0.4.20, (F_n) is equicontinuous, in the sense that if V is an arbitrary closed convex balanced set containing a neighbourhood of zero in H, there is a neighbourhood U of zero in K with $F_n(U) \subseteq V$ for all n. For $f \in U$,

$$F_0(f) = \lim_{n \to \infty} F_n(f) \in \bar{V} = V$$

and F_0 is continuous at the zero, and trivially linear, and so continuous everywhere. If X is a precompact set there are a_j and integers m_j $(j = 1, \ldots, k)$ with

$$X \subseteq \bigcup_{1 \leq j \leq k} (a_j + \tfrac{1}{3}U), \quad F_n(a_j) - F_0(a_j) \in \tfrac{1}{3}V \quad (n \geq m_j).$$

For $n_0 = \max(m_1, \ldots, m_k)$ and $n \geq n_0$, and each $f \in X$, there is some j with $f - a_j \in \tfrac{1}{3}U$, and

$$F_n(f) - F_0(f) = F_n(f - a_j) + F_n(a_j) - F_0(a_j)$$
$$- F_0(f - a_j) \in F_n(\tfrac{1}{3}U) + \tfrac{1}{3}V + F_0(\tfrac{1}{3}U) \subseteq V.$$

Thus the convergence on X is uniform. □

When $H = \mathbb{R}$ (real line), Theorem 0.4.20 (Banach and Steinhaus) holds for a normed K if and only if K is a barrel space. See Robertson and Robertson (1966), Proposition 1, Corollary 1, pp. 65–6. The proofs of Theorems 0.4.20, 0.4.21 are taken from that book. In these two theorems K can be a Banach space, a locally convex F-space, or a locally convex β-space; see Theorems 0.4.17, 0.4.18.

Exercise 0.4.1 Prove that the product topology contains the intersection of a finite number of its sets, and so satisfies all conditions for a topology.

Exercise 0.4.2 Prove that if $(T_j, \mathscr{G}_j)(j = 1, 2, 3)$ are topological spaces with $f: T_1 \to T_2$ and $g: T_2 \to T_3$ both continuous, then $g(f(.)): T_1 \to T_3$ is continuous.

Exercise 0.4.3 Let $(T, \mathscr{G})$ be a topological space, K the real line $\mathbb{R}$ or complex plane $\mathbb{C}$ with the modulus topology and let $f: T \to K, g: T \to K$ both be continuous. Given that $x \in K$, show that

$$|f|, \quad xf, \quad f+g, \quad fg$$

are continuous. Given that $K = \mathbb{R}$, show that

$$\max(f, g), \min(f, g)$$

are continuous.

Exercise 0.4.4 Prove that a convergent sequence and its limit form a compact set.

Example 0.4.5 In Theorem 0.4.8, let $K_2 = \mathbb{R}$, the real line with the modulus topology, and $K_1 = \mathbb{R} \otimes \mathbb{R}$. Then equality of $f(\bar{X})$ and $\overline{f(X)}$ does not always occur. For let $f(x, y) = x + y$, continuous in K_1. Let X be the set of all negative integers and Y the set of real numbers $n + \frac{1}{2} + 1/n$ ($n = 3, 4, \ldots$). Then

$$\bar{X} = X, \quad \bar{Y} = Y, \quad \tfrac{1}{2} \in \overline{X+Y}, \quad \tfrac{1}{2} \notin \bar{X} + \bar{Y}.$$

Exercise 0.4.6 Let $x.y$ be continuous in (x, y) with X, Y compact in $(K, \mathscr{G})$. Prove that $X.Y$ is compact in the product topology of $K \otimes K$. (*Hint*: Prove that $X \otimes Y$ is compact and then use Theorem 0.4.7.)

Exercise 0.4.7 Let $(T, \mathscr{G})$ be a topological space with C $\mathscr{G}$-compact and $(x_j) \subseteq C$. Prove that there is a point $v \in C$ such that if G is a $\mathscr{G}$-neighbourhood of v then $x_j \in G$ for an infinity of j.

(*Hint*: If F_j is the closure of the set of points $x_k (k \geq j)$, the sequence (F_j) has the finite intersection property relative to C. Hence there is a point v in the intersection of all the F_j. Let m be the greatest integer j for which x_j has been found to lie in G. As $v \in F_{m+1}$, then a value x_t, with $t \geq m+1$, lies in G.)

Exercise 0.4.8 Whatever the space K, define $\rho(x, y)$ to be 0 ($x = y$) and 1 ($x \neq y$). Given that (x_n) is a ρ-fundamental sequence show that for some N, $x_n = x_N$ (all $n \geq N$), so that (x_n) is convergent, and (K, ρ) is complete.

Exercise 0.4.9 For s the space of all sequences of complex numbers and $x = (x_n) \in s$, $y = (y_n) \in s$, let

$$\rho(x, y) = \sum_{n=1}^{\infty} 2^{-n} |x_n - y_n|/\{1 + |x_n - y_n|\}.$$

Show that ρ is a metric for which s is complete.

Exercise 0.4.10 Let m be the space of bounded sequences of complex numbers, with

$$\rho(x, y) = \sup_n |x_n - y_n|.$$

Show that ρ is a metric for which m is complete.

Exercise 0.4.11 Let c be the space of convergent sequences of complex numbers, which metric as in Exercise 0.4.10. Show that c is complete for this metric.

Exercise 0.4.12 Let K be a group and a locally compact Hausdorff space with continuous multiplication. Prove that K is a topological group (R. Ellis (1957)).

0.5 BASIC DEFINITIONS AND THEORY FOR ORDER

A set K is *partially ordered* if a relation $x \leq y$ between some x and y of K satisfies:

(0.5.1) $\qquad\qquad x \leq y$ and $y \leq z$ imply $x \leq z$;

(0.5.2) $\qquad\qquad x \leq x \quad$ (all $x \in K$).

If $X \subseteq K$, a $k \in K$ is an *upper bound of* X when $x \leq k$ for each $x \in X$. A $k \in K$ is a *supremum of* X if k is an upper bound and if $k \leq m$ for all upper bounds m of X. A $k \in K$ is a *lower bound of* X if $k \leq x$ for each $x \in X$, and k is an *infimum of* X if k is a lower bound and $m \leq k$ for all lower bounds m of X. The set K is *directed upwards* if every finite subset of K has an upper bound; K is *directed downwards* if every finite subset of K has a lower bound. In either case K is a *directed set*.

0.5 BASIC DEFINITIONS AND THEORY FOR ORDER

In this book we use two important examples of directed sets, the set $\mathscr{I}$ of all positive integers with the natural order, and the set $A|E$ given later, set inclusion being the order, and the direction being 'as $\mathscr{U}$ shrinks' (defined later). A *sequence* is a function $S: \mathscr{I} \to K$, and a *generalized sequence for a directed set* D is a function $S: D \to K$.

Using the same sign $\leq$ for D (directed set) and K (partially ordered set), if

(0.5.3) $\qquad S: D \to K,\ S(d_1) \leq S(d_2) \quad (d_1 \leq d_2 \text{ in } D),$

we say that S is *monotone increasing*. If we replace (0.5.3) by

(0.5.4) $\qquad S: D \to K,\ S(d_2) \leq S(d_1) \quad (d_1 \leq d_2 \text{ in } D),$

we say that S is *monotone decreasing*. If either occurs, we say that S is *monotone*. We often replace K by 2^K, the family of subsets of K, the partial order being set inclusion.

For $(K, \mathscr{G})$ a topological space, D a directed set, and $S: D \to K$, S is *D-convergent* with *D-limit* $k \in K$ if, given any $\mathscr{G}$-neighbourhood G of k, there is a $d_1 \in D$ depending on G, with $S(d) \in G$ when $d_1 \leq d$ ($d \leq d_1$) in D for D directed upwards (downwards, respectively). If $S: D \to 2^K$ we replace $S(d) \in G$ by $S(d) \subseteq G$ in the definition of S *D-convergent to* k. When S is monotone decreasing, S is *D-convergent to a set* $X \subseteq K$ if $X \subseteq \mathscr{G} S(d)$ for all $d \in D$, and if for $t \notin X$ there is a $d \in D$ with $t \notin \mathscr{G} S(d)$, Then $X = \bigcap_D \mathscr{G} S(d)$.

Theorem 0.5.1. *For* $(K, \mathscr{G})$ *a Hausdorff space and* D *a directed set, every D-convergent* $S: D \to K$ (*or* 2^K) *has only one D-limit* $k \in K$.

Proof. Two different D-limits have disjoint $\mathscr{G}$-neighbourhoods, say G_1, G_2. For D directed upwards and $S: D \to K$, there are $d_j \in D$ such that $S(d) \in G_j$ for all $d \in D$ with $d_j \leq d$ ($j = 1, 2$). For d' an upper bound of d_1, d_2, and $d \in D$ with $d' \leq d$, $S(d) \in G_1 \cap G_2$, the empty set. Hence if the D-limit exists it is unique. For D directed downwards, use a lower bound, and if $S: D \to 2^K$ use $S(d) \subseteq G_j$. □

Theorem 0.5.2. *For* $(K, \mathscr{G})$ *a topological space,* C $\mathscr{G}$-*compact in* K, D *directed,* $S: D \to C$, *and* $X(d)$ *the set of all* $S(d')$ *with* $d' \in D$ *and* $d \leq d'$ ($d' \leq d$) *for* D *directed upwards (downwards, respectively), then*

(0.5.5) $\qquad\qquad\qquad \bar{S} \equiv \bigcap_D \mathscr{G} X(d)$

is not empty, and if G *is a* $\mathscr{G}$-*neighbourhood of* $\bar{S}$, *there is a* $d \in D$ *with*

(0.5.6) $\qquad\qquad\qquad X(d) \subseteq G \cap C.$

If $S: D \to 2^K$ *with* D *directed downwards,* S *monotone decreasing in the direction of* D, *and* $S(d) \subseteq C$ *for some* $d \in D$, *then* $X(d') = S(d')$ (*all* $d' \in D$), $S(d') \subseteq C$ (*all* $d' \leq d$ *in* D), *and* (0.5.5), (0.5.6) *hold for* S.

(0.5.7) Further, if, with $(L, \mathscr{G}_1)$ a Hausdorff topological space, $f: K \to L$ is $\mathscr{G}$-$\mathscr{G}_1$-continuous, then $f(X(d))$ has D-limit $f(\bar{S})$.

Proof. As D is directed and by construction of $X(d)$, $X(.)$ and $\mathscr{G}X(.)$ have the finite intersection property relative to C. By Theorems 0.4.6, 0.4.5, $\bar{S} \cap C$ is not empty and $C \backslash G$ is $\mathscr{G}$-compact for $\bar{S} \subseteq G \in \mathscr{G}$. If $x \in C \backslash G$ then $x \notin \bar{S}$ and there are $d(x) \in D$ $x \in G(x) \in \mathscr{G}$ with $S(d) \cap G(x)$ empty for all $d \in D$ with $d(x) \leq d$ (for D directed upwards). The $G(x)$ cover $C \backslash G$, so that so do a finite number, say $G(x_1), \ldots, G(x_q)$. For d' an upper bound of $d(x_1), \ldots, d(x_q)$, $G(x_1) \cup \cdots \cup G(x_q)$ contains no points of $S(d)$ when $d' \leq d$ in D, and $\mathscr{G}X(d') \cap (C \backslash G)$ is empty and (0.5.6) true. Similarly for D directed downwards and for $S: D \to 2^K$. For (0.5.7) let $t \notin f(\bar{S})$. As $\mathscr{G}_1$ is Hausdorff and $\bar{S}$ compact, so that $f(\bar{S})$ is compact by Theorem 0.4.7(0.4.20), there are disjoint $\mathscr{G}_1$-neighbourhoods G_1, G_2, respectively, of t and $f(\bar{S})$ by Theorem 0.4.9. For G a $\mathscr{G}_1$-neighbourhood of $f(\bar{S})$, by $\mathscr{G}$-$\mathscr{G}_1$-continuity of f there is a $\mathscr{G}$-neighbourhood G_3 of $\bar{S}$ with $f(G_3) \subseteq G \cap G_2$. There is a $d \in D$ with $X(d) \subseteq G_3$, by (0.5.6), so that $f(X(d)) \subseteq G \cap G_2$ and $t \notin f(X(d))$. Hence (0.5.7). □

A *lattice* is a partially ordered set K for which

(0.5.8)   if $x \leq y$ and $y \leq x$, then $x = y$ $(x, y \in K)$;

(0.5.9) if $x, y \in K$, pair (x, y) has a *supremum* $x \vee y$ and an *infimum* $x \wedge y$.

A lattice K is *order-complete* if every non-empty subset of K with an upper bound has a supremum, and every non-empty subset of K with a lower bound has an infimum. A lattice K is *relatively order-compact* if every non-empty subset of K has a supremum and an infimum. An order-complete lattice can be made relatively order-compact by adjoining two elements analogous to $+\infty$ and $-\infty$.

Using o for an abbreviation for order, with D a set directed upwards and K a lattice, a function $S: D \to K$ is *o-convergent* to the *o-limit* $k \in K$, written $S(d) \overset{o}{\to} k$, or $k = \text{o-lim}\, S(d)$, if there are two sets D_1, D_2, directed upwards, and $T_j: D_j \to K$ $(j = 1, 2)$, T_1 monotone increasing, T_2 monotone decreasing, such that

(0.5.10) $$k = \sup T_1(p) = \inf T_2(q);$$

(0.5.11) given p, q, there is a $d \in D$ such that for all d' of D with $d \leq d'$,
$$T_1(p) \leq S(d') \leq T_2(q).$$

Temple (1971), p. 16, calls such an arrangement the 'bracketing of a wild function by two tame ones'.

Theorem 0.5.3. *For D a set directed upwards and K a lattice, the necessary and sufficient condition for a monotone increasing $S: D \to K$ to be o-convergent to k, is that $\sup S(d)$ exists and is k.*

Proof. Let $S(d) \xrightarrow{o} k$. Given q, a d exists for (0.5.11). D being directed upwards, given $d' \in D$, there is an upper bound d'' of d, d' with $S(d'') \leqslant T_2(q)$. By the monotonicity of S, $S(d') \leqslant T_2(q)$ and $S(d')$ is a lower bound of the $T_2(q)$ in (0.5.11). Thus

(0.5.12) $\qquad S(d') \leqslant \inf T_2(q) = k$ (all $d' \in D$), $\sup S(d) \leqslant k$.

If $S(d) \leqslant v$ (all $d \in D$) then by (0.5.11), $T_1(p) \leqslant v$ (all $p \in D_1$) and v is an upper bound of the $T_1(p)$, so that $k = \sup T_1(p) \leqslant v$. For $v = \sup S(d)$, $k \leqslant \sup S(d)$. By (0.5.12), (0.5.8), $k = \sup S(d)$. Conversely, for monotone increasing $S(d)$ with supremum k we take $D_1 = D_2 = D$, $T_2(q) = k$ (all $q \in D$), and $T_1(p) = S(p)$ (all $p \in D$), and (0.5.10), (0.5.11) are true. Hence the result. □

Theorem 0.5.4.

(0.5.13) *If $S: D \to K$, $S^*: D \to K$, $S(d) \leqslant S^*(d)(d \in D)$, $S(d) \xrightarrow{o} k$, $S^*(d) \xrightarrow{o} k^*$, then $k \leqslant k^*$.*

(0.5.14) *If $S(d)$ o-converges to a limit, it is unique.*

(0.5.15) *If $S'(d) \leqslant S(d) \leqslant S^*(d)(d \in D)$, o-lim $S'(d) = k =$ o-lim $S^*(d)$, then $S(d) \xrightarrow{o} k$.*

(0.5.16) *Like a subsequence of a sequence, if $D^* \subset D$ with the same order, if for each $d \in D$ there is a $d^* \in D^*$ with $d \leqslant d^*$, and if $S: D \to K$, $S^*: D^* \to K$ have $S^*(d) = S(d)(d \in D^*)$, then $S(d) \xrightarrow{o} k$ implies $S^*(d) \xrightarrow{o} k$.*

Proof. For (0.5.13) let T_2^* be the T_2 for S^*. For $p \in D_1$, $q \in D_2^*$ there are $d, d^* \in D$ with

$$T_1(p) \leqslant S(d'), \quad S^*(d'') \leqslant T_2^*(q) \quad (\text{all } d', d'' \in D \text{ with } d \leqslant d', d^* \leqslant d'').$$

Both inequalities hold with both d, d^* replaced by $d_1 = \sup(d, d^*)$, and with $d' = d'' \geqslant d_1$, so that

$$T_1(p) \leqslant S(d') \leqslant S^*(d') \leqslant T_2^*(q), \quad k = \sup T_1(p) \leqslant T_2^*(q), \quad k \leqslant \inf T_2^*(q) = k^*.$$

(0.5.14) uses (0.5.13) and (0.5.8). For (0.5.15) replace S', S^* by T'_1, T_2^*. For each p, q there are $d', d^* \in D$ with

$$T'_1(p) \leqslant S'(d_1)(d' \leqslant d_1 \text{ in } D), \quad S^*(d_2) \leqslant T_2^*(q)(d^* \leqslant d_2 \text{ in } D).$$

Use $\sup(d', d^*)$. For (0.5.16) D_1 and D_2 are independent of D and so of D^*. If (0.5.11) holds for D, then it holds when we restrict d, d' to lie in D^*, which is possible since for each d there is a d^* with $d \leqslant d^*$. Thus the first o-limit implies the second. □

The definitions of fundamental and convergent sequences in a metric space (see the remarks between (0.4.16) and Theorem 0.4.3) are straightforward and well known, as is the definition of order-convergence involving (0.5.10) and (0.5.11). Turning to fundamental (Cauchy) and order-fundamental (order-

Cauchy) sequences and functions in a space K with a multiplication operation $x.y$, possibly without a subtraction operation, it appears that we cannot use Weil's uniform spaces unless K is a group or something similar. McShane (1969), p. 3 and p. 5 (2.7), seems to be the first to give a suitable definition, followed by Henstock (1978), p. 72, and the following is a simplified version.

Let D be a set directed upwards. If D is directed downwards we replace $d \leq d'$ by $d' \leq d$. For a topology $\mathscr{G}$ in K and a multiplication operation $x.y$ on every pair x, y of points of K, a function $S: D \to K$ (respectively, 2^K) is *D-convergent* if there is a $t \in K$ such that for each $\mathscr{G}$-neighbourhood G of t there is a $d \in D$ with $S(d') \in G$ (respectively, $\subseteq G$) for all $d \leq d'$ in D. Then t is called the *D-limit of S*. The function S is *D-fundamental* if there is a $t \in K$ such that for each $\mathscr{G}$-neighbourhood G of t there are an $s \in K$ and a $d \in D$ with $s.S(d') \in G$ (respectively, $\subseteq G$) for all $d \leq d'$ in D. If for K, D, every such D-fundamental S is D-convergent we say that $(K, \mathscr{G})$ is *D-complete*.

In a topological group the definition of D-fundamental becomes more traditional. For $s.S(d') \in G$ gives $S(d') \in s^{-1}.G$, a $\mathscr{G}$-neighbourhood of $s^{-1}.t$. If also $S(d^*) \in s^{-1}.G$, then

$$S(d^*)^{-1}.S(d') \in G^{-1}.s.s^{-1}.G = G^{-1}.G \equiv G_0,$$

a $\mathscr{G}$-neighbourhood of the unit u, which is the traditional definition for a group with a topology. Conversely, if $S(d^*)^{-1}.S(d') \in G_0$, we take $s = S(d^*)^{-1}$, $t = u$; for the semigroup definition of D-fundamental.

For order-fundamental functions the $\mathscr{G}$-neighbourhood is replaced by two functions T_j as in (0.5.10) with t replacing k and with $s.S(d')$ replacing $S(d')$ in (0.5.11).

Theorem 0.5.5. *For some $k \in K$ let $k.x$ be continuous in x. Then a D-convergent S is D-fundamental.*

Proof. If $x = D$-limit S let G be an arbitrary $\mathscr{G}$-neighbourhood of $k.x$. By continuity there are a $G_1 \in \mathscr{G}$ with $x \in G_1$ and $k.G_1 \subseteq G$, and a $d \in D$ with $S(d') \in G_1$ (respectively, $\subseteq G_1$) for all $d \leq d'$ in D. Then $k.S(d') \in G$ (respectively, $\subseteq G$), as required by the definition of D-fundamental. □

Theorem 0.5.6. *For the positive integers $\mathscr{I}$ let an $N: D \to \mathscr{I}$ exist, monotone increasing and unbounded. Then a D-complete K is $\mathscr{I}$-complete. Conversely, let each point of K have a countable base of its $\mathscr{G}$-neighbourhoods and let $x.k$ be continuous in k for each $x \in K$. Let $S: D \to K$ be D-fundamental, so that for a $t \in K$ and a countable base (G_n) of $\mathscr{G}$-neighbourhoods of t that we can assume are monotone decreasing, we have*

(0.5.17) $(s_n) \subseteq K$, $(d_n) \subseteq D$, $s_n \cdot S(d) \in G_n$ *(all $d \in D$ with $d_n \leq d$), for $n = 1, 2, \ldots$.*

0.5 BASIC DEFINITIONS AND THEORY FOR ORDER 31

(0.5.18) *If for some point $t \in K$, each $\mathscr{G}$-neighbourhood G of t, and some $s \in K$, $d \in D$, both depending on G, we have $s.S(d') \in G$ (or $\subseteq G$) for all $d \leq d'$ in D, with $S(d_n) \to x$ as $n \to \infty$, then $S(d)$ is D-convergent, and so to x. If K is $\mathscr{I}$-complete, then it is also D-complete.*

Proof. Let $(x_n) \subseteq K$ be an $\mathscr{I}$-fundamental sequence with $V(d)$ the set of x_n with $n \geq N(d)$. Then a $t \in K$ is such that for each $\mathscr{G}$-neighbourhood G of t, there are an $s \in K$ and an integer p with $s.x_m \in G$ when $m \geq p$. If d has $N(d) \geq p$ then $s.V(d) \subseteq G$ and $V(d)$ is D-fundamental and so D-convergent to a D-limit, say x. By construction of $V(d)$, (x_n) is $\mathscr{I}$-convergent to x and K is $\mathscr{I}$-complete. Conversely, let K be $\mathscr{I}$-complete and let S, (s_n), (d_n), G_n be as in (0.5.17). The G_n being monotone decreasing, we can assume the d_n monotone increasing, replacing d_n by $\sup(d_1, d_2, \ldots, d_n)$, if necessary. Writing $x_n = S(d_n)$, $S_n.x_m \in G_n$ ($m \geq n$). As the G_n are a base for the $\mathscr{G}$-neighbourhoods of t, (x_n) is $\mathscr{I}$-fundamental and so $\mathscr{I}$-convergent, say to x. Thus $S(d_n)$ is $\mathscr{I}$-convergent to x, and so S satisfies the conditions of (0.5.18), and is D-convergent to x, so that K is D-complete. □

Requirement (0.5.18) is complicated but makes good sense when K is a topological group, taking $t = u$. For we have

$$S(d_n) \in x.G(n \geq m), S(d') \in s^{-1}G, S(d_n)^{-1}.S(d') \in G^{-1}ss^{-1}G = G^{-1}G,$$

$$S(d') \in S(d_n)G^{-1}G \in xGG^{-1}G.$$

A judicious use of Theorem 0.4.13 with G replaced by a G_j shows that $S(d)$ is D-convergent to x. Thus (0.5.18) is not just a hope but a reasonable requirement.

Finally we have a slight anomaly.

(0.5.19) *When $K = \mathbb{R}$ or $\mathbb{C}$ with $x.y$ denoting ordinary multiplication, then in the definition of a D-fundamental $S(d)$ we have to take $t \neq 0$.*

For if in the definition $t = 0 = s$, then $s.S(d') \in G$ for any $S(d')$, even for a divergent $S(d)$. However, if $t \neq 0$ we can take G so that $0 \notin G$ and then we cannot have $s = 0$.

Example 0.5.1 For an example in which intuition can lead us astray let D be the family of finite sets of real numbers such as d and d_1, and let a direction in D be given by $d \leq d_1$ meaning $d \subseteq d_1$. If

$$f(x, y) = \begin{cases} 0 & (x \neq y) \\ 1 & (x = y) \end{cases}, \quad f_d(x) \equiv \sup_{y \in d} f(x, y),$$

then $f_d = 0$ almost everywhere, yet for direction upwards, $\lim_d f_d = 1$ everywhere.

0.6 SHORT HISTORY OF INTEGRATION

In the simplest case the process of integration is the addition of areas of non-overlapping elementary figures, possibly followed by the taking of some kind of limit. As with so many other things the Greeks began the process, and eventually Eudoxus (c. 408–355 B.C.) produced the method of exhaustions in which the geometry of the figures was used to fit a sequence of non-overlapping triangles inside the main figure to finally exhaust the area. About 120 years afterwards Archimedes (c. 287–212 B.C.) detailed this first crude limit process, giving the areas of the circle and sections of the parabola and similar figures. A barrier to further progress was the lack of graphs, which were invented two thousand years after Eudoxus, in A.D. 1619, by René Descartes (1596–1650), and mentioned in a letter from Christiaan Huygens to his brother Lodewijk on 21 November 1669. At about that time the calculus was invented and used by I. Newton (1642–1727) and G. W. Leibnitz (1646–1716), and integrals were computed by inverting the differentiation of known functions. As shown in Section 0.1, the modern refinement of this process leads to the *gauge* or *Kurzweil–Henstock integral*.

The first constructive definition of the integral was developed by many people including Daniel Bernoulli (1700–1782), L. Euler (1707–1783), S. D. Poisson (1781–1840), A. L. Cauchy (1789–1867), G. F. B. Riemann (1826–1866), and J. G. Darboux (1842–1917), who finally gave the definition that elementary calculus books now use. See Cauchy (1821), Riemann (1868) (written 1854), H. J. S. Smith (1875), Darboux (1875), and Gillespie (1915), while a recent account can be found in Henstock (1988a), pp. 1–3.

Riemann (1868), art. 5, shows that the necessary and sufficient condition for his integral of a bounded function to exist is that the total length of the subintervals for which the oscillation is greater than any fixed positive number, is arbitrarily small. Those already knowing measure theory will recognize that here appears the result that a bounded Riemann-integrable function is continuous almost everywhere, though not expressed in the 'Lebesgue' notation. (Nowadays it is customary to refer results in Lebesgue integration back to Lebesgue, even though he did not produce everything.) Note that E. Borel (1871–1956) and H. Lebesgue (1875–1941) were both born after Riemann died. It could be that Riemann's condition inspired them to set up a theory of measure of sets of points on the real line, and then the Lebesgue integral, which we will consider in turn.

Given an interval $[a, b]$ on the real line with $b - a$ finite, Riemann used partitions (0.1.2), choosing an arbitrary point x_j in $u_{j-1} \leqslant x_j \leqslant u_j$. If $f:[a, b] \to R$ is the function to be integrated, Riemann calculated the sum

$$s = \sum_{j=1}^{n} f(x_j)(u_j - u_{j-1}).$$

If s tends to a limit as the mesh of the partition tends to 0, where the *mesh* is

0.6 SHORT HISTORY OF INTEGRATION

the greatest length of the subintervals, and so the maximum of $u_j - u_{j-1}$, we say that the *Riemann integral of f over* $[a, b]$ *exists*. Darboux (1875) Section 2, replaced $f(x_j)$ by the supremum of f in $[u_{j-1}, u_j]$ to obtain an *upper Darboux sum* U. For a *lower Darboux sum* L he replaced $f(x_j)$ by the infimum of f in $[u_{j-1}, u_j]$. Clearly $L \leqslant s \leqslant U$. If L and U have the same limit I as the mesh tends to 0, we say that the *Riemann–Darboux integral of f exists over* $[a, b]$ *with value I*, in which case s also tends to the same limit, which is therefore the value of the corresponding Riemann integral. Conversely, for a fixed partition of $[a, b]$, we can take $f(x_j)$ as near as we like to the supremum in $[u_{j-1}, u_j]$, so that s can be taken as near as we like to U. Similarly, for different choices of the x_j we can take s as near as we like to L, showing that if the Riemann integral exists, so does the Riemann–Darboux integral, with the same value, and for real-valued f the two integrals are equivalent. However, the Riemann–Darboux definition needs an order in the set of values and so can be used for lattice-valued functions, whereas for complex-valued functions the real and imaginary parts have to be integrated separately, and similarly for values in finite-dimensional Banach spaces. But the Riemann definition can be extended immediately to functions with these values and to functions with values in a topological space with addition and multiplication by scalars; see Graves (1927). In fact the Darboux method, useful in practice, turns out to be a barrier to further progress such as in the direction of the gauge integral.

On $\mathbb{R}$ a real-valued Riemann-integrable f has to be bounded, because of the Riemann–Darboux equivalent, but it can integrate some bounded simply discontinuous functions, see Example 0.1.1. But derivatives need not be bounded, see (0.1.8), though they satisfy Darboux's theorem on derivatives. Thus neither the Newton integral nor the Riemann integral includes the other. The common family of functions integrable to the same value by both methods on $[a, b]$, contains all continuous functions over $[a, b]$. (The indefinite Riemann integral of a continuous function over the finite interval can be differentiated everywhere in that interval to obtain the continuous function, so that each continuous function is a derivative.)

T. J. Stieltjes (1856–94), in his research (Stieltjes (1894)), on continued fractions, used two functions f and g, instead of the single f, replacing $f(t)(v - u)$ by $f(t)\{g(v) - g(u)\}$, and defining what is now known as the *Riemann–Stieltjes integral*. This gives the famous expression of F. Riesz (1880–1956), of the most general continuous linear functional on the space of continuous functions on a compact interval; see Riesz (1914). Also see Bliss (1917a), Carmichael (1919), Dunford (1938), p. 312, Vanderlijn (1941) (with g-values in a Banach space), and Vulih (1941) (B. Z. Vulih (1913–78)).

H. L. Smith (1925) integrated $\frac{1}{2}\{f(v) + f(u)\}\{g(v) - g(u)\}$ (*Stieltjes mean integrals*) used by Steffensen (1932) in actuarial mathematics. de Finetti and Jacob (1935) gave the integration by parts, with an expression for the most general continuous linear functional on the space of bounded functions

34 THE GENERAL THEORY OF INTEGRATION

having at most discontinuities of the first kind. Also see Fréchet (1936) (M. R. Fréchet (1878–1973)) and Kaltenborn (1934, 1937). A complicated expression was given by W. H. Young (1914) (W. H. Young (1863–1942)). Expressions

$$\{f(v) - f(u)\}^2/\{g(v) - g(u)\} \quad \text{and} \quad \{f(v) - f(u)\}\{h(v) - h(u)\}/\{g(v) - g(u)\}$$

were integrated by Hellinger (1907, 1909) (*Hellinger integrals*), Hobson (1920), and Kolmogorov (1930). Then Gowurin (1936) integrated a bilinear function of $f(x)$ and $g(v) - g(u)$. J. C. Burkill (1924a, b, c) systematized the theory, integrating *interval functions* $h(u, v)$ (*Burkill integral*). See also Saks (1927), R. C. Young (1928a), Getchell (1935), and Kempisty (1936b, 1939).

From the Riemann integral up to this point, the integrals are *mesh limits*, we take the limits of sums as the mesh tends to 0. If the limits exist, they give constructive integrals as an algorithm can be used. For example, for the u_j of (0.1.2) we can take $u_j = a + (b - a)j \cdot 2^{-n}$ ($j = 0, 1, 2, 3, \ldots, 2^n$) with $x_j = u_j$.

Moore (1915) seems to be the first to use *refinement limits* (σ-*limits, refinements of subdivisions*) followed by Pollard (1923), H. L. Smith (1925), Hyslop (1926), Dushnik (1931), Getchell (1935), L. C. Young (1936), and Glivenko (1936). Lebesgue (1904, 2nd edn 1928, p. 272) (H. Lebesgue (1875–1941)) and Ridder (1933c) gave integrals equivalent to the refinement integral. See also Shohat (1930), Fréchet (1936), Copeland (1937), Dienes (1947) (P. Dienes (1882–1952)), and Appling (1962a, b, c; 1963a, b, and many more). Henstock (1946, 1948) applied σ-limits to interval function integration, while Çesari (1962) used an abstractly defined mesh.

Real and complex numbers have two algebraic operations, addition and multiplication. In all of the above we have used sums. Replacing sums by products, we have *product integrals*. Birkhoff (1937b, pp. 104–24), Masani (1947, 1981), Neuberger (1958), MacNerney (1964), B. W. Helton (1966, 1969, 1973, 1976), Chatfield (1973, 1979), Martin (1973), Plant (1974), J. C. Helton (1975a, b, 1978), Kay (1975), and J. C. Helton and Stackwisch (1978) use the refinement definition, and so does part of the book, Dollard and Friedman (1979). Arley and Borchsenius (1945) give a special definition, and so does Hildebrandt (1959) (T. H. Hildebrandt (1888–1980)), that is equivalent to the refinement integral.

All such integrals share the major limitations of mesh limits, they cannot integrate all finite-valued derivatives (or the equivalent for product integration), nor the limits of all bounded convergent sequences of integrable functions over compact intervals. The Lebesgue integral removed the second of these limitations; see Lebesgue (1902, 1904, 1909a, 1910, and 2nd edn 1928). W. H. Young (1904b, 1905, 1910, 1914) used monotone sequences to extend Riemann's integral, W. H. Young (1904b) being independent of Lebesgue (1902) as Young's paper was written in 1903 and Young read Lebesgue's paper in 1904. Daniell (1917/1918) (P. J. Daniell (1889–1946)) appeared after Young's paper, so that W. H. Young has the priority for the method of using monotone sequences of functions. See also Daniell (1919/1920), Stone

(1948(I), 1949), and McShane (1949) (E. J. McShane (1904–89)). Hahn (1915) and Tonelli (1923) (L. Tonelli (1885–1946)) used open sets of small measure, on the complement of which the function is continuous, extended the function linearly over each interval of the open sets, then took the limit of the Riemann integral of the extended function as the open sets shrank. Riesz (1919) used limits of step-functions. Banach (1923, 1932, pp. 30–32) (S. Banach (1892–1945)) used an abstract linear functional 'Hahn–Banach' approach, with monotone sequences in Banach (1937), followed by Goldstine (1941) and Matsuyama (1942). All these methods were used to produce an absolute integral, for which, if f is integrable, so is $|f|$, and the same integral was produced in the end for real- and complex-valued functions f.

Integration of Banach-space-valued functions began with Hildebrandt (1927), then Bochner (1933), Gelfand (1936, 1938), Dunford (1936a, 1937) and Mikusiński (1964a, b). Or the values can lie in a lattice, see Izumi and Nakamura (1940), Orihara and Sunouchi (1942), Izumi, Matsuyama, and Orihara (1942), Izumi (1942a), and Nakano (1943). Izumi (1942a) showed that the functions f can be replaced by a linear σ-complete lattice having a sublattice of step-functions, following a method given by MacNeille (1941).

Product integrals using Lebesgue-type methods can be found in Birkhoff (1937b, pp. 124–30) and Dollard and Friedman (1978a, b, 1979).

Stieltjes-type integrals in this region began with Radon (1913) (J. Radon (1887–1956)) using a completely additive set function, now called a *Radon measure*, obtained from a point function g of bounded variation in n dimensions. But for $n = 1$ the Riemann–Stieltjes integral of 1 relative to g in $[a, b]$ always exists equal to $g(b) - g(a)$, whatever the g, whereas the Radon integral does not exist when g is not of bounded variation. If the Riemann–Stieltjes integral of f relative to a g of bounded variation exists, so does the corresponding Radon integral, the two integrals being equal if g is also continuous. The Radon integral integrates the limit function of a bounded sequence of Radon-integrable functions over a finite interval if g is fixed. The Lebesgue and Radon integrals over a Baire set M contained in a finite or infinite interval I follow on multiplying the integrand by the indicator of M and integrating over I. Fréchet (1915) generalized Radon's theory to a σ-field in a general set, using W. H. Young's upper and lower integrals in part of the theory. Dunford (1935a, b) defined the integral in a Banach space, using the L_1-norm, while Birkhoff (1935) extended Fréchet's theory to a Banach space, his integral including Dunford's. Next, the measure values were put in a Banach space, with a bilinear transformation from the two Banach spaces for the values of f and the measure, to a third Banach space; see Bochner and Taylor (1938), and Price (1940) who used ideas from Gowurin (1936), W. H. Young, and Birkhoff. Pettis (1938) defined a 'weak' integral using continuous linear functionals on a normed linear space with weak convergence, see Bourbaki (1959), Brooks (1969a), Uhl (1972), and Chatterji (1974). Phillips (1940) had values in locally convex linear topological spaces. Rickart (1942,

1944) generalized Phillips (1940), Gowurin (1936), and Price (1940). Bochner and Fan (1947) gave a Young–Daniell-type definition with measure values in certain ordered spaces, the point function being real-valued. Carathéodory (1938) replaced the subsets on which the measure function is defined, by a Boolean algebra, and was followed by Wecken (1939), Ridder (1941, 1946), Olmsted (1942), Gomes (1946), and Kappos (1949). See also Freudenthal (1936). Izumi (1941) replaced Lebesgue sums by abstract linear forms and applied Moore–Smith limits. Also see Bartle (1956), Bogdanowicz (1965), Dinculeanu (1966) and Brooks (1969a, b). Dobrakov (1970a, b) generalized Bartle (1956); both take f-values in a Banach space X and measure values as bounded linear operators from X to another Banach space Y, so that the measure value operates on the f-value. Kunugi (1954) defined a structure, an 'espace rangé' (ranked space), weaker than a topology, and used it to define an absolute integral; see Kunugi (1954, 1956), Enomoto (1954a, b, c, 1955a, b), where the name changed to Nakanishi (1956, 1957a, b, 1958, 1968a, b, c, 1969a, b, 1974, 1978a, b, 1978/1979, 1979, 1984), Nakanishi and Fujita (1970), and Okano (1957, 1958a, b, 1959a, b, c 1960, 1962a, b).

Previous attempts at a Riemann-type definition of Lebesgue integration were given by Lebesgue (1909a, pp. 30–33), Borel (1910a, b), Hahn (1914), Denjoy (1919, 1931), and Levi (1941), but little progress was made. Only in the 1960s was a successful theory developed, in the sense that it included at least the proof of some theorem of 'Lebesgue' power, the first being Henstock (1961b). But as the Riemann-type definition also applies to non-absolute integration, in which $|f|$ need not be integrable even if f is, we turn to non-absolute integration, the main theme of this book. It began implicitly with the calculus and was brought out by Cauchy's work on functions with asymptotes in the bounded interval $[a, b]$. If at a, the integral over $[a, b]$ is defined as the limit (if it exists) of the integral over $[c, b]$ as $c \to a +$, where the latter integral is a Riemann integral or has itself been defined by such an extra limit process. This is now known as a *Cauchy limit*, see Cauchy (1823, Lec. 25); and similarly if the asymptote is at b. If there are many such asymptotes the process becomes difficult. If a is replaced by $-\infty$, or b by $+\infty$, or both, we use the Cauchy limit process again, see Cauchy (1823, Lec. 24). For finite a, b, the integral over $(-\infty, b]$ is the limit of the integral over $[a, b]$ as $a \to -\infty$, the integral over $[a, +\infty)$ is the limit of the integral over $[a, b]$ as $b \to +\infty$, supposing the limits exist, while the integral over $(-\infty, +\infty)$ is the sum of the integrals over $(-\infty, 0]$ and $[0, +\infty)$, or the limit of the integral over $[a, b]$ as $-a$ and b tend *independently* to $+\infty$. These details are elementary but should be remembered in this context; the implications were studied by de la Vallée Poussin (1892a, b) before Lebesgue's famous paper. A similar integral occurs in complex variable theory, but there, $-a = b \to +\infty$, and a divergence on the left can be cancelled by a divergence on the right, and a lot of the theory fails. After 1902 it was seen that a Riemann-integrable f is

0.6 SHORT HISTORY OF INTEGRATION

Lebesgue-integrable, the converse being false, as the indicator of the rationals shows. But the modulus of the derivative of $x^2 \sin(x^{-2})$ is not Lebesgue integrable over $[-1, 1]$, so that Newton's integral is not included in Lebesgue's and it is natural to look for an integral that includes both. Beginning with Lebesgue integration, Denjoy (1912a, b) (A. Denjoy (1884–1974)) defined a process of *totalization* that others called the *special Denjoy integral*, using Cauchy limits and an extension of Harnack (1884), see Theorem 2.10.7, in a transfinite inductive process. Modifying the Harnack extension he defined the *general Denjoy integral*, Denjoy (1916a), and independently Hinčin (Khintchine) (1916, 1918). A systematic account is given in Denjoy (1916b). Luzin (1912c) gave a descriptive definition of the special Denjoy integral.

Lebesgue and Newton integrals are included in Denjoy's special integral, which is included in his general integral. The Denjoy integrals are not absolute, $|f|$ need not be integrable if f is. But if both f and $|f|$ are Denjoy integrable, they are also Lebesgue integrable. Lebesgue integrals are in some ways like absolutely convergent series, while the Denjoy integrals are analogous to conditionally convergent series, and so need to be studied, not just for their esoteric value, but for practical use. Thus there is a need of simpler definitions than those of Denjoy. His process was axiomatized, see Saks (S. Saks (1897–1942)) (1930, 1937 pp. 254–9), MacNeille (1941), Natanson (1961, pp. 169–78), and Solomon (1967, 1969a).

A chain of descriptive integrals began with the Perron integral of Perron (1914) (O. Perron (1880–1975)). Then Hake (1921) proved that if the special Denjoy integral exists, so does the Perron integral, with the same value, and Aleksandroff (1924a, b) and Looman (1925) proved independently that if the Perron integral exists, so does the special Denjoy integral. Thus the two integrals are equivalent, see Saks (1937, pp. 247–52). Then Tolstov (1940) gave a Perron-type integral equivalent to the general Denjoy integral.

A special Denjoy–Stieltjes integral can be defined if the integrator g is fairly smooth, and in particular if g is continuous and monotone; see Henstock (1960c). A Perron–Stieltjes integral can also be defined when g is strictly increasing. Otherwise definitions tend to be difficult. This led Ward (1936a) to replace derivatives by increments to define the Ward integral, equivalent to the Perron–Stieltjes and special Denjoy–Stieltjes integrals when they are defined. The variational integral, equivalent to the Ward integral, was defined in Henstock (1960a, 1963c).

The next step, travelling back to a Riemann-type integral, was taken independently by Kurzweil (1957, 1973a, b, 1978, 1980) and Henstock (1955b, pp. 277–8, 1963c, 1968a, 1988a). See also Henstock (1961b) which gives a Riemann-type integral for an axiomatic system that developed into the division systems and spaces of later papers and this book, see Henstock (1968b, 1969, 1973a, b, 1974, 1978, 1979, 1980a, b, 1982, 1983, 1988/1989).

For Denjoy-type integrals, more general derivatives were used in Hinčin (1916, 1917, 1918), Denjoy (1917), J. C. Burkill (1923), Looman (1923), Saks (1923), Ridder (1931b, 1932, 1933a, b, 1934a, b, 1935a), Izumi (1933, 1935), Kempisty (1934, 1936a), Krzyzański (1934), Verblunsky (1934), Jeffery (1939), Romanovski (1941a), and H. W. Ellis (1949). For Perron-type integrals see J. C. Burkill (1931, 1932, 1935, 1936b, 1951a); some of these can integrate all sum functions of convergent trigonometric series. So can the MZ-integral of Marcinkiewicz and Zygmund (1936), which exists when J. C. Burkill's (1951a) *symmetric Çesàro–Perron (SCP) integral* exists. But Skvorcov (1972) contradicts Bullen and Lee (1973a, p. 499) as he says that the MZ-integral can exist sometimes when the SCP-integral does not. These integrals cannot integrate all Abel sums of Abel-convergent trigonometric series, and the *Abel–Poisson–Perron integral* of Taylor (1955) was constructed for this purpose. These and many more integrals that use a convergence-factor to smooth the integrand's oscillation, and with a length function as integrator, were put on a general foundation by Jeffery and Miller (1945), and Bullen and Lee (1973a) went further. Between the two papers, Henstock (1960b) combined and generalized the approaches of Ward, and Jeffery and Miller, to give the N-integral, with the equivalent N-variational integral in Henstock (1960c). Being of Stieltjes type, the Ward, N-, and N-variational integrals cannot be included in Bullen and Lee's system, nor in Jeffery and Miller's system, while the N- and N-variational integrals include the other two. The N- and N-variational integrals are put into decomposable division spaces in Section 7.3 in the present book.

Almost every integral that does not use a convergence factor in its definition is equivalent to a generalized Riemann integral over a division system or space or a similar structure. Henstock (1968b, pp. 219–25) gave several examples, and these and many more are given in Chapter 1 here. This causes a great economy in proving and displaying the theory. One no longer has to prove a result for one integral, and then when another integral of the same kind is considered, to prove the result again for the new integral. One proof normally suffices for all integrals of the particular system, and usually the greater generality gives a deeper insight into the mathematics involved. The generalized Riemann and variational integration theory has developed considerably since the original papers. The book, Henstock (1963c), now out of print, can now be replaced in the main by Henstock (1988a), points not covered there being covered here. Many have written on the basic theory and have then developed other parts, and I hope to include as much as possible in the present book. Applications of the absolute part of the theory are given in McShane (1983), and for the moment I will not go over any part of the list of applications unless more of interest can be written. Neither will I deal with any non-standard theory using infinitesimals, such as can be found in Benninghofen (1984), Foglio (1985), and Mawhin (1986). Nor will any systematic theory of the trigonometric or other orthogonal series be given at

0.6 SHORT HISTORY OF INTEGRATION

present. There is enough theory involved with the generalized Riemann and variational and N- and N-variational integrations, with functional analytic questions, to fill the present medium-sized book.

Finally, this section shows that integration theory was not produced by a few individuals, but, on the contrary, it has been a co-operative effort of many people over a century (see the dates given). This section is an update of Henstock (1988*b*).

CHAPTER 1

DIVISION SYSTEMS AND DIVISION SPACES

1.1 DEFINITIONS

Generalized Riemann and variational integration theory connect a base space T of points of some kind, a space K of values, and three structures, (a) division systems and division spaces in T, (b) the algebra of K, (c) a topology or an order in K to define limits. As (a) is independent of (b) and (c) together, we can consider many combinations of the first structure with the other two. Two alternative theories of (a) differ trivially; here we begin with the theory using disjoint intervals, mentioning the one using non-overlapping intervals at the end of this section. The space K is the real line, the complex plane, or a Banach or more general space. Many integration processes have been devised for a variety of reasons. Almost every such process that does not use a convergence factor (a smoothing device) in its definition, can be given by a suitable division system or space—this is the point of the theory, as mentioned at the end of Section 0.6 on a partial history of integration theory.

In the non-empty base space T of points of some kind we use a non-empty collection $\mathcal{T}$ of some non-empty subsets I of T called *(generalized) intervals*. These I are our 'building bricks', like intervals and rectangles in Euclidean space. But we are not restricted to Euclidean-type spaces, for see Section 1.11 that deals with a general topological T_3-space in which generalized intervals are compact set differences that have interior points.

In integration theory we often associate points $t \in T$ with intervals $I \in \mathcal{T}$, using products $f(t)\mu(I)$, for example. Here we use a fixed family $\mathcal{U}^1$ of some interval-point pairs (I, t) ($I \in \mathcal{T}$, $t \in T$), saying that t is an *associated point of I* if $(I, t) \in \mathcal{U}^1$. In Hellinger (1907, 1909), J. C. Burkill (1924a, b, c), and Çesari (1962) the t are omitted. Burkill integration was the first process that embraced many integrations in one method, but by including the t instead of omitting it, we are able to deal with many more integration processes. Usually there are many associated points t of each $I \in \mathcal{T}$, and to each $t \in T$ there correspond many $I \in \mathcal{T}$ with t as associated point. Sometimes the t lie in one space, the I in another, with connections between the two; or several t could replace the single t, as in the *weighted refinement integral* of Wright and Baker (1969). The theory then proceeds up to the decomposability, where proofs for a collection of t might fail. For simplicity, here we keep to the single t in (I, t), with both t and I in T, noting that results in Sections 2.1, 2.3, 2.5, 2.7,

1.1 DEFINITIONS

and possibly other results, also apply to integrations such as the weighted refinement integral.

We use non-empty families **A** of some non-empty subsets $\mathcal{U} \subseteq \mathcal{U}^1$; these **A** are the vital components in the theory, differences between the families **A** being reflected in the differences between the corresponding integrals. See the other sections of this chapter.

We begin with the weakest basis for an integration theory. $(T, \mathcal{T}, \mathbf{A})$ is a *free division system* if **A** is directed downwards in the sense of set inclusion, i.e. given $\mathcal{U}_j \in \mathbf{A}$ $(j = 1, 2)$, there is a $\mathcal{U} \in \mathbf{A}$ (hence non-empty $\mathcal{U}$) with $\mathcal{U} \subseteq \mathcal{U}_1 \cap \mathcal{U}_2$. The convergence is a Moore–Smith convergence along this downwards direction, colloquially called 'as $\mathcal{U}$ shrinks', and this arrangement is sufficient for variational integration.

Next, for $X \subseteq T$ and $\mathcal{U} \subseteq \mathcal{U}^1$ we define

$$\mathcal{U}[X] \equiv \{(I, t): (I, t) \in \mathcal{U}, t \in X\}.$$

$(T, \mathcal{T}, \mathbf{A})$ is *freely fully decomposable* (respectively, *freely decomposable*, or *freely measurably decomposable* relative to a measure or measure space defined later) if to every family (respectively, countable family, or countable family of measurable sets) $\mathcal{X}$ of mutually disjoint subsets $X \subseteq T$, and every function $\mathcal{U}(\cdot): \mathcal{X} \to \mathbf{A}$, there is a $\mathcal{U} \in \mathbf{A}$ with

(1.1.1) $\qquad \mathcal{U}[X] \subseteq \mathcal{U}(X) \quad (X \in \mathcal{X}).$

There is no need for the union of the $X \in \mathcal{X}$ to be T. If, for the given X,

$$\mathcal{U}[X] = \mathcal{U}(X)[X] \quad (X \in \mathcal{X})$$

we call $\mathcal{U}$ the *diagonal* of the $(\mathcal{U}(X), \mathcal{X})$.

To construct the generalized Riemann integral we need divisions of fixed sets in T, and so we move beyond the free division system. First, an *elementary set* E is a non-empty set that is an interval or a union of a finite number of mutually disjoint intervals. A subfamily $\mathcal{U} \subseteq \mathcal{U}^1$ *divides* E, if, for a finite subfamily $\mathscr{E} \subseteq \mathcal{U}$, called a *division of E from $\mathcal{U}$*, the $(I, t) \in \mathscr{E}$ have mutually disjoint I with union E. The corresponding collection of the I alone is called a *partition* $\mathscr{P}$ of E. These I are called *partial intervals of E from $\mathcal{U}$, $\mathscr{E}$, and $\mathscr{P}$*. A non-empty subset of $\mathscr{E}$, including $\mathscr{E}$ itself, is called a *partial division of E from $\mathcal{U}$ and $\mathscr{E}$*, and the union of I from the partial division is called a *partial set* P of E that comes from $\mathscr{E}$ and $\mathcal{U}$, while P is called *proper* if also $P \neq E$. If there are partial sets $P_1, \ldots, P_n$ from the same $\mathscr{E}$, the P_j are called *co-partitional* and their union is also a partial set. The subfamily $\mathcal{U}$ is called *infinitely divisible* if each elementary set divided by $\mathcal{U}$ contains a proper partial set also divided by $\mathcal{U}$. A division $\mathscr{E}_0$ of E is called a *refinement of* (or, *refines*) a division $\mathscr{E}$ of E, if there is a division of each I with $(I, t) \in \mathscr{E}$, denoted by $\mathscr{E}_0.I$, formed of those $(J, x) \in \mathscr{E}_0$ with $J \subseteq I$. We write $\mathscr{E}_0 \leq \mathscr{E}$ since $J \subseteq I$.

Just one choice of t occurs in $\mathscr{E}$ for each I with $(I, t) \in \mathscr{E}$; two or more t with the same I would give non-disjoint I. (This is a trivial place where the arrangement for the weighted refinement integral breaks down, and the point can be disregarded.) But a t might have several I with $(I, t) \in \mathscr{E}$. The concept of a partition is near to the concept of a compact set covered by a finite number of open sets; the intrinsic topology of Section 2.9 develops this. But open sets usually overlap, while intervals of a partition are disjoint, making some proofs more difficult than those in topology. $\mathscr{E}$ is a finite collection, which is sometimes crucial in proofs.

Let $\mathscr{U}.E$ be the set of all $(I, t) \in \mathscr{U}$ with I a partial interval of E, and $\mathbf{A}|E$ the set of all $\mathscr{U}.E$ dividing E with $\mathscr{U} \in \mathbf{A}$. The notation $\mathbf{A}|E$ is borrowed from number theory, where for integers a, b, $a|b$ denotes that a divides b. If $\mathbf{A}|E$ is not empty we say that $\mathbf{A}$ *divides* E. Also we say that $\mathbf{A}$ is *directed for divisions of E* if, given $\mathscr{U}_1$ and $\mathscr{U}_2$ in $\mathbf{A}|E$, there is a $\mathscr{U}_3 \in \mathbf{A}|E$ with $\mathscr{U}_3 \subseteq \mathscr{U}_1 \cap \mathscr{U}_2$; here this is the direction 'as $\mathscr{U}$ shrinks'. If $\mathbf{A}$ is directed for divisions of E, and divides E, we call $(T, \mathscr{T}, \mathbf{A})$ a *division system for* E. This is the minimal arrangement of conditions for defining the generalized Riemann integral over E.

Intuitively, if the integral exists over E, it ought to exist over every partial set P of E, in order to have great use. To this end, a *restriction of $\mathscr{U}$ to P* is a family $\mathscr{U}_1 \subseteq \mathscr{U}.P$. The family $\mathbf{A}$ *has the restriction property* if, for each elementary set E, each partial set P, and each $\mathscr{U} \in \mathbf{A}|E$, there is in $\mathbf{A}|P$ a restriction of $\mathscr{U}$ to P. If this holds, with $(T, \mathscr{T}, \mathbf{A})$ a division system for all elementary sets, and if $P_1, \ldots, P_n$ are co-partitional whenever the P_j are partial sets of the same E, we call $(T, \mathscr{T}, \mathbf{A})$ *a division space*. We need the P_j co-partitional to avoid the pathology of Example 2.3.7, and the condition also turns out to be the condition needed in order to construct an additive division space (see below) from a division space.

The integral for a division space has many useful properties, but does not always behave well for integrability over the union of two disjoint elementary sets, given the integrability over the separate sets. To ensure good behaviour we say that $\mathbf{A}$ is *additive*, if, given disjoint elementary sets E_j and $\mathscr{U}_j \in \mathbf{A}|E_j$ ($j = 1, 2$), there is a $\mathscr{U} \in \mathbf{A}|E_1 \cup E_2$ with $\mathscr{U} \subseteq \mathscr{U}_1 \cup \mathscr{U}_2$. If $\mathbf{A}$ is additive in a division space $(T, \mathscr{T}, \mathbf{A})$, we call it an *additive division space*. This concept works well in Euclidean-type and similar spaces, such as Cartesian products of the real line. But in topological T_3-spaces in which 'intervals' are compact set differences with non-empty interior, it has not been obvious till recently how to have an additive $\mathbf{A}$. See the end of Section 2.7, which uses an old construction.

More definitions occur out of the mainstream. In a division space the link between divisions $\mathscr{E}$ over E and divisions over the separate I with $(I, t) \in \mathscr{E}$, need only go one way, from E to the I. We need the opposite way to prove results for E when results for the I are known. We say that the space is *linked* if two more properties hold. First, for every partial set P of E and every $\mathscr{U} \in \mathbf{A}|P$, there is a $\mathscr{U}_1 \in \mathbf{A}|E$ with $\mathscr{U}_1.P \subseteq \mathscr{U}$. Secondly, if I is a partial interval of E, P a partial set of E, and $I \subseteq P$, then I is a partial interval of P.

1.1 DEFINITIONS

Note that if P is a proper partial set of E, then so is $E\setminus P$. Thus if $\mathcal{U}_2 \in \mathbf{A}|E\setminus P$ there is a $\mathcal{U}_3 \in \mathbf{A}|E$ with $\mathcal{U}_3.(E\setminus P) \subseteq \mathcal{U}_2$, when the space is linked. If $\mathcal{U}_4 \in \mathbf{A}|E$ with $\mathcal{U}_4 \subseteq \mathcal{U}_1 \cap \mathcal{U}_3$ then $\mathcal{U}_4.P \subseteq \mathcal{U}$, $\mathcal{U}_4.(E\setminus P) \subseteq \mathcal{U}_2$. But some $(I, t) \in \mathcal{U}_4$ might not be in $(\mathcal{U}_4.P) \cup (\mathcal{U}_4.(E\setminus P))$ and $\mathbf{A}$ need not be additive.

Alternatively we use star-sets, the complements of open sets of the intrinsic topology of Section 2.9. For E an elementary set let $E^*_{\mathcal{U}}$ be the set of all t with $(I, t) \in \mathcal{U}.E$ for some $I \in \mathcal{T}$. Then the *star-set* E^* is the intersection of $E^*_{\mathcal{U}}$ for all $\mathcal{U} \in \mathbf{A}|E$. For P a proper partial set of E, the frontier star-set $\mathcal{F}(E; P)$ is $P^* \cap (E\setminus P)^*$. If there is a $\mathcal{U}_E \in \mathbf{A}|E$ such that every $\mathcal{U} \in \mathbf{A}|E$ with $\mathcal{U} \subseteq \mathcal{U}_E$ has $E^*_{\mathcal{U}} = E^*$, we say that $(T, \mathcal{T}, \mathbf{A})$ *is stable for* E. If true for all elementary sets E we say that $(T, \mathcal{T}, \mathbf{A})$ *is stable*.

For $\mathcal{T}_0$ a special non-empty subset of $\mathcal{T}$, e.g. $\mathcal{T}_0 = \mathcal{T}$, if E is an elementary set and P a proper partial set of E from $\mathcal{T}_0$, i.e. the union of a finite number of disjoint partial intervals of E from $\mathcal{T}_0$, we say that the division space $(T, \mathcal{T}, \mathbf{A})$ is *weakly $\mathcal{T}_0$-compatible with P in E* when there is a $\mathcal{U}_P \in \mathbf{A}|E$ such that $(I, t) \in \mathcal{U}_P$ and $t \notin P^*$ imply $I \subseteq E\setminus P$, and I, P are co-partitional. If also $I^* \subseteq E^*\setminus P^*$ for some $\mathcal{U}_P \in \mathbf{A}|E$, we say that $(T, \mathcal{T}, \mathbf{A})$ is *strongly $\mathcal{T}_0$-compatible with P in E*. More details on star-sets and the intrinsic topology are in Section 2.9.

For $\mathcal{G}$ a topology in T, $(T, \mathcal{T}, \mathbf{A})$ *is compatible with $\mathcal{G}$* if for each $G \in \mathcal{G}$ and each elementary set E, there is a $\mathcal{U} \in \mathbf{A}|E$ such that $I \subseteq G \cap E$ when $(I, t) \in \mathcal{U}$ with $t \in G$.

Returning to mainstream definitions, a division space such as in Section 1.2 for Riemann integration gives enough structure for Riemann-type results, as will be seen. But Riemann integration has no Lebesgue-type limit theorems, so that the latter theorems need more structure, and we use the decomposability in free division systems, modified for divisions of elementary sets. Thus $(T, \mathcal{T}, \mathbf{A})$ *is fully decomposable* (respectively, *decomposable*, or *measurably decomposable relative to a measure or measure space* defined later) if to every family (respectively, countable family, or countable family of measurable sets) $\mathcal{X}$ of mutually disjoint subsets $X \subseteq T$, every elementary set E, and every function $\mathcal{U}(.) : \mathcal{X} \to \mathbf{A}|E$, there is a $\mathcal{U} \in \mathbf{A}|E$ with $\mathcal{U}[X] \subseteq \mathcal{U}(X) (X \in \mathcal{X})$. There is no need for the union of the $X \in \mathcal{X}$ to be T or E^*. Again we have the definition of the *diagonal* of the $(\mathcal{U}(X), \mathcal{X})$.

In the definition of stability for E we note that for every $t \notin E^*$ there is a $\mathcal{U} = \mathcal{U}(t) \in \mathbf{A}|E$ with $t \notin E^*_{\mathcal{U}*}$. If $(T, \mathcal{T}, \mathbf{A})$ is fully decomposable there is a $\mathcal{U} \in \mathbf{A}|E$, independent of t, that lies in the diagonal of the $\mathcal{U}(t)$, and $\mathcal{U}$ can serve as the $\mathcal{U}_E$ in the definition of stability. Thus:

(1.1.2) *If a division system for E is fully decomposable, it is also stable.*

Turning to product spaces, let $T^z = T^x \otimes T^y$, the Cartesian product space of the spaces T^x and T^y, and let $\mathcal{T}^u$ be the family of u-intervals $I^u \subseteq T^u$, with $\mathcal{U}^{1u}$ the corresponding family of (I^u, u), all for $u = x, y$. Then $\mathcal{T}^z$, $\mathcal{U}^{1z}$ are taken to be the respective families of all $I^z = I^x \otimes I^y$ and all $(I^z, z) = (I^x \otimes I^y, (x, y))$, written $(I^x, x) \otimes (I^y, y)$, for all $I^u \in \mathcal{T}^u$ and all $(I^u, u) \in \mathcal{U}^{1u}$ ($u = x, y$). Let $\mathbf{A}^u$ be a family of subsets of $\mathcal{U}^{1u}$, for $u = x, y, z$. Then we say that $\mathbf{A}^x, \mathbf{A}^y, \mathbf{A}^z$ *have the Fubini property in common*, if two properties hold. For E^u an arbitrary elementary set in T^u ($u = x, y$) and $E^z = E^x \otimes E^y$, with $\mathcal{U}^z$ an arbitrary member of $\mathbf{A}^z | E^z$, then to each $x \in (E^x)^*$ there is a $\mathcal{U}^y(x) \in \mathbf{A}^y | E^y$, and to each collection of divisions $\mathcal{E}^y(x)$ of E^y from $\mathcal{U}^y(x)$, one division for each such x, there corresponds a $\mathcal{U}^x \in \mathbf{A}^x | E^x$ such that if $(I^x, x) \in \mathcal{U}^x$, $(I^y, y) \in \mathcal{E}^y(x)$, then $(I^x, x) \otimes (I^y, y) \in \mathcal{U}^z$. As this property is unsymmetrical in x, y, we usually assume also the property in which x and y are interchanged, keeping the product space as $T^x \otimes T^y$. If the property is true only one way, we can say that $\mathbf{A}^x, \mathbf{A}^y, \mathbf{A}^z$ *have an unsymmetrical Fubini property in common*. If $\mathbf{A}^u$ ($u = x, y, z$) have the Fubini property in common and if $(T^z, \mathcal{T}^z, \mathbf{A}^z)$ is a division system (space) we call $(T^z, \mathcal{T}^z, \mathbf{A}^z)$ a *Fubini division system (space)*.

Note the use of star-sets to locate points, and note that the two division systems (spaces) in x, y could be totally different.

To obtain a suitable $(T^z, \mathcal{T}^z, \mathbf{A}^z)$ from the separate $(T^u, \mathcal{T}^u, \mathbf{A}^u)$ ($u = x, y$) we construct $\mathbf{A}^z$. Let E^u be an elementary set in T^u and, for each $z \in T^z$ let $\mathcal{U}^u(z) \in \mathbf{A}^u | E^u$ ($u = x, y$). Let $\mathcal{U}$ be the family of all $(I^x, x) \otimes (I^y, y)$ with I^u a partial interval of E^u and $(I^u, u) \in \mathcal{U}^u(z)$ ($u = x, y$). Let $\mathbf{A}^z$ be the family of finite unions of such $\mathcal{U}^z$ for all finite unions of disjoint products $E^x \otimes E^y$. Then we call $(T^z, \mathcal{T}^z, \mathbf{A}^z)$ the *product division system (space) of the* $(T^u, \mathcal{T}^u, \mathbf{A}^u)$ ($u = x, y$).

To define the product division system (space) that uses a finite set C of more than two coordinate variables we define $T^C, \mathcal{T}^C, \mathcal{U}^{1C}$ as the respective Cartesian products of $T^u, \mathcal{T}^u, \mathcal{U}^{1u}$ ($u \in C$) and then define $\mathbf{A}^C$ in the following way. Let E^u be an elementary set in T^u and, for each $t \in T^C$, let $\mathcal{U}^u(t) \in \mathbf{A}^u | E^u$ ($u \in C$). Let $\mathbf{A}^C$ be the family of finite unions of such $\mathcal{U}^C$ for all finite unions of disjoint products $\bigotimes_C E^u$. Then $(T^C, \mathcal{T}^C, \mathbf{A}^C)$ is the *product division system (space) of the* $(T^u, \mathcal{T}^u, \mathbf{A}^u)$. Product division systems (spaces) with a finite number of coordinate variables are studied in Chapter 5, while Chapter 6 deals with infinite-dimensional spaces, in which we need further definitions.

Thus, for each b of an index set B let us have a stable fully decomposable division space $(T(b), \mathcal{T}(b), \mathbf{A}(b))$. For the Cartesian product T of $T(b)$ ($b \in B$) we define a suitable $(T, \mathcal{T}, \mathbf{A})$. For each $C \subseteq B$ let $\prod(X(b); C)$ be the Cartesian product of $X(b)$ where

$$X(b) \subseteq T(b) \, (b \in C), \qquad X(b) = T(b) \, (b \in B \setminus C).$$

Let $X(C)$ be the Cartesian product of $X(b)$ for $b \in C$ alone, so that $T(C)$ comes from $T(b)$. We take the intervals of T to be $I = \prod(I(b); C)$ for all finite sets $C \subseteq B$ and all $I(b) \in \mathcal{T}(b)$ ($b \in C$), while $\mathcal{U}^1(C), \mathcal{U}^1$ are the Cartesian products of $(I(b), f(b))$ for all $b \in C$, all $b \in B$, respectively. To construct suitable families $\mathcal{U}$ that divide elementary sets E, we associate a finite set $C_1(\mathcal{U}, f) \subseteq B$ with

1.1 DEFINITIONS
45

each f with $f(b) \in T(b)^* (b \in B)$ and a $\mathcal{U}_1(C; \mathcal{U}; f) \in \mathbf{A}(C)$ for each finite $C \subseteq B$, the $(T(C), \mathcal{T}(C), \mathbf{A}(C))$ being as constructed previously. Then $\mathcal{U}$ is the family of all (I, f) with

(1.1.3) $\quad\quad I = \prod (I(b); C) \subseteq E, C \quad \text{finite}, \quad C_1(\mathcal{U}; f) \subseteq C \subseteq B;$

(1.1.4) $\quad\quad\quad\quad\quad \bigotimes_C (I(b), f(b)) \in \mathcal{U}_1(C; \mathcal{U}; f);$

(1.1.5) $\quad\quad\quad\quad\quad\quad f(b) \in T(b)^* (b \in B \setminus C).$

For $\mathbf{A}$ the family of all such $\mathcal{U}$, $(T, \mathcal{T}, \mathbf{A})$ is the *product division space of* $(T(b), \mathcal{T}(b), \mathbf{A}(b)) (b \in B)$.

Note that there may not be any order in the index set B, in which case the $b \in B$ can be rearranged without affecting the construction. For example, B can be the set of positive integers, so that we are dealing with a sequence space, or the positive real axis, so that we are dealing with a function space and, for example, the functions could be random functions in a stochastic process. Here B has an order. Or B could be the complex plane, with no reasonable order in the sense we need.

In the alternative theory of divisions, two intervals are *non-overlapping* if no $I \in \mathcal{T}$ lies in both. An *elementary set* E is an interval or a union of a finite number of mutually non-overlapping intervals. Two elementary sets are *non-overlapping* if no $I \in \mathcal{T}$ lies in both. A *division* of an elementary set E from $\mathcal{U} \subseteq \mathcal{U}^1$, is a finite subset $\mathcal{E}$ of $\mathcal{U}$, such that the I from the $(I, t) \in \mathcal{E}$ are mutually non-overlapping with union E. Other definitions repeat the previous pattern, replacing 'disjoint' by 'non-overlapping', except for the definition of an additive $\mathbf{A}$. Here, for an arbitrary positive integer n we take mutually non-overlapping elementary sets $E_1, \ldots, E_n$, to avoid the pathology of Example 1.1.1. Another pathology is shown in Example 1.1.2.

The non-additive division spaces and division spaces of the original theory are now written as division spaces and additive divisions spaces, respectively. B. S. Thomson (1971a, b, 1972a, b) examined the early theory and showed that variational integration only needed what he called 'division systems' and what in this book I have called 'free division systems', reserving the name 'division system' for the arrangement where divisions are over some fixed elementary set. Thomson also pointed out that for divisions of E only the (I, t) were used with $I \subseteq E$, and all other (I, t) could be forgotten. Thus I borrowed a number theory notation and defined the $\mathbf{A}|E$. Originally I used E^* when $(T, \mathcal{T}, \mathbf{A})$ was stable, but Thomson pointed out that E^* could be defined always. For such clarifications I am very grateful.

Example 1.1.1 In Euclidean two-dimensional space let 'intervals' be closed rectangles with sides parallel to two given axes at right angles, and also be segments of two lines at right angles, the segments having a common endpoint, just like the letter L in various orientations. If a rectangular interval is bisected twice by two lines at right angles, giving rectangular intervals

$I_1, \ldots, I_4$, then I_1 does not overlap with the other three, but I_1 has an L in common with the union of the other three. Thus an elementary set E can be non-overlapping with each of the mutually nonoverlapping $E_1, \ldots, E_n$ while E overlaps with the union.

Example 1.1.2 On the real line let $a < b < c$ and let $[a, b]$, $[a, b)$, $[b, c]$, and $(b, c]$, be four 'intervals' with

$$[a, b) \cup [b, c] = [a, c] = [a, b] \cup (b, c] = [a, b] \cup [b, c],$$

so that $[b, c]$ is a partial set of $[a, c]$. But $(b, c]$ does not form a partition of $[b, c]$, and

$$[a, b) \cup (b, c] \neq [a, c].$$

Example 1.1.3 Thomson (1971a, b, c, 1972a, b, 1975) altered the theory, obtaining a less general but possibly more simple arrangement. His division system $(T, \mathcal{T}, \mathbf{A})$ satisfies: given $\mathcal{U}_1, \mathcal{U}_2 \in \mathbf{A}$, there is a $\mathcal{U}_3 \in \mathbf{A}$ with $\mathcal{U}_3 \subseteq \mathcal{U}_1 \cap \mathcal{U}_2$, and $(\varnothing, t) \in \mathcal{U}$ for all $\mathcal{U} \in \mathbf{A}$, all $t \in T$, where $\varnothing$ is the empty set. It is a division space if also every $\mathcal{U} \in \mathbf{A}$ divides every elementary set. Instead of additivity, Thomson uses the set

$$\mathcal{U}(X) = \{(I, t) : (I, t) \in \mathcal{U}, I \subseteq X\}$$

and assumes that for every $\mathcal{U} \in \mathbf{A}$ and every elementary set E with complement $T \setminus E = \setminus E$, there is a $\mathcal{U}_1 \in \mathbf{A}$ with $\mathcal{U}_1 \subseteq \mathcal{U}(E) \cup \mathcal{U}(\setminus E)$. This is very near to the property of additivity, which can use a division instead of E and $\setminus E$. Many needs are satisfied, but it cannot deal with situations in which divisions are taken in several directions simultaneously, as in rotations of a plane. For example, as in Henstock (1973b), pp. 320, 321, let $\mathcal{T}$ be the set of all finite rectangles on the x, y plane with sides parallel to the axes, and all finite rectangles on the same plane with sides at angles of $45°$ to the axes, and let $(T, \mathcal{T}, \mathbf{A})$ satisfy Thomson's conditions. For S the square with centre the origin, sides of unit length and parallel to the axes, and Q a square with one side a diagonal d of S, and a $\mathcal{U} \in \mathbf{A}$, let $\mathcal{U}_1, \mathcal{U}_2 \in \mathbf{A}$ satisfy

$$\mathcal{U}_1 \subseteq \mathcal{U}(S) \cup \mathcal{U}(\setminus S), \quad \mathcal{U}_2 \subseteq \mathcal{U}_1(Q) \cup \mathcal{U}_1(\setminus Q).$$

Then each rectangle I with $(I, (x, y)) \in \mathcal{U}_2$, is in one of $S \cap Q$, $S \setminus Q$, $Q \setminus S$, $\setminus(Q \cup S)$. If $(I, (x, y))$ is in a division $\mathcal{E}$ of S from $\mathcal{U}_2$, the sides of I have to be parallel to the axes, or else we would need triangles with points to complete $\mathcal{E}$, and triangles are not in $\mathcal{T}$. The union of such I from $\mathcal{E}$ is S and so includes the diagonal d. As I is in $S \cap Q$ or $S \setminus Q$, $I \cap d$ is empty or a single point. Hence an at most finite number of points of d is covered by those I. By this contradiction $\mathcal{U}_2$ cannot divide S and the example cannot satisfy Thomson's conditions. This example follows since each $\mathcal{U} \in \mathbf{A}$ has to divide each elementary set.

Example 1.1.4 McShane (1969) generalized Henstock (1968a) to produce a division space included in the arrangement of this book. His partition of a set $Y \subseteq T$ is a collection of mutually disjoint intervals $I_j (1 \leq j \leq k)$ whose union contains Y, each I_j being associated with a point t_j. Our $\mathscr{E}$ corresponds to the collections of $(I_j \cap Y, t_j)$, and $\mathscr{T}$ is such that if $J_1, J_2 \in T$ then $J_1 \cap J_2 \in \mathscr{T}$ or is empty. For a non-empty set Δ of labels and each $\delta \in \Delta$, $P(\delta)$ is a set of $(I, t)(I \in \mathscr{T}, t \in T)$ with the property that if $\delta_1, \delta_2 \in \Delta$, there is a $\delta_3 \in \Delta$ such that

$$P(\delta_3) \subseteq P(\delta_1) \cap P(\delta_2).$$

Thus the **A** constructed from the $P(\delta)(\delta \in \Delta)$ is directed for divisions. As the empty set is not in $\mathscr{T}$, McShane (1969, p. 7, (3.1)(v)(vi)(a)) makes (p. 7, (iv)(b)) unnecessary. To go from intervals Y to subintervals, McShane (1969) used p. 11, (4.4), so that **A** has the restriction property, and $(T, \mathscr{T}, \mathbf{A})$ is a division space. McShane integrates functions $h(I \cap Y, t)$ that are finitely additive in Y, to compensate for the lack of additivity in **A**.. Property McShane (1969, p. 20, (7.6)(v)) is a restricted form of decomposability. Thus his system is included in the arrangements of this book. He gives a simpler version of the theory in McShane (1973).

Some of the following sections are useful illustrations of the definitions of the present section and show that many integrals can be defined as division space integrals. Also see Chapter 6, which deals with integration in sequence and function spaces.

1.2 THE NORM INTEGRAL
(HENSTOCK (1968b, pp. 219–20, Ex. 43.1))

Here, intervals are n-dimensional bricks I with edges parallel to the coordinate axes,

$$I = \bigotimes_{j=1}^{n} [a_j, b_j), \bar{I} = \bigotimes_{j=1}^{n} [a_j, b_j], I^0 = \bigotimes_{j=1}^{n} (a_j, b_j) \ (a_j < b_j, j = 1, 2, \ldots, n),$$

$\bar{I}$ is the *closure* and I^0 the *interior* of I. An *edge* is the locus $a_k \leq x_k \leq b_k$ for a fixed k in $1 \leq k \leq n$, with $x_j = a_j$ or b_j (all $j \neq k$), and a *face* is the locus $x_k = a_k$ or b_k with $a_j \leq x_j \leq b_j$ ($j \neq k$). The *vertices* are the $(x_1, \ldots, x_n)$ with $x_j = a_j$ or b_j for each j.

Burkill integration of functions of I does not use associated points, while in Riemann and Riemann–Stieltjes integration the associated points lie arbitrarily in $\bar{I}$. If f, g are functions of $t = (t_1, \ldots t_n)$, the brick-point functions are $f(t)\Delta_1 \ldots \Delta_n g$,

$$\Delta_j g \equiv g(x_1, \ldots, x_{j-1}, b_j, x_{j+1}, \ldots, x_n) - g(x_1, \ldots, x_{j-1}, a_j, x_{j+1}, \ldots, x_n)$$

The norm of I, norm (I), is either

$$N_1 = \max_{1 \leq j \leq n} (b_j - a_j) \text{ or } N_2 = \left\{ \sum_{j=1}^{n} (b_j - a_j)^2 \right\}^{1/2},$$

the same function for each I. As $N_1 < N_2 \leq n^{1/2} N_1$, either can be used for the norm. Then the *norm of a division* $\mathscr{E}$ is the greatest norm(I) for $(I, t) \in \mathscr{E}$, and we have a limit process as the norm tends to 0. Thus for Riemann, Riemann–Stieltjes, Hellinger, and Burkill integration we include in $\mathbf{A}$ those $\mathscr{U} \subseteq \mathscr{U}^1$ for which there are some elementary set E, some $\varepsilon > 0$, with $(I, t) \in \mathscr{U}$ if and only if $I \subseteq E$, $t \in \bar{I}$, and norm$(I) < \varepsilon$. The family $\mathbf{A}|E$ is not empty, for there are disjoint bricks with union E and we can bisect each such brick in the direction of the $(n - 1)$-dimensional plane of all axes but that of x_j, and do this for $j = 1, 2, \ldots, n$, giving 2^n smaller bricks, and repeat till the norms are less than ε. The family $\mathbf{A}$ is directed for divisions of E, for if $\mathscr{U}_j$ has $\varepsilon = \varepsilon_j > 0$ ($j = 1, 2$), we take the $\mathscr{U}_3$ defined by $\min(\varepsilon_1, \varepsilon_2) > 0$. If the elementary set $P \subset E$, and if, given $\varepsilon > 0$, the faces of the bricks forming a partition of P of norm $< \varepsilon$ are continued to meet the faces of the bricks making up E, we get a finite number of bricks forming a partition of E, some of whom form a partition of P, showing that P is a partial set of E. Similarly, when the disjoint P_1, P_2 are partial sets of E, so is $P_1 \cup P_2$. For $\mathscr{U} . P$ we add $I \subseteq P$ to the other conditions on $\mathscr{U}$, so showing that $\mathscr{U} . P \in \mathbf{A}|P$ when $\mathscr{U} \in \mathbf{A}|E$. Thus $(T, \mathscr{T}, \mathbf{A})$ is a division space. $\mathbf{A}$ is not additive, for if elementary sets E_1, E_2 are disjoint but with closures containing a common $(n - 1)$-dimensional brick B with $(n - 1)$-dimensional interior B^0, there is a brick $I \subseteq E_1 \cup E_2$ with $I \not\subseteq E_j$ ($j = 1, 2$), $I^0 \cap B^0$ not empty, and norm(I) arbitrarily small. For $\mathscr{U}_j \in \mathbf{A}|E_j$ ($j = 1, 2$), $\mathscr{U} \in \mathbf{A}|E_1 \cup E_2$, $t \in \bar{I}$, then $(I, t) \in \mathscr{U}$ for norm(I) small enough, but $(I, t) \notin \mathscr{U}_j$ ($j = 1, 2$). However, the space is linked. For if P is a proper partial set of E and $\varepsilon > 0$ defines $\mathscr{U} \in \mathbf{A}|P$, the same ε defines a $\mathscr{U}_1 \in \mathbf{A}|E$ with $\mathscr{U}_1 . P = \mathscr{U}$. If I is a brick in E and $I \subseteq P$, then I is a partial interval of E and P. Next, if I is fixed and $(I, t) \in \mathscr{U} \in \mathbf{A}|E$, then t ranges over the whole of $\bar{I}$, so that $E^*_\mathscr{U} = \bar{E}$, $E^* = \bar{E}$, and $(T, \mathscr{T}, \mathbf{A})$ is stable. Thus the frontier star-set $\mathscr{F}(E; P) = \bar{P} \cap \overline{E \setminus P}$. Further, $(T, \mathscr{T}, \mathbf{A})$ is not weakly $\mathscr{T}$-compatible with P. For if $t \notin P^*$, then $t \in \setminus \bar{P}$, an open set, and, given $\varepsilon > 0$, t can be nearer to P than ε. Thus if $(I, t) \in \mathscr{U}$ defined by ε, the interior of I can meet P in an n-dimensional brick, and I, P are not disjoint. $(T, \mathscr{T}, \mathbf{A})$ is not decomposable, nor measurably decomposable if closed bricks are measurable in the sense used. For let $(\bar{I}_j)$ be a sequence of mutually disjoint closed bricks in E with norm tending to 0. If

$$0 < \varepsilon_j < \text{norm}(I_j) \quad (j = 1, 2, \ldots)$$

then $\varepsilon_j \to 0$ as $j \to \infty$ and, given $\varepsilon > 0$, an integer j has $\varepsilon_j < \text{norm}(I_j) < \varepsilon$. If $t \in \bar{I}_j$, (I_j, t) is in the $\mathscr{U}$ defined by ε, but not in the $\mathscr{U}_j$ defined by ε_j.

When $n > 1$, $(T, \mathcal{T}, \mathbf{A})$ is a Fubini division space. In fact, if $\mathbf{A}^u$ ($u = x, y, z$) are respectively for m-, n-, and $(m + n)$-dimensional spaces of the foregoing type, they have the Fubini property in common. For if E^u is an arbitrary elementary set in T^u ($u = x, y$) and if $E^z = E^x \otimes E^y$, let $\mathcal{U}^z \in \mathbf{A}^z | E^z$ be defined by $\varepsilon > 0$, with $\mathcal{U}^y(x) \in \mathbf{A}^y | E^y$ defined by $2^{-1/2}\varepsilon$, and a division $\mathcal{E}^y(x)$ of E^y-from $\mathcal{U}^y(x)$. If norm$(I^x) < 2^{-1/2}\varepsilon$, $x \in \bar{I}^x$, $(I^y, y) \in \mathcal{E}^y(x)$, then norm$(I^x \otimes I^y) < \varepsilon$, using either N_1 or N_2, and so $(I^x, x) \otimes (I^y, y) \in \mathcal{U}^z$. Interchanging x, y, we have the proof of the other Fubini property.

Several examples are contained in the book, Henstock (1988a), Sections 1, 2, pp. 1–29, and many can be found in books on Riemann integration.

1.3 THE REFINEMENT INTEGRAL
(HENSTOCK (1968b, p. 220, Ex. 43.2))

For T a set of points and $\mathcal{T}$ a finitely additive family (i.e. if $I_1, I_2 \in \mathcal{T}$, then $I_1 \cup I_2 \in \mathcal{T}$ when I_1, I_2 are disjoint, and otherwise $I_1 \cap I_2 \in \mathcal{T}$) of non-empty subsets, called *intervals*, of $\mathcal{T}$, so that elementary sets are intervals, let $\mathbf{A}$ be the collection of exactly those $\mathcal{U}$ for which there are an interval I and a partition $\mathcal{P}$ of I, such that all partitions $\mathcal{P}'$ of I with $\mathcal{P}' \leq \mathcal{P}$ (i.e. to each $J \in \mathcal{P}'$ there corresponds an $I \in \mathcal{P}$ with $J \subseteq I$) come from $\mathcal{U}$. This leads to Moore–Pollard (refinement) integration, Moore (1900, 1915, 1939), Moore and Smith (1922), and Pollard (1923). Clearly $\mathbf{A} | I$ is not empty if $I \in \mathcal{T}$. Using (J, t) with $J \subseteq I$ and $t \in J$, or $t \in \bar{J}$ for a reasonable $\bar{J}$, then $\mathbf{A}$ is directed for divisions of I. For if $\mathcal{P}, \mathcal{P}'$ are partitions of I with $\mathcal{P}''$ the family of all non-empty $J \cap J'$ with $J \in \mathcal{P}$, $J' \in \mathcal{P}'$, then $J \cap J' \in \mathcal{T}$ and $\mathcal{P}''$ is a partition of I with $\mathcal{P}'' \leq \mathcal{P}$, $\mathcal{P}'' \leq \mathcal{P}'$. Each partial set P of I is in $\mathcal{T}$, and if $P \neq I$, so is $I \setminus P$, and $\mathbf{A}$ has the restriction property. If I_1, I_2 are disjoint with $\mathcal{U}_j$ defined by a partition $\mathcal{P}_j$ of I_j ($j = 1, 2$), we define $\mathcal{U}$ for $I_1 \cup I_2$ by the partition $\mathcal{P}_1 \cup \mathcal{P}_2$. Thus $\mathbf{A}$ is additive and $(T, \mathcal{T}, \mathbf{A})$ is an *additive division space*.

For T a Cartesian product of two such sets T^u ($u = x, y$) we define $\mathcal{T}$ to be the set of Cartesian products of intervals. Assuming where necessary that $(T^u, \mathcal{T}^u, \mathbf{A}^u)$ is stable ($u = x, y$) we follow the pattern of this example to define $\mathbf{A}$, and we have that $(T, \mathcal{T}, \mathbf{A})$ is an additive Fubini division space. But in general $(T, \mathcal{T}, \mathbf{A})$ is not decomposable nor measurably decomposable, particularly when the following construction can take place.

Let $J \in \mathcal{T}$, let (J_j) be a sequence of mutually disjoint intervals in J with union J and with $J_j^* \subseteq J_j$ ($j = 1, 2, \ldots$). If $\mathcal{U}$ is decomposable relative to these J_j^* and if $\mathcal{E}$ is a division of J, then only a finite number of points t in (I, t) are used to form $\mathcal{E}$. Let K be the largest integer such that J_K^* contains one of these t. Then $\bigcup_{j > K} J_j^*$ cannot be covered by $\mathcal{E}$, and this contradiction stops decomposability.

1.4 THE GAUGE (KURZWEIL–HENSTOCK) INTEGRAL
(HENSTOCK (1968b, pp. 220, 221, Ex. 43.3; and 1988a))

We modify the arrangement of Section 1.2, going as far as but not including the norm, and not taking all $t \in \bar{I}$. We include in **A** those $\mathscr{U} \subseteq \mathscr{U}^1$ for which there are a positive function δ on T and an elementary set E, such that $(I, t) \in \mathscr{U}$ if $I \subseteq E$, $t \in \bar{I}$, and $I \subseteq S(t, \delta(t))$, the open sphere centre t and radius $\delta(t)$. Following McShane (1969) we then say that (I, t) is δ-fine. The calculus, Riemann, refinement, Lebesgue, special Denjoy, Perron, Ward, and gauge (Kurzweil–Henstock) integrals are all included in the integral produced by this $(T, \mathscr{T}, \mathbf{A})$.

To define $\mathscr{U} \in \mathbf{A}$ and a division $\mathscr{E}$ of E from $\mathscr{U}$, it is enough for δ to be a positive function on $\bar{E}$. To prove that $\mathbf{A} | E$ is not empty we have

Theorem 1.4.1. *Given an elementary set E and a positive function δ on $\bar{E}$, there is a δ-fine division $\mathscr{E}$ of E.*

Proof. E is a finite union of bricks I; if a δ-fine division of each such I exists, the union of the divisions is a δ-fine division of $\mathscr{E}$. Thus we can take $E = I$. Bisecting each edge of I we have 2^n smaller bricks J. If the theorem is false, by the same argument at least one J has no δ-fine division. There is thus a sequence $I = I_1 \supset I_2 \supset \cdots$ of bricks having no δ-fine division, with $\mathrm{diam}(I_{j+1}) = \frac{1}{2}\mathrm{diam}(I_j)$ for $j = 1, 2, \ldots$. Let (x_j) be the sequence of centres of the bricks. Then for $k > j$, $x_k \in I_j$, and so for metric $\|x - y\|$,

$$\|x_k - x_j\| \leqslant \sup \|x - y\| (x, y \in I_j) = \mathrm{diam}(I_j)$$
$$= 2^{1-j}\mathrm{diam}(I_1) \to 0 \quad (j \to \infty).$$

Thus (x_j) is fundamental and so convergent to some limit x. As $x_j \in I \subseteq E$, $x \in \bar{E}$, and $\delta(x) > 0$ is defined. For $k > j$, $x_k \in I_j$, so $x \in \bar{I}_j$. As $\mathrm{diam}(I_j) \to 0$ $(j \to \infty)$, eventually $I_j \subseteq S(x; \delta(x))$ and (I_j, x) forms a δ-fine division of I_j, contradicting its definition. Hence the theorem. $\square$

In one dimension Borel's covering theorem gives a proof. Again we can assume that $E = [a, b)$ with a, b finite. Each $x \in [a, b]$ has a symmetrical open neighbourhood $I(x) \equiv (x - \delta(x), x + \delta(x))$. As a, b are finite, by Borel's covering theorem a finite number, say $I(u_1), \ldots, I(u_m)$, cover $[a, b]$ (i.e. their union contains $[a, b]$.) We now look for the most economical cover.

(1.4.1) *We can arrange that each point of $[a, b]$ lies in at least one and at most two of the $I(u_k)$.*

For $a \leqslant u < v < w \leqslant b$ let $I(u), I(v), I(w)$ have a common point. We remove one of the three. If $y - \delta(v) \leqslant u - \delta(u)$, then

$$\delta(v) \geqslant \delta(u) + v - u > \delta(u), \ v + \delta(v) > v + \delta(u) > u + \delta(u), \ I(u) \subseteq I(v),$$

1.4 THE GAUGE (KURZWEIL–HENSTOCK) INTEGRAL

and we omit $I(u)$. If $v + \delta(v) \geq w + \delta(w)$, similarly $I(w) \subseteq I(v)$ and we omit $I(w)$. Otherwise, as all three intervals have a common point,

$$u - \delta(u) < v - \delta(v) < v + \delta(v) < w + \delta(w), \qquad I(v) \subseteq I(u) \cup I(w),$$

and we omit $I(v)$. (1.4.1) follows, from which we can assume that

$$a \leq u_1 < u_2 < \cdots < u_m \leq b, \qquad a \in I(u_1), \quad b \in I(u_m)$$

the $I(u_k)$ ($k = 1, 2, \ldots, m$) being consecutive. The intersection

(1.4.2) $$(u_j, u_{j+1}) \cap I(u_j) \cap I(u_{j+1}) = G,$$

say, is not empty and so is an open interval, in which we choose a point y_j ($j = 1, \ldots, m - 1$). The a, b, u_j, y_j are suitable points of partition for a δ-fine division of $[a, b]$. We have proved more than what is required.

Returning to n dimensions, $\mathbf{A}$ is directed for divisions of E, for by Theorem 1.4.1, if $\delta_j > 0$ defines $\mathscr{U}_j$ ($j = 1, 2$), $\delta \equiv \min(\delta_1, \delta_2) > 0$ defines $\mathscr{U}$ with $\mathscr{U} \subseteq \mathscr{U}_1 \cap \mathscr{U}_2$. As in Section 1.2, the $\mathbf{A}$ here has the restriction property, so that $(T, \mathscr{T}, \mathbf{A})$ is a division space. $\mathbf{A}$ is not additive, since if disjoint elementary sets E_j have a common point x in their closures that is not at a vertex, as in Section 1.2, and the $(n-1)$-dimensional brick B, we can find n-dimensional bricks J in $E_1 \cup E_2$ with (J, x) δ-fine, but with J not entirely in E_1 nor in E_2.

As in Section 1.2, if we have elementary sets P, E with $P \subset E$, continuing the faces of P to meet the faces of E, we show that $E \setminus P$ is an elementary set. Divisions of P and $E \setminus P$ from Theorem 1.4.1 using $\delta > 0$ in $\bar{E}$, then show that P is a partial set of E. This is true trivially if $P = E$. If (I, t) is δ-fine and $I \subseteq E$, similarly I is a partial interval of E. If P is a partial set of E and the δ-fine $I \subseteq P$, then I is a partial interval of P. If with the above, $\mathscr{U} \in \mathbf{A}|P$, defined by $\delta > 0$ on $\bar{E}$, we can extend $\mathscr{U}$ to $\mathscr{U}_0 \in \mathbf{A}|E$ by using δ, and then $\mathscr{U}_0 \cdot P = \mathscr{U}$, and $(T, \mathscr{T}, \mathbf{A})$ is linked. For E an elementary set, $E_{\mathscr{U}}^* = \bar{E}$ and $E^* = \bar{E}$, and the space is stable. It is strongly $\mathscr{T}$-compatible with P in E, where P, E are elementary sets and $P \subset E$, for if $t \notin P^* = \bar{P}$ then t lies in the open set $\setminus \bar{P}$ and so is the centre of a sphere contained in $\setminus \bar{P}$. The only restriction on δ is that $\delta(x) > 0$ ($x \in \bar{E}$), so that $(T, \mathscr{T}, \mathbf{A})$ is fully decomposable.

Now let E^z be the Cartesian product of elementary sets E^u in T^u ($u = x, y$), with δ a positive function on $\bar{E}^z$. Using the Pythagorean metric, to each fixed point $x \in \bar{E}^x$ with $\delta_{1x}(y) = 2^{-1/2}\delta(x, y)$, let there be a δ_{1x}-fine division $\mathscr{E}(x)$ of E^y, and $\delta_2(x)$ the least of the $\delta_{1x}(y)$ for which $(I^y, y) \in \mathscr{E}(x)$. If (I^x, x) is δ_2-fine and $(I^y, y) \in \mathscr{E}(x)$, then

$$\sqrt{\{\delta_2(x)^2 + \tfrac{1}{2}\delta(x, y)^2\}} \leq \sqrt{\{\tfrac{1}{2}\delta(x, y)^2 + \tfrac{1}{2}\delta(x, y)^2\}} = \delta(x, y),$$

so that $I^x \otimes I^y$ lies in the sphere centre (x, y) and radius $\delta(x, y)$, and $(I^x, x) \otimes (I^y, y)$ is δ-fine. Thus the first Fubini property holds. As usual, the second only needs x and y interchanging. So $(T^z, \mathscr{T}^z, \mathbf{A}^z)$ is a Fubini division space. Note that T^x and T^y can have different dimensions.

The first appearance of the gauge integral was implicit in Henstock (1955b, pp. 277–8), and then in Kurzweil (1957, pp. 424–8). The book, Henstock (1963c), due to be published in 1962, was held up with production difficulties of further books in a series. It was a pioneering book with a few errors found subsequently by students at lectures, and is now out of print. (The book, Henstock (1988a), contains the bulk of the material, with many improvements and simplified proofs.) On 3 October 1963, K. Kartàk informed me of J. Kurzweil's paper, so that until then, Kurzweil and myself were working independently. Subsequently Henstock (1968a) appeared, a simplified approach to the theory. Then J. Mawhin found a lemma of Cousin (1895) that gives a two-dimensional version of Theorem 1.4.1. Luzin (1915) used the one-dimensional case for trigonometric series, and it was mentioned by W. H. and G. C. Young (1915). The first incomplete proofs of Theorem 1.4.1 are in Kurzweil (1957, p. 423, Lemma 1.1.1) and independently in Henstock (1961a, pp. 129–30, Theorem 16). M. McCrudden found the error in my proof for $n = 1$. When the unsymmetrical defining interval at x is

$$(x - \delta_1(x), x + \delta_2(x))$$

the proof breaks down for the case when, for some j,

$$u_{j-1} < u_{j+1} - \delta_1(u_{j+1}) < u_{j-1} + \delta_2(u_{j-1}) < u_j - \delta_1(u_j) < u_j < u_{j+1}$$
$$< u_{j+1} + \delta_2(u_{j+1}) < u_{j+2} - \delta_1(u_{j+2}) < u_j + \delta_2(u_j) < u_{j+2}$$

The intervals are so interlocked that the removal of one uncovers part of the main interval. However, the proof is sound for symmetrical intervals with $\delta_1(x) = \delta_2(x)$.

1.5 THE GAUGE INTEGRAL, ASSOCIATED POINTS AT VERTICES, AND INFINITE INTERVALS

A simple change in Section 1.4 gives an additive division space. We include in **A** those $\mathcal{U} \subseteq \mathcal{U}^1$ for which there are an elementary set E and a positive function δ defined on T (or, equivalently, on $\bar{E}$) such that $(I, t) \in \mathcal{U}$ if $I \subseteq E$, $I \subseteq S(t, \delta(t))$, and t a vertex of I. The only change is that involving a vertex. A division of E from $\mathcal{U}$ is called a *restricted division*. As in Section 1.4, the proofs show that $(T, \mathcal{T}, \mathbf{A})$ is a division space. The proof of Theorem 1.4.1 needs a slight alteration, since at the end the limit x need not be at a vertex of I_j. However, splitting I_j by hyperplanes through x and parallel to all but one coordinate axis, we have a δ-fine division of I_j with x as associated point at various vertices.

The best proof of additivity seems to be in Henstock (1988a, Theorem 4.6(4.7), p. 46). Perhaps a shorter proof will be found.

1.5 THE GAUGE INTEGRAL

Let E_1, E_2 be disjoint elementary sets with $\bar{E}_1 \cap \bar{E}_2$ not empty. (Otherwise the proof is trivial.) For E^0 the interior of E and $E^b = \bar{E}\setminus E^0$ its boundary or frontier, by geometry E_j^b is a finite union of parts of faces of bricks that make up E_j, the parts being bricks of dimension one less than the dimension of T. If $t \in E_1^0 \cup E_2^0$ let $2\delta(t)$ be the distance from t to $E_1^b \cup E_2^b$. If $t \in E_1^b \cup E_2^b$ let $2\delta(t)$ be the least distance from t to the parts of faces in $E_1^b \cup E_2^b$ on which t does not lie. Then $\delta > 0$ is defined on $\overline{E_1 \cup E_2}$. If (I, t) is δ-fine and $\bar{I}$ has no points in common with $E_1^b \cup E_2^b$ then by construction $t \in E_1^0$ and $I \subseteq E_1$, or $t \in E_2^0$ and $I \subseteq E_2$. If $\bar{I}$ intersects $E_1^b \cup E_2^b$ then $t \in E_1^b \cup E_2^b$ and $\bar{I}$ does not intersect any part of a face in $E_1^b \cup E_2^b$ on which t does not lie. If t lies on face F, then as t is a vertex of I, no interior point of I lies on F. As $I \subseteq E_1 \cup E_2$, I cannot have interior points in E_1 and other interior points in E_2, or else a face would have interior points of I. Hence either $I \subseteq E_1$ or $I \subseteq E_2$, finishing the proof.

The elementary sets we have considered have been bounded. In order to have arrangements for unbounded intervals such as $(-\infty, a)$, $[b, +\infty)$, $(-\infty, +\infty)$ in one dimension and similar intervals in higher dimensions we consider conventional vectors x with $x_j = -\infty$ or $x_j = +\infty$, for one or more j. Instead of spheres centre these conventional x, we use Cartesian products of (a_j, b_j), where $a_j < x_j < b_j$ and $2x_j = a_j + b_j$ (when x_j is finite) or $a_j = x_j = -\infty$, or $x_j = b_j = +\infty$. Then we can deal directly with generalized Riemann integrals over infinite intervals, avoiding an extra limit process. See Section 2.10.

We have assumed that the gauge function δ is completely arbitrary, except that it is positive everywhere. In Section 0.1 it is pointed out that if δ is continuous on the real line then we can take δ constant on compact intervals, which leads to ordinary Riemann integration on those intervals. Thus we need discontinuous δ. We can assume that δ takes values in the set of reciprocals, $\{1/n: n = 1, 2, 3, \ldots\}$, which in this case implies that decomposability and full decomposability are the same property, Examples 2.4.1–3 look further into the question of decomposability and full decomposability.

At the 10th Summer Conference in Vancouver, B.C., Canada, in 1986, P. S. Bullen asked whether δ could be taken measurable (see Bullen (1986/1987)). Liu (1987/1988) proved this on the real line, while Pfeffer (1988b) went further, showing that on the real line, δ can be taken upper semicontinuous on a subset Z with $\setminus Z$ of measure zero, i.e. of L-variation zero, where L is the length function on the real line. Buczolich (1987/1988) extended this to $\mathbb{R}^m$, i.e. finite-dimensional Euclidean space. Foran and Meinershagen (1987/1988) showed in Example 1 that on the real line there is a Lebesgue integrable f (i.e. f and $|f|$ are generalized Riemann integrable) for which there is no Borel measurable δ to give Riemann sums within $\varepsilon > 0$ of the integral, and also in various theorems that if Z_f is the set where the derivative of the integral exists and is f, and if Z is an F_σ subset contained in Z_f with complement of measure zero (of L-variation zero) then δ can be chosen to be Baire 2 in Z. If either (a) F is ACG* and $f = F'$ wherever F' exists, and $f = 0$ otherwise, or (b) if $|f|$

is dominated by a Baire function, or (c) f is bounded and measurable, then δ can be Baire 2 everywhere. Can these results be extended to $\mathbb{R}^m$?

Examples of gauges are given in Henstock (1988a), pp. 48–50, Exx. 4.n ($1 \leq n \leq 9$).

1.6 McSHANE'S MODIFICATION, GIVING ABSOLUTE INTEGRALS

McShane (1969, 1973, 1983) modified the constructions in Sections 1.4, 1.5, omitting '$t \in \bar{\varGamma}$' and 't a vertex of $\varGamma$', respectively, and only imposing the condition $t \in \bar{E}$, with the δ-fine condition, which increases the number of δ-fine divisions, potentially imposing a greater restriction on the integral. It turns out that the integral is absolute, equivalent to that of Lebesgue and Radon, and cannot integrate all derivatives. The different geometrical arrangement of McShane's from those of Sections 1.4, 1.5 produces the differences between McShane's and the gauge integrals and there therefore arises the question: what is the essential geometric difference?

The (T, $\mathscr{T}$, $\mathbf{A}$) follow roughly the same pattern for McShane's arrangement as the gauge integral pattern in Section 1.4, additivity being missing.

1.7 SYMMETRIC INTERVALS

McGrotty (1962) modified the constructions of Sections 1.4, 1.5, for one dimension, in another way, taking associated points at the centres of intervals, with a gauge δ. He proved that $[-a, a)$ can be divided if $\delta > 0$ is defined there, provided that in place of the symmetric intervals at $-a$ and a we use $[-a, -a+h)$ and $[a-h, a)$ together, for $0 < h < \delta(-a)$. Thus we are effectively dealing with intervals mod($2a$) or with arcs of a circle of circumference $2a$. Thomson (1980/1981a), pp. 86–7, omitted the $[-a, -a+h)$ and $[a-h, a)$, proving that there is a set $D \subseteq (0, a)$ with the closure of $(0, a) \backslash D$ countable, such that every interval $[-x, x)$, with $x \in D$, is divided. (Actually he translated the main interval to $[a, b)$ with centre c in place of 0.) Then Preiss and Thomson (1988–89) proved that if the symmetric intervals are given for every real x, there is a countable set N on the real line such that every interval is divided if neither of its end-points is in N. Clearly these results can be extended to finite-dimensional Cartesian products of such arrangements that are independent in each dimension, e.g. see Henstock (1968b), p. 222, Ex. 43.6. A simple argument will suffice for a proof of the first two papers' results.

Theorem 1.7.1. *Let the gauge $\delta > 0$ be defined at each point of (a, b), but not necessarily at the ends. For c the midpoint $\frac{1}{2}(a + b)$, $[c - x, c + x)$ can be*

1.8 THE DIVERGENCE THEOREM

divided, for all x except possibly a finite number, in $(0, d)$, for each d in $0 < d < \frac{1}{2}(b - a)$. If intervals $[a, a + h)$ and $[b - h, b)$ can be used together, for some $\eta > 0$ and all h in $0 < h < \eta$, then $[a, b)$ can also be divided.

Proof. Let s be the supremum of numbers $y > 0$ such that $(c - x, c + x)$ can be divided, for all x in $(0, y - c)$ except possibly for a finite number of x, where it can easily be arranged that $\delta(c - x) = \delta(c + x)$ for all x in $(0, b - c)$. If $s < b$ then also $s > a$ and $\delta(s) > 0$ exists. Using intervals $(s - x, s + x)$ with $0 < x < \delta(s)$ and $[a + b + x - s, s - x)$ divided, and $[a + b - x - s, a + b + x - s)$, $[a + b - x - s, s + x)$ is divided, for all but a finite number of $x < b - s$ and $x < \delta(s)$. This contradicts the definition of s, hence $s = b$, and the first result follows. Note that the $s < b$ in the proof can join the exceptional points. For the second result we only need one h in $0 < h < \eta$ for which $[a + h, b - h)$ is divided, and the two extra intervals fill the gaps. □

Next, in two dimensions we have squares with sides parallel to the axes and with associated points at their centres. These were mentioned in Henstock (1961a) as a way of dealing with twice integrated trigonometric series $F(x)$ on putting $G(x, y) = F(x + y)$. G turns out to be a variational integral, using squares, of the sum-function of the original trigonometric series. Unfortunately the gauge integral of this arrangement does not exist, and it might be necessary to modify the gauges used for Henstock (1961a), pp. 109, 110.

For let S be the main square $[-\pi, \pi) \otimes [-\pi, \pi)$ and let D be either the diagonal connecting the points $(-\pi, -\pi)$ and (π, π), or the diagonal connecting $(-\pi, \pi)$ and $(\pi, -\pi)$. For each $(x, y) \notin D$ let $d(x, y)$ be the least distance from (x, y) to D. If $\delta(x, y) > 0$ is arbitrary on D and satisfies $\delta(x, y) \leq \frac{1}{2}d(x, y)$ at other points, then any possible division of S will cover D. By construction the only squares whose closure contains a point of D, and that are δ-fine, must have their centres, the associated points, on D. As these closures do not overlap, two adjacent closures must have one common point, on D, leaving two V-shaped regions between them. By construction no finite collection of off-diagonal squares can completely fill these V-shaped regions, so that no division is possible.

If D is $x + y = 0$ then we can have $0 < \delta(x, y) < \frac{1}{2}|x + y|$ off the diagonal, so that even if $\delta(x, y)$ is a function of $x + y$ alone, we still get no division.

1.8 THE DIVERGENCE THEOREM

After the integration of derivatives in one dimension, the next stage is the divergence theorem, which amounts to the same result in more than one dimension. Many papers have been written on the subject and work continues on what is the best integral to tackle the divergence theorem on manifolds. Until the theory clarifies it seems best to give a list of papers: Verblunsky

(1949), Mařik (1956), Mařik and Matyska (1965), Shapiro (1958), Mawhin (1981a,b), Pfeffer (1982, 1984a,b, 1986, 1987a, 1988a, 1988/1989b), Pfeffer and Yang (1988/1989, 1989/1990), Jarnik and Kurzweil (1985a, 1988), and Jarnik, Kurzweil, and Schwabik (1983).

1.9 THE GENERAL DENJOY INTEGRAL

After constructing the first of his extensions of Lebesgue integration (now known as *Denjoy–Perron* or *special Denjoy integration*), Denjoy (1916a, b) gave an even deeper construction, now known as the *general Denjoy* or *Denjoy–Hinčin integral* (see Hinčin (1916, 1918). Later, Tolstov (1939b) gave a Perron-type construction of an integral equivalent to the general Denjoy integral, and in turn we can construct the corresponding generalized Riemann integral (see Henstock (1968b, pp. 222–3, Ex. 43.9; 1979, pp. 2, 3)).

Let the real interval $[a, b]$ be the union of a sequence (P_j) of perfect sets and let δ be a gauge on $[a, b]$. Then the corresponding $\mathcal{U}$ is the family of all (I, t) with $I = [u, v)$, $t = u$ or $t = v$, u and v in the same P_j, and $v - u < \delta(t)$, except that when that P_j is isolated at t on the side on which the other end-point lies, we can let that end-point take all values distant not more than $\delta(t)$ away from t.

To prove that $[a, b)$ and all subintervals $[u, v)$ are divided, let us say that an interval $I = [u, v) \subseteq [a, b)$ is *admissible* if every $[w, x) \subseteq I$ is divided. If two admissible intervals abut or overlap, their union is admissible. If H is the union of interiors of all admissible intervals and if the closed interval $I \subseteq H$, then every $t \in I$ lies in the interior of an admissible interval. Hence by Borel's covering theorem I is covered by a finite number of such interiors and so is admissible. If $P = [a, b] \setminus H$ then H is open and so P closed. Apart possibly from a, b, if P has an isolated point t then for some integer j, $t \in P_j$. If P_j is isolated at t, then on the isolated side all intervals from t to w (say) are included that have $0 < |t - w| < \delta(t)$. On the other side, or on both sides, t is approached by other points of P_j, giving intervals that link together divisions on both sides of t. Hence t is an interior point of an admissible interval, contradicting its definition. Hence P is closed with no isolated points and so is perfect (except possibly for a, b). If $P \cap (a, b)$ is not empty, then by Baire's density theorem, P is of the second category since $[a, b]$ is complete. Hence there are a closed interval J and an integer j, such that $J^\circ \cap P$ is not empty and that P_j is dense in $J \cap P$. Hence $J \cap P \subseteq P_j$, so that each $t \in J \cap P$ is the centre of an interval $K(t) = (t - \delta(t), t + \delta(t))$ such that every interval $[u, t)$, $[t, u) \subseteq K(t)$ with $u \in P$, is divided. By Young's covering theorem, which says that $J \cap P$ is compact, a finite number of $K(t)$, and so of $[t, u)$, $[u, t)$, cover $J \cap P$. The intervals of J outside the cover are admissible and finite in number, so that J is admissible and $J \cap P$ empty. Hence at most, $a, b \in P$. But

1.10 BURKILL'S APPROXIMATE PERRON INTEGRAL

there are suitable intervals from a, b to show that $[a, b)$ is admissible, completing the proof.

Exercise 1.9.1 Show that for some integer j, P_j contains an interval over which we have an arrangement involving a gauge, and so a special Denjoy integral is definable.

1.10 BURKILL'S APPROXIMATE PERRON INTEGRAL

The approximate Perron integral of Burkill (1931) has a generalized Riemann integral form that is easily obtained. See Henstock (1961a; 1968b, p. 223, Ex. 43.11; 1979, p. 3).

Let M be a continuous non-atomic measure on the Borel sets of $[a, b]$, with $M((x, y)) > 0$ for all $(x, y) \subseteq (a, b)$. For each $t \in (a, b)$ let $U_-(t)$ be the family of all $[u, t] \subseteq [a, b]$ with u in a set of left lower M-density $\geq d_-$ at t, the set varying with t, and let $U_+(t)$ be the family of all $[t, u] \subseteq [a, b]$ with u in a set of right lower M-density $\geq d_+$ at t. If d_-, d_+ are independent of t, with $d_- + d_+ > 1$, then this arrangement divides every interval of $[a, b]$. If necessary, we can put

$$U = \bigcup_{a \leq t \leq b} U_-^*(t) \cup U_+^*(t),$$

we take a one-sided arrangement at a and at b, where the asterisk means that t is also included.

The proof is taken from Henstock (1961a), p. 131. Taking

$$\varepsilon = (d_- + d_+ - 1)/3$$

then $\varepsilon > 0$, and to each $t \in (a, b)$ there is a gauge $\delta(t)$ such that if $h \in (0, \delta(t))$ the set of points y in $[t - h, t)$ with $[y, t) \in U_-(t)$, has M-measure greater than $(d_- - \varepsilon)M((t - h, t))$, and the set of points y in $[t, t + h)$ with $[t, y) \in U_+(t)$, has M-measure greater than $(d_+ - \varepsilon)M((x - h, x))$. Let

(1.10.1) $$t - \delta(t) < u < t < u + \delta(u).$$

Then $[u, t)$ is in $[u, u + \delta(u))$ and in $[t - \delta(t), t)$, so that the set of points y in the open interval (u, t) with $[u, y) \in U_+(u)$ or $[y, t) \in U_-(t)$, have respective M-measures greater than $(d_+ - \varepsilon)M((u, t))$, $(d_- - \varepsilon)M((u, t))$. As

$$d_+ - \varepsilon + d_- - \varepsilon = d_+ + d_- - 2(d_- + d_+ - 1)/3$$
$$= (2 + d_- + d_+)/3 > 1,$$

the two sets have a y in common, and $[u, y) \in U_+(u)$, $[y, t) \in U_-(t)$. For X_n the set of t with $\delta(t) \geq n^{-1}$, clearly $\bigcup X_n = [a, b]$.

Let us say that $[u, v)$ is admissible if every $[w, x) \subseteq [u, v)$ is divided. We then proceed as in Section 1.9, taking H the union of interiors of all admissible intervals, and $P = [a, b] \setminus H$. If $t \in (a, b)$ is an isolated point of P there are suitable $[u, t)$, $[t, v)$ with u and v in appropriate sets, and the u, v link up with admissible intervals on the left and right, so that t is an interior point of an admissible interval, giving a contradiction. Hence P is perfect or empty. If P is not empty, there are an integer n and a portion $P \cap J$ of P such that X_n is everywhere dense in $P \cap J$. If $[w, v]$ is any interval containing points of $P \cap J$, then $P \cap J \cap [w, v]$ can be covered by a finite number of intervals (u, t) satisfying (1.10.1), and these intervals link up with the finite number of intervals of the open set $(w, v) \setminus (P \cap J)$, $P \cap J$ is empty, and the theorem is proved.

The case $d_- = 1 = d_+$ gives the approximate Perron integral by using the theory of Section 7.1

1.11 DIVISION SYSTEMS AND SPACES IN A TOPOLOGY

See Henstock (1968b, pp. 224–5, Ex. 43.14; 1980a, pp. 397–8). Let T be a topological T_3-space (i.e. every point is a closed set, and for each $t \in T$ and each neighbourhood G of t, there is a neighbourhood G_1 of t whose closure is contained in G). In T we can construct a fully decomposable stable division space, as follows. Let $\bar{X}$, X° be the closure and interior of $X \subseteq T$, respectively, with $\mathscr{H}$ the family of non-empty compact sets. The generalized intervals are those $I = X \setminus Y (X, Y \in \mathscr{H})$ with I° non-empty. Let $\mathbf{A}$ be the family of all $\mathscr{U}$ defined by an elementary set E and a function $J: T \to T$ with $t \in J(t)^\circ$, such that $\mathscr{U}$ contains all (I, t) with $t \in \bar{E} \cap \bar{I}$ and $I \subseteq J(t) \cap E$. Then $\mathbf{A}$ divides each E. It is enough to show that $\mathbf{A}$ divides each interval $I = X \setminus Y$. As $\bar{I} \subseteq \bar{X} = X \in \mathscr{H}$, $\bar{I}$ is compact. Thus for a finite number of points $t_1, \ldots, t_n$ in $\bar{I}$, the union of the $J(t_j)^\circ$ ($1 \leq j \leq n$) contains $\bar{I}$. As T is a T_3-space there are open sets G_j containing the t_j, with disjoint $\bar{G}_j \subseteq J(t_j)^\circ$. Let

$$G_j^+ = J(t_j)^\circ \setminus \bigcup_{k \neq j}^n \bar{G}_k \quad (j = 1, \ldots, n).$$

Then G_j^+ is a neighbourhood of t_j, and the union of the G_j^+ contains $\bar{I}$, while $G_j^+ \cap \bar{G}_k$ ($k \neq j$) is empty. The following mutually disjoint I_j form a division of $I = X \setminus Y$.

$$I_1 = X \cap G_1^+ \setminus Y, \quad I_j = X \cap G_j^+ \setminus \bigcup_{k=1}^{j-1} G_k^+ \cup Y \quad (j = 2, \ldots, n).$$

For each $I_j \in T$ and $I_j \subseteq X \setminus Y$, and the union of the I_j contains $X \setminus Y$. The other properties are now easily shown. The construction of Section 2.7 gives an additive division space.

CHAPTER 2

GENERALIZED RIEMANN AND VARIATIONAL INTEGRATION IN DIVISION SYSTEMS AND DIVISION SPACES

2.1 FREE DIVISION SYSTEMS, THE FREE p-VARIATION, FREE NORM VARIATION, AND CORRESPONDING VARIATIONAL INTEGRAL, AND THE FREE VARIATION SET

In the base space T of points we use a collection $\mathcal{T}$ of some non-empty subsets $I \subseteq T$ called (*generalized*) *intervals*, a non-empty collection $\mathcal{U}^1$ of some interval-point pairs (I, t) ($I \in \mathcal{T}, t \in T$), and a non-empty family $\mathbf{A}$ of some non-empty subcollections $\mathcal{U} \subseteq \mathcal{U}^1$. $(T, \mathcal{T}, \mathbf{A})$ is a *free division system* if $\mathbf{A}$ is directed downwards in the sense of set inclusion (i.e. given $\mathcal{U}_j \in \mathbf{A}$ ($j = 1, 2$), there is a $\mathcal{U} \in \mathbf{A}$ with $\mathcal{U} \subseteq \mathcal{U}_1 \cap \mathcal{U}_2$.) Colloquially, this results in convergence 'as $\mathcal{U}$ shrinks'.

The minimal requirement of the space K of values is that it has a multiplication (or addition) operation $x \cdot y \in K$ for all $x, y \in K$. If also it has a function $p: K \to \mathbb{R}^+$ (non-negative real line) such that

(2.1.1) $p(x \cdot y) \leq p(x) + p(y)$ $(x, y \in K)$ and $p(u) = 0$

for the unit u in K, if it exists, we call $p(y)$ the *p-modulus of y*. From (2.1.1),

(2.1.2)   if $p(x) = 0 = p(y)$ then $p(x \cdot y) = 0$.

Given $h: \mathcal{U}^1 \to K$ we define

$$FV_p(h; \mathcal{U}) \equiv \sup(\mathcal{E}) \sum p(h(I, t))$$

over all arbitrary *finite* collections $\mathcal{E}$ of some $(I, t) \in \mathcal{U}$ with mutually disjoint I, where $\mathcal{E}$ and $\mathcal{U}$ are non-empty, $\mathcal{E} \subseteq \mathcal{U} \subseteq \mathcal{U}^1$, and where $(\mathcal{E})\sum$ denotes summation over $\mathcal{E}$. If a finite supremum does not exist, by convention we write $FV_p(h; \mathcal{U}) = +\infty$. Clearly

(2.1.3) $FV_p(h; \mathcal{U}) \leq FV_p(h; \mathcal{U}_j)$ $(\mathcal{U} \subseteq \mathcal{U}_1 \cap \mathcal{U}_2, \mathcal{U}_j \subseteq \mathcal{U}^1, j = 1, 2, \mathcal{U}$ non-empty), even allowing for the convention. Thus it is reasonable to define the *free p-variation of h relative to $\mathbf{A}$* to be

$$\overline{FV}_p(h; \mathbf{A}) \equiv \inf FV_p(h; \mathcal{U}) \quad (\mathcal{U} \in \mathbf{A}), \qquad = \limsup(\mathcal{E}) \sum p(h(I, t))$$

as $\mathscr{U}$ shrinks. Here, conventionally, $\overline{FV}_p(h; \mathbf{A}) = +\infty$ if and only if $FV_p(h; \mathscr{U}) = +\infty$ for all $\mathscr{U} \in \mathbf{A}$. If $\overline{FV}_p(h; \mathbf{A})$ is finite, we say that *h is of free bounded p-variation relative to* $\mathbf{A}$, and if $\overline{FV}_p(h; \mathbf{A}) = 0$, we say that *h is of free p-variation zero relative to* $\mathbf{A}$.

Sometimes the topology of K needs an infinity of functions p to define it. But if $p(x)$ is a norm $\|x\|$ in K we change the term 'free p-variation' to 'free norm variation'; or 'free variation' if K lies in $\mathbb{R}$ or $\mathbb{C}$ with the modulus as norm. We also omit p from FV_p and $\overline{FV}_p$.

Using the indicator $\chi(X; \cdot)$ of the set X let $p(h(I, t), X)$ be $p(h(I, t))\chi(X; t)$,

$$FV_p(h; \mathscr{U}; X) \equiv \sup(\mathscr{E}) \sum p(h(I, t); X),$$

$$\overline{FV}_p(h; \mathbf{A}; X) \equiv \inf FV_p(h; \mathscr{U}; X) = \limsup(\mathscr{E}) \sum p(h(I, t))\chi(X; t)$$

as $\mathscr{U}$ shrinks, with the same requirements for $\mathscr{E}$, $\mathscr{U}$ and the same conventions for $+\infty$. Clearly, even with the convention,

(2.1.4)

$$\overline{FV}_p(h; \mathbf{A}; X) \leqslant \overline{FV}_p(h; \mathbf{A}; Y) \leqslant \overline{FV}_p(h; \mathbf{A}; T) = \overline{FV}_p(h; \mathbf{A}) \quad (X \subseteq Y \subseteq T).$$

If $\overline{FV}_p(h; \mathbf{A}; X) = 0$, we say that X *is of free p-h-variation zero*, and if a property holds except in such an X, we say that the *property holds free p-h-almost everywhere*. If $h = h^* . h_1$, $h^* = h . h_2$, and h_1, h_2 are of free p-variation zero, we say that *h and h* are free p-variationally equivalent*. If K is a group, $h_2 = h_1^{-1}$.

Theorem 2.1.1.

(2.1.5) $\qquad FV_p(h_1 . h_2; \mathscr{U}) \leqslant FV_p(h_1; \mathscr{U}) + FV_p(h_2; \mathscr{U}),$

(2.1.6) $\qquad \overline{FV}_p(h_1 . h_2; \mathbf{A}) \leqslant \overline{FV}_p(h_1; \mathbf{A}) + \overline{FV}_p(h_2; \mathbf{A}).$

If h and h* are free p-variationally equivalent then

(2.1.7) $\qquad \overline{FV}_p(h; \mathbf{A}) = \overline{FV}_p(h^*; \mathbf{A}).$

(2.1.8) *If K is a semigroup, free p-variational equivalence is an equivalence relation.*

Proof. From (2.1.1), when $\mathscr{E}$ comes from $\mathscr{U}$,

$$(\mathscr{E})\sum p(h_1 . h_2) \leqslant (\mathscr{E})\sum p(h_1) + (\mathscr{E})\sum p(h_2) \leqslant FV_p(h_1; \mathscr{U}) + FV_p(h_2; \mathscr{U}).$$

Hence (2.1.5). Given $\varepsilon > 0$, there are $\mathscr{U}_j \in \mathbf{A}$ for which

$$FV_p(h_j; \mathscr{U}_j) < \overline{FV}_p(h_j; \mathbf{A}) + \tfrac{1}{2}\varepsilon \quad (j = 1, 2).$$

2.1 FREE DIVISION SYSTEMS

Let $\mathscr{U} \subseteq \mathscr{U}_1 \cap \mathscr{U}_2$, $\mathscr{U} \in \mathbf{A}$. Then by (2.1.5), (2.1.3),

$$\overline{FV}_p(h_1.h_2; \mathbf{A}) \leq FV_p(h_1.h_2; \mathscr{U}) \leq FV_p(h_1; \mathscr{U}) + FV_p(h_2; \mathscr{U})$$

$$< \overline{FV}_p(h_1; \mathbf{A}) + \overline{FV}_p(h_2; \mathbf{A}) + \varepsilon,$$

giving (2.1.6). Using this for (2.1.7),

$$\overline{FV}_p(h; \mathbf{A}) \leq \overline{FV}_p(h^*; \mathbf{A}), \quad \overline{FV}_p(h^*; \mathbf{A}) \leq \overline{FV}_p(h; \mathbf{A}).$$

When K is a semigroup, if h, h^* and h^*, h^+ are free p-variationally equivalent there are h_j ($j = 1, 2, 3, 4$) of free p-variation zero, such that

$$h = h^*.h_1, \quad h^* = h.h_2, \quad h^* = h^+.h_3, \quad h^+ = h^*.h_4, \quad h = h^+.(h_3.h_1),$$

$$h^+ = h.(h_2.h_4),$$

and by (2.1.6), $h_3.h_1$ and $h_2.h_4$ are of free p-variation zero. Hence h and h^+ are free p-variationally equivalent. The other properties of an equivalence relation are trivially true. □

We can multiply together n values $k_1, \ldots, k_n \in K$ by taking them in pairs, e.g. for $n = 4$ we have $((k_2.k_3).k_1).k_4$ or $(k_1.k_3).(k_2.k_4)$ or ... Given the n values, we write the set of all such products as $\prod_{j=1}^n k_j$. A finite union of mutually disjoint intervals is called an *elementary set*. A function H of elementary sets is finitely multiplicative if

$$\prod_{j=1}^n H(I_j) = \text{sing}(H(E))$$

when E is the finite union of mutually disjoint intervals I_j ($j = 1, \ldots, n$). In this case it is immaterial where the brackets are put in the multiplication, and the values of H lie in a commutative subset of K.

If the function H of elementary sets is finitely multiplicative and if H is free p-variationally equivalent to h where $h: \mathscr{U}^1 \to K$, we say that H is *a free p-variational integral of h*, writing

$$H(E) = (\mathbf{A}) \int_E dh,$$

and omitting $(\mathbf{A})$ if $\mathbf{A}$ is understood.

Theorem 2.1.2.

(2.1.9) *If h is of free p-variation zero, if the unit u exists in K, if $h^{-1}: \mathscr{U}^1 \to K$ exists with*

$$h(I, t).h^{-1}(I, t) = u((I, t) \in \mathscr{U} \in \mathbf{A})$$

and if h^{-1} is also of free p-variation zero, then the unit function $H(E) = u$ (all elementary sets E) is an integral of h. We write u for the unit function also.

(2.1.10) Let $\mathscr{V}$ be the family of all finitely multiplicative functions H of elementary sets of free p-variation zero, and let K be a commutative group. If $h: \mathscr{U}^1 \to K$ with free p-variational integrals H_1, H_2, then $H_1 = H_2.H_3$, $H_2 = H_1.H_4$ where H_3, $H_4 \in \mathscr{V}$. If $\mathscr{V} = \text{sing}(u)$ then $H_1 = H_2$ and the integral is unique.

Proof. Taking the u in pairs, if n is a positive integer, $u^n = u$ and the unit function is finitely multiplicative. As

$$h = u.h = H.h \quad \text{and} \quad H = h.h^{-1}$$

then by definition h is p-variationally integrable over E to $H(E) = u$, giving (2.1.9)

If K is a semigroup, Theorem 2.1.1(2.1.8) is true. Hence in (2.1.10), H_1 and H_2 are free p-variationally equivalent and so $H_1 = H_2.H_3$, $H_2 = H_1.H_4$, where H_3, H_4 are of free p-variation zero and are functions of elementary sets. To show that H_3, H_4 are finitely multiplicative, K needs to be a commutative group, and then H_3, H_4 are in $\mathscr{V}$. □

Theorem 2.1.3. *Let $X_1, \ldots, X_n$ be sets in T with union X. Then*

$$\overline{FV}_p(h; A; X) \leq \sum_{j=1}^{n} \overline{FV}_p(h; A; X_j).$$

Proof. By (2.1.4) we can assume the right side finite and the X_j mutually disjoint. Given $\varepsilon > 0$, let $\mathscr{U}_j \in \mathbf{A}$ satisfy

(2.1.11) $FV_p(h; \mathscr{U}_j; X_j) < \overline{FV}_p(h; A; X_j) + \varepsilon.2^{-j} \quad (j = 1, \ldots, n).$

By direction in $\mathbf{A}$ there is a $\mathscr{U} \in \mathbf{A}$ with

$$\mathscr{U} \subseteq \bigcap_{j=1}^{n} \mathscr{U}_j.$$

If $\mathscr{E}$ comes from $\mathscr{U}$, (2.1.11) gives

$$(\mathscr{E})\sum p(h; X) = \sum_{j=1}^{n} (\mathscr{E})\sum p(h; X_j) \leq \sum_{j=1}^{n} FV_p(h; \mathscr{U}_j; X_j)$$

$$< \sum_{j=1}^{n} \overline{FV}_p(h; A; X_j) + \varepsilon,$$

$$\overline{FV}_p(h; A; X) \leq FV_p(h; \mathscr{U}; X) \leq \sum_{j=1}^{n} \overline{FV}_p(h; A; X_j) + \varepsilon.$$

As $\varepsilon > 0$ is arbitrary, this gives the result. It cannot be extended to an infinite sequence of sets unless decomposability or something similar holds.

2.1 FREE DIVISION SYSTEMS

We say that a set $X \subseteq T$ is *(Carathéodory) free p-h-measurable* if, for each $Y \subseteq T$,

(2.1.12) $\quad \overline{FV}_p(h; \mathbf{A}; Y \cap X) + \overline{FV}_p(h; \mathbf{A}; Y \setminus X) \leq \overline{FV}_p(h; \mathbf{A}; Y).$

Then by Theorem 2.1.3, equality occurs in (2.1.12).

Theorem 2.1.4. *For X the union of mutually disjoint free p-h measurable sets $X_1, \ldots, X_n$, then X is free p-h-measurable, and for each $Y \subseteq T$,*

(2.1.13) $\quad \overline{FV}_p(h; \mathbf{A}; Y) = \sum_{j=1}^{n} \overline{FV}_p(h; \mathbf{A}; Y \cap X_j) + \overline{FV}_p(h; \mathbf{A}; Y \setminus X),$

(2.1.14) $\quad \overline{FV}_p(h; \mathbf{A}; X) = \sum_{j=1}^{n} \overline{FV}_p(h; \mathbf{A}; X_j).$

Proof by induction. Trivially true for X_1, suppose true for $X_2, \ldots, X_m$ ($m < n$) with union Z. By inductive hypothesis, Z is free p-h-measurable and $Z \cap X_{m+1}$ is empty. Hence by definition, even when the first term is $+\infty$,

$$\overline{FV}_p(h; \mathbf{A}; Y) = \overline{FV}_p(h; \mathbf{A}; Y \cap X_{m+1}) + \overline{FV}(h; \mathbf{A}; Y \setminus X_{m+1})$$
$$= \overline{FV}_p(h; \mathbf{A}; Y \cap X_{m+1}) + \overline{FV}_p(h; \mathbf{A}; (Y \setminus X_{m+1}) \cap Z)$$
$$+ \overline{FV}_p(h; \mathbf{A}; (Y \setminus X_{m+1}) \setminus Z)$$
$$= \overline{FV}_p(h; \mathbf{A}; Y \cap X_{m+1}) + \overline{FV}_p(h; \mathbf{A}; Y \cap Z)$$
$$+ \overline{FV}_p(h; \mathbf{A}; Y \setminus (X_{m+1} \cup Z))$$
$$= \sum_{j=1}^{m+1} \overline{FV}_p(h; \mathbf{A}; Y \cap X_j) + \overline{FV}_p(h; \mathbf{A}; Y \setminus (X_{m+1} \cup Z)),$$

using (2.1.14) for $n = m$. This gives (2.1.13) for $n = m + 1$. By Theorem 2.1.3,

$$\overline{FV}_p(h; \mathbf{A}; Y) \geq \overline{FV}_p(h; \mathbf{A}; Y \cap (X_{m+1} \cup Z)) + \overline{FV}_p(h; \mathbf{A}; Y \setminus (X_{m+1} \cup Z)).$$

Thus by definition $X_{m+1} \cup Z$ is free p-h-measurable. By induction we have (2.1.13), and (2.1.14) when $Y = X$. □

Thus the finite union of p-h-measurable sets is p-h-measurable, which leads to the following definition. A class S of sets in T is called *finitely additive* if (i) $\varnothing \in S$ (ii) $\setminus X \in S$ when $X \in S$ (iii) *the union of a finite number of sets of S, is also in S.*

Theorem 2.1.5.

(2.1.15) *If X and W are free p-h-measurable, so are $X \cap W$, $\setminus X$, $X \setminus W$.*

(2.1.16) *If also $W \subseteq X$,*

$$\overline{FV}_p(h; \mathbf{A}; X \setminus W) = \overline{FV}_p(h; \mathbf{A}; X) - \overline{FV}_p(h; \mathbf{A}; W).$$

(2.1.17) *The class S of free p-h-measurable sets is finitely additive.*

(2.1.18) *If each X of a family $\mathscr{X}$ of subsets of T is free p-h-measurable with $\overline{FV}_p(h; \mathbf{A}; X)$ finite, if $k(X)$ is finitely additive over the algebra $A(\mathscr{X})$ of finite unions, intersections, and differences of sets of $\mathscr{X}$, and if*

$$\overline{FV}_p(h; \mathbf{A}; X) = k(X) \quad (X \in \mathscr{X})$$

the same is true for $X \in A(\mathscr{X})$.

Proof. For $Y \subseteq T$ and X, W free p-h-measurable, Theorem 2.1.3 gives

$$\overline{FV}_p(h; \mathbf{A}; Y) = \overline{FV}_p(h; \mathbf{A}; Y \cap X) + \overline{FV}_p(h; \mathbf{A}; Y \setminus X)$$
$$= \overline{FV}_p(h; \mathbf{A}; Y \cap X \cap W) + \overline{FV}_p(h; \mathbf{A}; (Y \cap X) \setminus W)$$
$$+ \overline{FV}_p(h; \mathbf{A}; Y \setminus X)$$
$$\geq \overline{FV}_p(h; \mathbf{A}; Y \cap X \cap W) + \overline{FV}_p(h; \mathbf{A}; Y \setminus (X \cap W)).$$

Hence $X \cap W$ is free p-h-measurable. As definition (2.1.12) is symmetrical in X and $\setminus X$, $\setminus X$ is also free p-h-measurable. Then $X \setminus W = X \cap (\setminus W)$ is also free p-h-measurable. In (2.1.16), $X \cap W = W$, so that the result follows from (2.1.12) on putting W for X and then X for Y. For (2.1.17), (i) is trivially true, (ii) comes from (2.1.15), and then for (iii) we take the sets disjoint and use Theorem 2.1.4. Then (2.1.18) follows from (2.1.14), (2.1.16). □

The topology $\mathscr{G}$ of T is *normal* if, for each pair F_1, F_2 of disjoint closed sets, there are disjoint neighbourhoods of F_1, F_2. We say that $(T, \mathscr{T}, \mathbf{A})$ *is freely compatible with* $\mathscr{G}$ if, for each $G \in \mathscr{G}$, there is a $\mathscr{U} \in \mathbf{A}$ with $I \subseteq G$ for each $(I, t) \in \mathscr{U}$ with $t \in G$. These definitions lead to conditions when the variation is finitely additive without the sets having a measurability condition.

Theorem 2.1.6. *Let $(T, \mathscr{T}, \mathbf{A})$ be freely compatible with the topology $\mathscr{G}$ in T, let X_1, X_2 be subsets of T and G_1, G_2 be disjoint open sets with $X_j \subseteq G_j$ ($j = 1, 2$). Then*

(2.1.19) $\quad \overline{FV}_p(h; \mathbf{A}; X_1 \cup X_2) = \overline{FV}_p(h; \mathbf{A}; X_1) + \overline{FV}_p(h; \mathbf{A}; X_2).$

(2.1.20) *If $\mathscr{G}$ is also normal, (2.1.9) holds when the closures of X_1, X_2 are disjoint.*

Proof. By compatibility, corresponding to G_j there is a $\mathscr{U}_j \in \mathbf{A}$ ($j = 1, 2$) and by direction there is a $\mathscr{U} \in \mathbf{A}$ with $\mathscr{U} \subseteq \mathscr{U}_1 \cap \mathscr{U}_2$. If $(I_j, t_j) \in \mathscr{U}$ and $t_j \in X_j$ then $I_j \subseteq G_j$ ($j = 1, 2$) and I_1, I_2 are disjoint. Thus, given $\varepsilon > 0$, we can choose $\mathscr{E}_j \subseteq \mathscr{U}$ such that if $(I, t) \in \mathscr{E}_j$ then $t \in X_j$, $I \subseteq G_j$,

$$(\mathscr{E}_j) \sum p(h) \chi(X_j; t) > \overline{FV}_p(h; \mathbf{A}; X_j) - \tfrac{1}{2}\varepsilon \quad (j = 1, 2).$$

$$\overline{FV}_p(h; \mathbf{A}; X_1) + \overline{FV}_p(h; \mathbf{A}; X_2) - \varepsilon < (\mathscr{E}_1)\sum p(h)\chi(X_1; t) + (\mathscr{E}_2)\sum p(h)\chi(X_2; t)$$
$$= (\mathscr{E}_1 \cup \mathscr{E}_2)\sum p(h)\chi(X_1 \cup X_2; t) < FV_p(h; \mathscr{U}; X_1 \cup X_2),$$
$$\overline{FV}_p(h; \mathbf{A}; X_1) + \overline{FV}(h; \mathbf{A}; X_2) - \varepsilon \leqslant \overline{FV}_p(h; \mathbf{A}; X_1 \cup X_2).$$

Being true for all $\varepsilon > 0$, and using Theorem 2.1.3, we have (2.1.19). Then (2.1.20) follows from the definition of normal. □

For X_1 closed, it is sometimes possible to prove that for the open set $G = \setminus X_1$ there are closed sets $X_2 \subseteq G$ for which

(2.1.21) $$\overline{FV}_p(h; \mathbf{A}; X_2) \to \overline{FV}_p(h; \mathbf{A}; G)$$

and so prove the p-h-measurability of G.

There is another indication of the variability of h without using p. Given h: $\mathscr{U}^1 \to K$ and $\mathscr{U} \in \mathbf{A}$, the *free variation set* $FVS(h; \mathscr{U})$ for $\mathscr{U}$, is the set of values of all products $\sum_{j=1}^{n} h(I_j, t_j)$, for all finite collections $(I_j, t_j) \in \mathscr{U}$ ($j = 1, \ldots, n$) that have mutually disjoint I_j. There is no product when the collection is empty. However,

(2.1.22) $$FVS(h; \mathscr{U}) \subseteq FVS(h; \mathscr{U}_1) \quad (\mathscr{U} \subseteq \mathscr{U}_1 \subseteq \mathscr{U}^1),$$

except when $\mathscr{U}$ is empty. When K has a unit u, by convention we say that an empty collection has product value u, so that $u \in FVS(h; \mathscr{U})(\mathscr{U} \in \mathbf{A})$, and then (2.1.22) is true even for an empty $\mathscr{U}$, though such a $\mathscr{U}$ would not be in $\mathbf{A}$. Also, for $X \subseteq T$ put

$$h(I, t; X) \equiv \begin{cases} h(I, t) & (t \in X), \\ u & (t \notin X), \end{cases} \quad FVS(h; \mathscr{U}; X) = FVS(h(I, t; X); \mathscr{U}).$$

When K has a topology $\mathscr{G}$ let the *limiting free variation set* $\overline{FVS}(h; \mathbf{A})$ be the intersection, for all $\mathscr{U} \in \mathbf{A}$, of the closure $\mathscr{G}FVS(h; \mathscr{U})$, and similarly $\overline{FVS}(h; \mathbf{A}; X)$ the intersection of $\mathscr{G}FVS(h; \mathscr{U}; X)$. We can now ask the question: how many properties of the free p-variation are copied in some sense by free variation sets?

First, if for some $\mathscr{U} \in \mathbf{A}$, $\mathscr{G}FVS(h; \mathscr{U})$ is compact, we say that h *is of free bounded variation (relative to* $\mathbf{A}$), and then, if $\overline{FVS}(h; \mathbf{A})$ only contains u, we say that h *is of free variation zero (relative to* $\mathbf{A}$). If $\mathscr{G}FVS(h; \mathscr{U}; X)$ is compact for some $\mathscr{U} \in \mathbf{A}$, with $\overline{FVS}(h; \mathbf{A}; X) = \text{sing}(u)$, we say that X *is of free h-variation zero*. If a property holds except in such a set X, we say that *the property holds free h-almost everywhere*. If $h = h^* \cdot h_1$, $h^* = h \cdot h_2$ with h_1, h_2 of free variation zero, we say that h and h^* are *free variationally equivalent*. As will be seen, no clash occurs with earlier definitions.

First we have an approximation result.

Theorem 2.1.7. *Let $\mathscr{G}FVS(h; \mathscr{U})$ be compact for a $\mathscr{U} \in \mathbf{A}$. Then $\overline{FVS}(h; \mathbf{A})$ is not empty even if no unit exists, and for G an open neighbourhood of $\overline{FVS}(h; \mathbf{A})$, there is a $\mathscr{U}_1 \in \mathbf{A}$ with $FVS(h; \mathscr{U}_1) \subseteq G$.*

Proof. A being a free division system, the $\mathscr{U} \in \mathbf{A}$ are directed downwards by set inclusion, making **A** a downwards directed set for Theorem 0.5.2, and $FVS(h; \mathscr{U})$ is an $X(d)$ ($d = \mathscr{U}$). Hence the results. □

Theorem 2.1.8. *Let $h_j : \mathscr{U}^1 \to K$ (an additive semigroup). Then*

(2.1.23) $\quad FVS(h_1 + h_2; \mathscr{U}) \subseteq FVS(h_1; \mathscr{U}) + FVS(h_2; \mathscr{U}) \quad (\mathscr{U} \in \mathbf{A})$.

(2.1.24) *If also $\mathscr{G}$ is Hausdorff, $x + y$ continuous in (x, y), $\mathscr{U}_j \in \mathbf{A}$, $\mathscr{G}FVS(h_j; \mathscr{U}_j)$ compact $(j = 1, 2)$, then for $\mathscr{U} \in \mathbf{A}$, $\mathscr{U} \subseteq \mathscr{U}_1 \cap \mathscr{U}_2$, $\mathscr{G}FVS(h_1 + h_2; \mathscr{U})$ is compact and*

(2.1.25) $\quad \overline{FVS}(h_1 + h_2; \mathbf{A}) \subseteq \overline{FVS}(h_1; \mathbf{A}) + \overline{FVS}(h_2; \mathbf{A})$.

(2.1.26) *If also h_1, h_2 are of free variation zero, then $h_1 + h_2$ is of free variation zero.*

(2.1.27) *If also h, h^* are free variationally equivalent, then,*

$$\overline{FVS}(h; \mathbf{A}) = \overline{FVS}(h^*; \mathbf{A})$$

and free variational equivalence is an equivalence relation.

Proof. (2.1.23) follows from commutativity, associativity, $\sum$ replacing $\prod$, and

$$(\mathscr{E})\sum \{h_1 + h_2\} = (\mathscr{E})\sum h_1 + (\mathscr{E})\sum h_2 \in FVS(h_1; \mathscr{U}) + FVS(h_2; \mathscr{U})(\mathscr{E} \subseteq \mathscr{U}).$$

In (2.1.24), Exercise 0.4.6 and continuity of $x + y$ in (x, y) give $X + Y$ compact when X, Y are compact. (2.1.23) and direction in **A** imply that the closure of the right side of (2.1.23) and so the closure of the left side are compact for the $\mathscr{U}$ of (2.1.24), and all $\overline{FVS}$ in (2.1.25) are non-empty. Using Theorem 0.5.2(0.5.7) and Exercise 0.4.6, the closure of the right side of (2.1.23) tends to the right side of (2.1.25) as $\mathscr{U}$ shrinks, and (2.1.25) follows. Thus (2.1.26) follows, and (2.1.27) also, just as (2.1.7), (2.1.8) follow from (2.1.6). □

Theorem 2.1.9

(2.1.28) $\quad FVS(h; \mathscr{U}; X) \subseteq FVS(h; \mathscr{U}; Y), \overline{FVS}(h; \mathbf{A}; X) \subseteq \overline{FVS}(h; \mathbf{A}; Y)$
$$(X \subseteq Y \subseteq T).$$

For K an additive semigroup, $\mathscr{G}$ Hausdorff, $x + y$ continuous in (x, y), let $X_1, \ldots, X_n$ be n sets in T with union X and $\mathscr{G}FVS(h; \mathscr{U}_j; X_j)$ compact for some $\mathscr{U}_j \in \mathbf{A}$ $(j = 1, \ldots, n)$.

2.1 FREE DIVISION SYSTEMS

(2.1.29) Then $\mathcal{G}FVS(h; \mathcal{U}; X)$ is compact if $\mathcal{U} \in \mathbf{A}$, $\mathcal{U} \subseteq \bigcap_{j=1}^{n} \mathcal{U}_j$, and

(2.1.30) $\overline{FVS}(h; \mathbf{A}; X) \subseteq \sum_{j=1}^{n} \overline{FVS}(h; \mathbf{A}; X_j)$.

Proof. By the trivial (2.1.28) we can in (2.1.30) assume the X_j mutually disjoint. Then

$$h(I, t; X) = \sum_{j=1}^{n} h(I, t; X_j),$$

giving the results from Theorem 2.1.8(2.1.24),(2.1.25) and mathematical induction. They correspond to the results of Theorem 2.1.3. □

We now attempt to tie the variation sets with the p-variation. For each non-negative real number M let $S_p(M)$ be the set of $k \in K$ with $p(k) \leq M$.

Theorem 2.1.10.

(2.1.31) For p continuous relative to $\mathcal{G}$, $\overline{FVS}(h; \mathbf{A}) \subseteq S_p(\overline{FV_p}(h; \mathbf{A}))$.

(2.1.32) Further, if $\overline{FV_p}(h; \mathbf{A}) = 0$, $\mathcal{G}FVS(h; \mathcal{U})$ compact for some $\mathcal{U} \in \mathbf{A}$, and $p(k) > 0$ ($k \in K$, $k \neq u$, the unit), then h is of free variation zero.

(2.1.33) Further, for K an additive semigroup, if, for some fixed integer $q > 0$, the $k \in K$ are sums of at most q components $k^{(m)}$ ($m = 1, \ldots, q$) such that

$$p\left(\sum_{j=1}^{n} k_j^{(m)}\right) = \sum_{j=1}^{n} p(k_j^{(m)}), \; p(k^{(m)}) \leq p(k) \leq \sum_{m=1}^{q} p(k^{(m)}), \; \sum_{j=1}^{n} k_j^{(m)} = \left(\sum_{j=1}^{n} k_j\right)^{(m)}$$

($m = 1, \ldots, q$), and if $\mathcal{G}S_p(M)$ is compact for each $M \geq 0$, then some point $k \in \overline{FVS}(h; \mathbf{A})$ has $p(k) \geq \overline{FV_p}(h; \mathbf{A})/q$ (if this is finite), or, for all $\mathcal{U} \in \mathbf{A}$, $\mathcal{G}FVS(h; \mathcal{U})$ is not compact (if $\overline{FV_p}(h; \mathbf{A}) = +\infty$).

(2.1.34) If in (2.1.33) h has free variation zero, then $\overline{FV_p}(h; \mathbf{A}) = 0$.

(If p is the modulus in $K = \mathbb{R}$ let $k^{(1)} = \max(k, 0)$, $k^{(2)} = \min(k, 0)$, giving two components satisfying the conditions in (2.1.33). If K is s-dimensional Euclidean we can use $2s$ components. Are there other interesting cases?)

Proof. For $X \subseteq K$ let $p(X) = \sup \{p(k) : k \in X\}$. By (2.1.1), finite $\mathcal{E} \subseteq \mathcal{U}$, and p continuous,

$$p((\mathcal{E}) \sum h(I, t)) \leq (\mathcal{E}) \sum p(h(I, t)) \leq FV_p(h; \mathcal{U}), \; p(\overline{FVS}(h; \mathbf{A}))$$

$$\leq p(\mathcal{G}FVS(h; \mathcal{U})) \leq FV_p(h; \mathcal{U}),$$

giving (2.1.31) as $\mathcal{U}$ shrinks. From this, in (2.1.32),

$$p(\overline{FVS}(h; \mathbf{A})) = 0, \; \overline{FVS}(h; \mathbf{A}) = \text{sing}(u).$$

For (2.1.33) let $p(FVS(h; \mathscr{U})) = M$, finite. Then for finite $\mathscr{E} \subseteq \mathscr{U}$,

(2.1.35) $(\mathscr{E})\sum p(h(I, t)^{(m)}) = p((\mathscr{E})\sum h(I, t)^{(m)}) = p(\{(\mathscr{E})\sum h(I, t)\}^{(m)})$
$$\leq p((\mathscr{E})\sum h(I, t)) \leq M,$$
$$(\mathscr{E})\sum p(h(I, t)) \leq qM, \overline{FV}_p(h; A) \leq FV_p(h; \mathscr{U}) \leq qM,$$

and $\overline{FV}_p(h; A)$ is finite, a converse to (2.1.31), partial because $q > 1$. Further, given $\varepsilon > 0$ (even if $M = 0$), there is a $k \in FVS(h; \mathscr{U})$ with

$$qM/(q + \varepsilon) \leq p(k) \leq M.$$

As $S_p(M')$ is compact for all $M' \geq 0$, there is a $k \in \mathscr{G}FVS(h; \mathscr{U})$ with $p(k) = M = p(FVS(h; U))$. Then from (2.1.31), (2.1.35),

$$p(k) \leq FV_p(h; \mathscr{U}) \leq qp(k).$$

Again using the compactness of $S_p(M')$, there is a $k \in \overline{FVS}(h; A)$ with

$$p(k) = p(\overline{FVS}(h; A)) \leq \overline{FV}_p(h; A) \leq qp(k),$$

giving the first part of (2.1.33). However, if $p(FVS(h; \mathscr{U})) = +\infty$ for all $\mathscr{U} \in A$, (2.1.31) shows that $\overline{FV}_p(h; A) = +\infty$. As a continuous function is bounded on a compact set, $FVS(h; \mathscr{U})$ cannot lie in a compact set. In (2.1.34), by choice of $\mathscr{U} \in A$, $M = p(FVS(h; \mathscr{U}))$ is as small as we please, and (2.1.35) gives $\overline{FV}_p(h; A) = 0$.

A set $X \subseteq T$ is called *free h-measurable* if, for each $Y \subseteq T$ with $\mathscr{G}FVS(h; \mathscr{U}; Y)$ compact for some $\mathscr{U} \in A$,

(2.1.36) $\overline{FVS}(h; A; Y \cap X) + \overline{FVS}(h; A; Y \setminus X) \subseteq \overline{FVS}(h; A; Y).$

This looks like (2.1.12), but the $\overline{FVS}$ are sets, and the sum of two sets is the set of all sums of two values, one from each set. By (2.1.28), $\mathscr{G}FVS(h; \mathscr{U}; Z)$ is compact for $Z = Y \cap X$ and $Y \setminus X$, and by (2.1.30) there is equality in (2.1.36).

Corresponding to Theorem 2.1.4, we have the following results.

Theorem 2.1.11. *For K an additive semigroup, $\mathscr{G}$ Hausdorff, $x + y$ continuous in (x, y), let X be the union of mutually disjoint free h-measurable sets X_j ($j = 1, \ldots, n$), with $\mathscr{U}_j \in A$ and $\mathscr{G}FVS(h; \mathscr{U}_j; X_j)$ compact ($j = 1, \ldots, n$). Then X is free h-measurable and for each set Y with $\mathscr{G}FVS(h; \mathscr{U}; Y)$ compact for some $\mathscr{U} \in A$,*

(2.1.37) $\overline{FVS}(h; A; Y) = \sum_{j=1}^{n} \overline{FVS}(h; A; Y \cap X_j) + \overline{FVS}(h; A; Y \setminus X),$

(2.1.38) $\overline{FVS}(h; A; X) = \sum_{j=1}^{n} \overline{FVS}(h; A; X_j).$

Proof. Follow the proof of Theorem 2.1.4 with $\supseteq$ replacing $\geq$. □

2.1 FREE DIVISION SYSTEMS

Theorem 2.1.12. *For the hypotheses on K as in Theorem 2.1.11, if X, W are free h-measurable, so is $X \cap W$, and the class of free h-measurable sets is finitely additive.*

Proof. Follow the proof of Theorem 2.1.5. □

Theorem 2.1.13. *For $(T, \mathcal{T}, \mathbf{A})$ freely compatible with the topology $\mathcal{G}$ of T, and $G_j \in \mathcal{G}$, $X_j \subseteq G_j$ ($j = 1, 2$), G_1, G_2 disjoint, then*

(2.1.39) $\overline{FVS}(h; \mathbf{A}; X_1) + \overline{FVS}(h; \mathbf{A}; X_2) = \overline{FVS}(h; \mathbf{A}; X_1 \cup X_2).$

Proof. See that of Theorem 2.1.6. □

My thanks are due to Thomson (1971a, b, c, 1972a, b, 1975), who examined the theory and showed that it would work with fewer hypotheses, coining the name, *division system*, which in this book is called a free division system. Here I have called a division system one that considers divisions of a fixed elementary set. See Example 1.1.3 for some details of Thomson's system.

While writing the book, Henstock (1968b), it became clear that the use of axioms would be most cumbersome, and J. J. McGrotty suggested the use of names to correspond to measure spaces. Thus I invented the name division space, and the original nomenclature can be found in that book, Section 43. Slight changes have occured from time to time, and this book uses the definitive nomenclature.

It appears that McShane (1969) pioneered the use of semigroup spaces of values, Henstock (1969) only using groups.

The very clever definition of measurability in (2.1.12) is of course due to Carathéodory, the 'embroidery' of this section being added on.

The variation is a generalization of that in, say, Saks (1937), see Henstock (1979), while the variation set first appeared in Henstock (1961b), p. 404. The N-variation first appeared in Henstock (1961a).

It turns out that the version of the free variation set for division systems (so that 'free' and 'F' are dropped) is the most natural way of defining the variational integral without use of a p-modulus, and of linking this integral to the generalized Riemann integral to be defined in Section 2.3. See Theorem 2.5.4.

Exercise 2.1.1 Let $(\mathbb{R}, \mathcal{T}, \mathbf{A})$ be the gauge (Kurzweil–Henstock) arrangement on the real line $\mathbb{R}$ of Section 1.4. Let $h([0, n^{-1}), t) = n$ ($n = 1, 2, \ldots$), and otherwise let $h = 0$. Show that $\overline{FV}(h; \mathbf{A}; [0, 1]) = +\infty$, $\overline{FVS}(h; \mathbf{A}; [0, 1]) = \text{sing}(0)$, and $FVS(h; \mathcal{U}; [0, 1))$ is unbounded for each $\mathcal{U} \in \mathbf{A}$. This illustrates the need in Theorem 2.1.7 and later theorems of an FVS set whose closure is compact, and not just a compact $\overline{FVS}$ set.

Exercise 2.1.2 If h and h^* are free p-variationally equivalent and if $X \subseteq T$, then show that

$$\overline{FV}_p(h; \mathbf{A}; X) = \overline{FV}_p(h^*; \mathbf{A}; X).$$

(If $\overline{FV}_p(k; \mathbf{A}) = 0$ then $\overline{FV}_p(k; \mathbf{A}; X) = 0$, as all terms in sums are omitted except those with $t \in X$, the only difference between the two kinds of sum. As h and h^* are free p-variationally equivalent there are h_1, h_2 with the property of k, such that $h = h^* \cdot h_1$, $h^* = h \cdot h_2$. Hence $h\chi(X;\cdot)$ and $h^*\chi(X;\cdot)$ are free p-variationally equivalent and Theorem 2.1.1(2.1.7) gives the result.)

Exercise 2.1.3 If $h: \mathcal{U}^1 \to \mathbb{R}$, $h^*: \mathcal{U}^1 \to \mathbb{R}$, $f: T \to \mathbb{R}$, and h, fh^* are free p-variationally equivalent, and if $|f| \leq C$ in the set $X \subseteq T$, then for p the modulus, show that

$$\overline{FV}(h; \mathbf{A}; X) \leq C . \overline{FV}(h^*; \mathbf{A}; X).$$

Changing $|f| \leq C$ to $|f| \geq C$ in X, then

$$\overline{FV}(h; \mathbf{A}; X) \geq C . \overline{FV}(h^*; \mathbf{A}; X).$$

Exercise 2.1.4 For a fixed positive integer n let K be the set of vectors k with n real components $k_1, \ldots, k_n$, and with $p(k) = \{k_1^2 + \cdots + k_n^2\}^{1/2}$. Show that $q = 2n$.

Exercise 2.1.5 Let K be the set of all bounded real-valued sequences $k = (k_n)$ with

$$p(k) = \sup_n |k_n|.$$

Then q is infinite. For $h: \mathcal{U}^1 \to K$ we have $h(I, t) = (h_n(I, t))$, a bounded sequence for each fixed (I, t). Let

$$h_n(I, t) = \begin{cases} 1 & (t = n^{-1}, I = [n^{-1}, u), n^{-1} < u < (n-1)^{-1}), \\ 0 & \text{(otherwise)}. \end{cases}$$

For the division system of Section 1.4 with suitable $\varepsilon > 0$, show that $FVS(h; \mathcal{U})$ is the set of (k_n) with $k_n = 0$ or 1 ($n = 1, 2, \ldots$) while $FV_p(h; \mathcal{U}) = +\infty$ (all $\mathcal{U} \in \mathbf{A}$).

Exercise 2.1.6 Given that the group operation in $\mathbb{R}$ or $\mathbb{C}$ is multiplication and $p(z) = |z|$, show that $q = 1$ with $p(wz) = p(w)p(z)$ in (2.1.1), and $p(\text{unit}) = 0$ becomes $p(1) = 1$ because of multiplication instead of addition. In (2.1.33), using (2.1.31), there is a $k \in \overline{FVS}(h; \mathbf{A})$ with $p(k) = \overline{FV}_p(h; \mathbf{A})$, and $p(\overline{FVS}(h; \mathbf{A})) = \overline{FV}_p(h; \mathbf{A})$.

2.2 FREELY DECOMPOSABLE DIVISION SYSTEMS

To the definitions of Section 2.1. we add the following: $(T, \mathcal{T}, \mathbf{A})$ is *freely fully decomposable* (respectively, *freely decomposable*, or *freely measurably decomposable relative to a measure* {see (2.1.12)} or measure space—defined later) if to every family (respectively, countable family, or countable family of measurable sets) $\mathcal{X}$ of mutually disjoint subsets $X \subseteq T$, and every function $\mathcal{U}(.)$: $\mathcal{X} \to \mathbf{A}$, there is a $\mathcal{U} \in \mathbf{A}$ with

$$\mathcal{U}[X] \equiv \{(I, t): (I, t) \in \mathcal{U}, t \in X\} \subseteq \mathcal{U}(X) \quad (X \in \mathcal{X}).$$

There is no need for the union of the $X \in \mathcal{X}$ to be T. If, for the given X,

$$\mathcal{U}[X] = \mathcal{U}(X)[X] \quad (X \in \mathcal{X})$$

we call $\mathcal{U}$ the *diagonal of the* $(\mathcal{U}(X), \mathcal{X})$.

Here we assume decomposability. The FV_p, FVS, and other functions and sets are as in Section 2.1, and we can now add to the results given there.

Theorem 2.2.1. *Let (X_j) be a sequence of subsets of T with union X. Then*

(2.2.1) $$\overline{FV}_p(h; \mathbf{A}; X) \leq \sum_{j=1}^{\infty} \overline{FV}_p(h; \mathbf{A}; X_j).$$

If the X_j are mutually disjoint and free p-h-measurable, then X is free p-h-measurable, and for each $Y \subseteq T$,

(2.2.2) $$\overline{FV}_p(h; \mathbf{A}; Y) = \sum_{j=1}^{\infty} \overline{FV}_p(h; \mathbf{A}; Y \cap X_j) + \overline{FV}_p(h; \mathbf{A}; Y \backslash X),$$

(2.2.3) $$\overline{FV}_p(h; \mathbf{A}; X) = \sum_{j=1}^{\infty} \overline{FV}_p(h; \mathbf{A}; X_j).$$

(The last result is the definition of the countable additivity of $\overline{FV}_p(h; \mathbf{A}; X)$ in the family of *p-h*-measurable sets.)

Let K also have a scalar multiplication by real or complex numbers, with $p(fk) \leq |f| p(k)$ ($k \in K$, f the scalar).

(2.2.4) *If $\overline{FV}_p(h; \mathbf{A}; X) = 0$, with f a scalar-valued point function, then $\overline{FV}_p(fh; \mathbf{A}; X) = 0$.*

(2.2.5) *If $\overline{FV}_p(fh; \mathbf{A}; X) = 0$ with $f \neq 0$ in $X_0 \subseteq X$, then $\overline{FV}_p(h; \mathbf{A}; X_0) = 0$.*

Proof. If in (2.2.1) all but a finite number of X_j are empty, we use Theorem 2.1.3. Otherwise, in the proof of that theorem we replace direction in $\mathbf{A}$ by free decomposability. By (2.1.4) we can take the X_j mutually disjoint, and there is a $\mathcal{U} \in \mathbf{A}$ with $\mathcal{U}[X_j] \subseteq U_j$ ($j \geq 1$). If $\mathscr{E}$ comes from $\mathcal{U}$ and if $\mathscr{P}$, $\mathscr{P}_j$ are the subsets of $\mathscr{E}$ for which the points t lie in X, X_j, respectively, as $\mathscr{E}$ is a finite set there is a greatest integer m depending on $\mathscr{E}$, with $\mathscr{P}_m$ not empty. Also we can assume the series in (2.2.1) convergent, or else there is nothing to prove. Then

$$(2.2.6) \quad (\mathscr{E})\sum p(h(I, t); X) = (\mathscr{P})\sum p(h) = \sum_{j=1}^{m} (\mathscr{P}_j)\sum p(h)$$

$$\leq \sum_{j=1}^{m} FV_p(h; \mathscr{U}_j; X_j)$$

$$< \sum_{j=1}^{m} (\overline{FV}_p(h; \mathbf{A}; X_j) + \varepsilon 2^{-j})$$

$$< \sum_{j=1}^{\infty} \overline{FV}_p(h; \mathbf{A}; X_j) + \varepsilon,$$

$$\overline{FV}_p(h; \mathbf{A}; X) \leq \overline{FV}_p(h; \mathscr{U}; X) \leq \sum_{j=1}^{\infty} \overline{FV}_p(h; \mathbf{A}; X_j) + \varepsilon,$$

and $\varepsilon > 0$ is arbitrary. Hence (2.2.1) is true. If in (2.2.1) the X_j are free p-h-measurable and mutually disjoint, then by (2.1.4) and Theorem 2.1.4(2.1.13),

$$\overline{FV}_p(h; \mathbf{A}; Y) \geq \sum_{j=1}^{n} \overline{FV}_p(h; \mathbf{A}; Y \cap X_j) + \overline{FV}_p(h; \mathbf{A}; Y \setminus X).$$

Being true for all positive integers n, we can replace n by $+\infty$. The opposite inequality follows from (2.2.1), giving (2.2.2), and (2.2.3) on putting $Y = X$. Then (2.1.12) follows for X on using (2.2.2) and (2.2.1), so that X is free p-h-measurable.

Next, from $p(fk) \leq |f|p(k)$ there is equality when $f = 0$, and when $f \neq 0$ we have $|f|p(f^{-1}(fk)) \leq |f||f^{-1}|p(fk) = p(fk)$, giving the opposite inequality, so that

$$(2.2.7) \qquad\qquad p(fk) = |f|p(k).$$

For (2.2.4) let X_j be the subset of X with $|f| \leq j$ ($j = 1, 2, \ldots$). Then by (2.2.1), (2.1.4), (2.2.7),

$$\overline{FV}_p(fh; \mathbf{A}; X) \leq \sum_{j=1}^{\infty} \overline{FV}_p(fh; \mathbf{A}; X_j) \leq \sum_{j=1}^{\infty} j \cdot \overline{FV}(h; \mathbf{A}; X_j)$$

$$\leq \sum_{j=1}^{\infty} j \cdot \overline{FV}_p(h; \mathbf{A}; X) = 0.$$

For (2.2.5) replace X and f by X_0 and $1/f$ in (2.2.4). □

Note that property (2.2.1) is the definition of an outer measure in Lebesgue theory.

We now turn to corresponding results concerning variation sets. Scalar multiplication seems difficult to handle, and is omitted.

2.2 FREELY DECOMPOSABLE DIVISION SYSTEMS

Theorem 2.2.2. *Let (X_j) be a sequence of subsets of T with union X, let K be an additive topological group, and for some $\mathcal{U}_j \in \mathbf{A}$ let $\mathcal{G}FVS(h; \mathcal{U}_j; X_j)$ be compact. Then*

(2.2.8) $$\overline{FVS}(h; \mathbf{A}; X) \subseteq \sum_{j=1}^{\infty} \overline{FVS}(h; \mathbf{A}; X_j).$$

Here the symbolism covers an operation that is not as simple as it seems. If $K_j \subseteq K$ $(j = 1, 2, \ldots)$ the symbol $\sum_{j=1}^{\infty} K_j$ denotes the closure of the set

$$\sum_{j=1}^{<\infty} K_j \equiv \left\{ \sum_{j=1}^{n} k_j : k_j \in K_j, \ j = 1, \ldots, n, \text{ for } n = 1, 2, \ldots \right\}.$$

If also the X_j are mutually disjoint and free h-measurable, then X is free h-measurable and for each $Y \subseteq T$,

(2.2.9) $\overline{FVS}(h; \mathbf{A}; Y) = \sum_{j=1}^{\infty} \overline{FVS}(h; \mathbf{A}; Y \cap X_j) + \overline{FVS}(h; \mathbf{A}; Y \setminus X),$

(2.2.10) $\overline{FVS}(h; \mathbf{A}; X) = \sum_{j=1}^{\infty} \overline{FVS}(h; \mathbf{A}; X_j).$

Thus $\overline{FVS}(h; \mathbf{A}; X)$ is countably additive in the given extended sense for addition of sets, for X in the family of all free h-measurable sets.

Proof. Theorem 2.1.9(2.1.28) shows that we can assume the X_j mutually disjoint, and if then all but a finite number of X_j are empty, Theorem 2.1.9(2.1.30) and Theorem 2.1.12 give the results. Otherwise we begin with Theorem 0.4.13, an open neighbourhood G of the zero 0, and a sequence (G_j) of open neighbourhoods of 0, such that

(2.2.11) $G_1 = G, \quad G_j - G_j \subseteq G_{j-1} \ (j \geq 2),$

$$G_{N+1} + G_{N+2} + \ldots + G_q \subseteq G_N \quad (q > N \geq 1).$$

By the approximation result, Theorem 2.1.7, there is a $\mathcal{U}_j \in \mathbf{A}$ with

(2.2.12) $$FVS(h; \mathcal{U}_j; X_j) \subseteq \overline{FVS}(h; \mathbf{A}; X_j) + G_j,$$

the right side being an open neighbourhood of $\overline{FVS}(h; \mathbf{A}; X_j)$. In the proof of Theorem 2.2.1 we now replace $p(h)$ by h, $\leq$ by $\in$ or $\subseteq$, and use the analogue of (2.2.6).

$$(\mathcal{E})\sum h(I, t; X) = (\mathcal{P})\sum h = \sum_{j=1}^{m} (\mathcal{P}_j)\sum h \in \sum_{j=1}^{m} FVS(h; \mathcal{U}_j; X_j)$$

$$\subseteq \sum_{j=1}^{m} \left\{ \overline{FVS}(h; \mathbf{A}; X_j) + G_j \right\}$$

$$= \sum_{j=1}^{m} \overline{FVS}(h; \mathbf{A}; X_j) + \sum_{j=1}^{m} G_j \subseteq \sum_{j=1}^{m} \overline{FVS}(h; \mathbf{A}; X_j) + G,$$

$$FVS(h; \mathcal{U}; X) \subseteq \sum_{j=1}^{<\infty} \overline{FVS}(h; \mathbf{A}; X_j) + G,$$

$$\overline{FVS}(h; \mathbf{A}; X) \subseteq \sum_{j=1}^{\infty} \overline{FVS}(h; \mathbf{A}; X_j) + \bar{G}.$$

As K is a topological group, Theorems 0.4.11, 0.4.14 now give (2.2.8).

Note that $\sum_{j=1}^{m}(A_j + B_j)$ is the set of $\sum_{j=1}^{m}(a_j + b_j) = \sum_{j=1}^{m} a_j + \sum_{j=1}^{m} b_j$, and so is $\sum_{j=1}^{m} A_j + \sum_{j=1}^{m} B_j$, and we can reverse the argument by pairing off a_j with b_j for each j.

To prove (2.2.9), (2.2.10) we follow the proofs of (2.2.2), (2.2.3), using Theorem 2.1.11(2.1.37). □

Decomposability first appeared as axiom (A4) in Henstock (1961a), p. 117, and was used to prove Theorem 8, p. 123, the N-variational form of Theorem 2.2.1 in the present book. It was named in Henstock (1968b), pp. 216, 217 as was *full decomposability*.

2.3 DIVISION SYSTEMS, THE GENERALIZED RIEMANN AND VARIATIONAL INTEGRALS, THE VARIATION, AND THE VARIATION SET

There are at least three possible ways of defining variational integrals of functions with values in a Banach space B, and two ways of defining generalized Riemann integrals of such functions. We can use the norm of B directly through the norm variation to obtain the strong variational integral, as in Section 2.1 before Theorem 2.1.2. Or we can use a common device of integration theory, namely, a family of continuous linear functionals $\mathscr{F}$ on B. If an **A**-*integral* $\mathbf{A}(k)$ is defined for real functions k, if $h: \mathcal{U}^1 \to B$, and if, for some $H \in B$ and each continuous linear functional $\mathscr{F}$ on B, $\mathbf{A}(\mathscr{F}(h))$ exists and is equal to $\mathscr{F}(H)$, then we say that *h is weakly **A**-integrable with weak **A**-integral H*. For example, Dunford (1936b, c) and Pettis (1938) define the *weak Lebesgue integral*, and we could define the *weak generalized Riemann* and *weak variational integrals*. Theorems for $\mathbf{A}(h)$ follow from theorems for $\mathbf{A}(\mathscr{F}(h))$ by standard ways, and for a unique H we need enough continuous linear functionals to separate all points of B. At times this method is attractive, but when K does not have enough continuous linear functionals we need some other method. Also, on moving to spaces K without a norm, it is difficult to extend, say, the majorized or dominated convergence theorem from $K = \mathbb{R}$ to general K, except for an equivalence found recently that uses Riemann sums. So here we first define the generalized Riemann integral and the variation directly, the latter giving a better variational integral.

In this section $(T, \mathcal{T}, \mathbf{A})$ is a division system for an elementary set E. For convenience we recite some of the definitions again. Using the collection $\mathcal{T}$ of

2.3 GENERALIZED RIEMANN AND VARIATIONAL INTEGRALS

generalized intervals I and $\mathcal{U}^1$ of interval-point pairs (I, t), the *elementary set* E is an interval or a finite union of mutually disjoint intervals. A subfamily $\mathcal{U} \subseteq \mathcal{U}^1$ *divides* E if, for a *finite* subfamily $\mathcal{E} \subseteq \mathcal{U}$, called a *division of E from $\mathcal{U}$*, the $(I, t) \in \mathcal{E}$ have mutually disjoint I with union E. The collection of these I alone is called a *partition of E from $\mathcal{U}$*, and the separate I are called *partial intervals of E from $\mathcal{U}$*, any $\mathcal{P} \subseteq \mathcal{E}$ is called a *partial division of E from $\mathcal{U}$*, and $\mathcal{U}.E$ is the set of all $(I, t) \in \mathcal{U}$ with I a partial interval of E. We use a family **A** of some non-empty subsets $\mathcal{U} \subseteq \mathcal{U}^1$, and $\mathbf{A}|E$ is the set of all $\mathcal{U}.E$ with $\mathcal{U} \in \mathbf{A}$ dividing E. $(T, \mathcal{T}, \mathbf{A})$ is called a *division system for E* if $\mathbf{A}|E$ is not empty and if **A** is *directed for divisions of E*, i.e. given $\mathcal{U}_j \in \mathbf{A}|E$ ($j = 1, 2$), there is a $\mathcal{U} \in \mathbf{A}|E$ with $\mathcal{U} \subseteq \mathcal{U}_1 \cap \mathcal{U}_2$.

Let the space K of values have a Hausdorff topology and a multiplication operation $x.y$ for all $x, y \in K$, and let $h: \mathcal{U}^1 \to K$. If $\mathcal{U} \in \mathbf{A}|E$ let $S(h; \mathcal{U}; E)$ be the set of values of all products $(\mathcal{E})\prod h(I, t)$ for the $(I, t) \in \mathcal{E}$ in any order and any grouping of pairs of terms in brackets separated by the multiplication dot, and for all divisions $\mathcal{E}$ of E from $\mathcal{U}$. In a semigroup K there is no need of brackets. If $x.y$ is commutative we often write it as $x + y$, and $(\mathcal{E})\sum$ for $(\mathcal{E})\prod$; then the order of terms in brackets is immaterial. As $\mathbf{A}|E$ is directed by set inclusion, then 'as $\mathcal{U}$ shrinks', $S(h; \mathcal{U}; E)$ is a monotone decreasing set-valued function of the $\mathcal{U} \in \mathbf{A}|E$, even in the general case. We say that $S(h; \mathcal{U}; E)$ is *convergent* $(\mathbf{A}; E)$ with limit $H \in K$ if, given an arbitrary $G \in \mathcal{G}$ with $H \in G$, there is a $\mathcal{U} \in \mathbf{A}|E$ depending on G, such that $S(h; \mathcal{U}; E) \subseteq G$. The limit H is called the *generalized Riemann integral of h over E (relative to $\mathbf{A}, \mathcal{G}$)*, written

$$H = H(E) = (\mathbf{A}) \int_E dh = \int_E dh,$$

omitting **A** when it is understood. For additive K with another (possibly scalar) multiplication $h(I, t) = f(t)g(I, t)$ and h integrable, we often write the integral as

$$(\mathbf{A}) \int_E f \, dg = \int_E f \, dg.$$

Theorem 2.3.1. *The generalized Riemann integral is uniquely defined.*

Proof. If $J \in K$ is also a limit, then given arbitrary G_1 in $J \in G_1 \in \mathcal{G}$, there is a $\mathcal{U}_1 \in \mathbf{A}|E$ with $S(h; \mathcal{U}_1; E) \subseteq G_1$. $\mathbf{A}|E$ being directed, there is a $\mathcal{U}_2 \in \mathbf{A}|E$ with $\mathcal{U}_2 \subseteq \mathcal{U} \cap \mathcal{U}_1$, and $S(h; \mathcal{U}_2; E)$ lies in $G \cap G_1$. As $\mathcal{G}$ is Hausdorff this can occur only when $J = H$. □

We say that $S(h; \mathcal{U}; E)$ is *fundamental* $(\mathbf{A}; E)$ if there is a point $k \in K$ such that to each neighbourhood G of k there correspond an $s \in K$ and a $\mathcal{U} \in \mathbf{A}|E$ such that $s.S(h; \mathcal{U}; E) \subseteq G$. This corresponds to the definitions given just before Theorem 0.5.5; that theorem and the next are vital here.

When $S(h; \mathcal{U}; E)$ does not have a limit we can use

$$\bar{S}(h; \mathbf{A}; E) = \bigcap \mathscr{G} S(h; \mathcal{U}; E) \quad (\text{all } \mathcal{U} \in \mathbf{A}|E).$$

Theorem 2.3.2. *Let $\mathscr{G}S(h; \mathcal{U}; E)$ be compact for some $\mathcal{U} \in \mathbf{A}|E$. Then $\bar{S}(h; \mathbf{A}; E)$ is not empty, and if G is a $\mathscr{G}$-neighbourhood of $\bar{S}(h; \mathbf{A}; E)$, there is a $\mathcal{U}_1 \in \mathbf{A}|E$ with*

$$(2.3.1) \qquad S(h; \mathcal{U}_1; E) \subseteq \mathscr{G}.$$

Proof. Follow the proof of Theorem 0.5.2(0.5.5), (0.5.6). □

Sometimes two disjoint partial intervals from different partitions of E cannot together be part of another partition of E, see Example 2.3.7. Thus we cannot transfer the definitions of FV_p and FVS to here, except by analogy. We need new definitions.

Given $h; \mathcal{U}^1 \to K$, the p-modulus $p(k)$ ($k \in K$), the elementary set E, and $\mathcal{U} \in \mathbf{A}|E$, define

$$V_p(h; \mathcal{U}; E) \equiv \sup(\mathscr{E}) \sum p(h(I, t)), \quad \bar{V}_p(h; \mathbf{A}; E) \equiv \inf V_p(h; \mathcal{U}; E)$$
$$= \limsup(\mathscr{E}) \sum p(h),$$

first over all divisions $\mathscr{E}$ of E from $\mathcal{U}$, and then over all $\mathcal{U} \in \mathbf{A}|E$. If $\bar{V}_p(h; \mathbf{A}; E)$ is finite we say that h is of *bounded p-variation* over E, and if $\bar{V}_p(h; \mathbf{A}; E) = 0$ we say that h is of *p-variation zero* over E. When p is the norm we omit the symbol p and say, *norm variation*.

The *variation set* $VS(h; \mathcal{U}; E)$ for a $\mathcal{U} \in \mathbf{A}|E$ is the set of values $(\mathscr{P}) \prod h(I, t)$ for all partial divisions $\mathscr{P}$ of E from $\mathcal{U}$, the values being taken in all orders, with all arrangements of brackets around pairs of elements of K, separated by the multiplication sign. If K has a unit u and if P is empty, then by convention we take the product to be u. The *limiting variation set* is defined to be

$$\overline{VS}(h; \mathbf{A}; E) \equiv \bigcap \mathscr{G} VS(h; \mathcal{U}; E) \quad (\text{all } \mathcal{U} \in \mathbf{A}|E).$$

If for some $\mathcal{U} \in \mathbf{A}|E$, $\mathscr{G} VS(h; \mathcal{U}; E)$ is compact, we say that h is *of bounded variation in E*; if also $\overline{VS}(h; \mathbf{A}; E) = \text{sing}(u)$, we say that h is *of variation zero in E*.

As a division of E from $\mathcal{U}$, is also a partial division, we have

$$(2.3.2) \qquad S(h; \mathcal{U}; E) \subseteq VS(h; \mathcal{U}; E), \quad \bar{S}(h; \mathbf{A}; E) \subseteq \overline{VS}(h; \mathbf{A}; E).$$

If h has variation zero in E then $S(h; \mathcal{U}; E)$ also has limit u as $\mathcal{U}$ shrinks,

$$(2.3.3) \qquad \int_E dh = u, \quad \bar{S}(h; \mathbf{A}; E) = \overline{VS}(h; \mathbf{A}; E) = \text{sing}(u).$$

2.3 GENERALIZED RIEMANN AND VARIATIONAL INTEGRALS

For a set $X \subseteq T$ we define

$$h(I, t; X) \equiv \begin{cases} h(I, t) & (t \in X), \\ u & (t \notin X), \end{cases}$$

and write $V_p(h(I, t; X); \mathcal{U}; E)$ as $V_p(h; \mathcal{U}; E; X)$. Similarly for $\bar{V}_p$, VS, $\overline{VS}$. If for some $\mathcal{U} \in \mathbf{A}|E$, $\mathcal{G}\ VS(h; \mathcal{U}; E; X)$ is compact and tends to a limit, which has to be u, as $\mathcal{U}$ shrinks, we say that X is of h-*variation zero, relative to* E, $\mathbf{A}$. Clearly

(2.3.4) $\qquad VS(h; \mathcal{U}; E; X) \subseteq VS(h; \mathcal{U}; E; Y) \subseteq VS(h; \mathcal{U}; E; T)$

$\qquad\qquad = VS(h; \mathcal{U}; E) \quad (X \subseteq Y \subseteq T),$

(2.3.5) if Y is of h-variation zero, then X is of h-variation zero for any $X \subseteq Y$ and

$$\int_E dh(I, t; X) = u.$$

Theorem 2.3.3. *For* K *an additive semigroup let* $h_j: \mathcal{U}^1 \to K (j = 1, 2)$. *Then*

(2.3.6) $\qquad S(h_1 + h_2; \mathcal{U}; E) \subseteq S(h_1; \mathcal{U}; E) + S(h_2; \mathcal{U}; E).$

For $x + y$ *continuous in* (x, y) *and* h_1, h_2 *integrable over* E, *the first integral below exists and*

(2.3.7) $\qquad \int_E d(h_1 + h_2) = \int_E dh_1 + \int_E dh_2.$

(2.3.8) *For* $K = \mathbb{C}$ *with its modulus topology,* $h_j: \mathcal{U}^1 \to \mathbb{R}$ *and numbers* $H_j \in \mathbb{R}$ $(j = 1, 2)$, *if* $h_1 + ih_2$ *is integrable to* $H_1 + iH_2$, *then* h_j *is integrable to* H_j $(j = 1, 2)$.

Proof.

$$(\mathcal{E})\sum\{h_1(I, t) + h_2(I, t)\} = (\mathcal{E})\sum h_1(I, t) + (\mathcal{E})\sum h_2(I, t)$$

$$\in S(h_1; \mathcal{U}; E) + S(h_2; \mathcal{U}; E)$$

gives (2.3.6). If the integrals H_j of h_j $(j = 1, 2)$ exist, then by continuity of $x + y$ in (x, y), if G is a given $\mathcal{G}$-neighbourhood of $H_1 + H_2$ there are $\mathcal{G}$-neighbourhoods G_j of H_j $(j = 1, 2)$ such that $G_1 + G_2 \subseteq G$, and $\mathcal{U}_j \in \mathbf{A}|E$ such that $S(h_j; \mathcal{U}_j; E) \subseteq G_j$ $(j = 1, 2)$. By direction a $\mathcal{U} \in \mathbf{A}|E$ exists with $\mathcal{U} \subseteq \mathcal{U}_1 \cap \mathcal{U}_2$. By (2.3.6),

$$S(h_1 + h_2; \mathcal{U}; E) \subseteq S(h_1; \mathcal{U}_1; E) + S(h_2; \mathcal{U}_2; E) \subseteq G_1 + G_2 \subseteq G,$$

$h_1 + h_2$ is integrable over E to $H_1 + H_2$, and (2.3.7) follows. For (2.3.8), a partial converse, given $\varepsilon > 0$, there is a $\mathcal{U} \in \mathbf{A}|E$ such that every division $\mathcal{E}$ of

E from $\mathscr{U}$ satisfies

$$\varepsilon > |(\mathscr{E})\sum(h_1 + ih_2) - (H_1 + iH_2)|$$
$$= |(\mathscr{E})\sum h_1 - H_1 + i\{(\mathscr{E})\sum h_2 - H_2\}| \geq |(\mathscr{E})\sum h_j - H_j|$$

($j = 1, 2$), so that as $\varepsilon > 0$ is arbitrary, H_j is the integral of h_j ($j = 1, 2$).

Instead of using a topology $\mathscr{G}$ in K, we now suppose that K is a complete lattice (see Section 0.5), and can then define the integral as the *o-limit* (*order limit*) of $S(h; \mathscr{U}; E)$ for the downwards directed system of sets $\mathscr{U} \in A|E$. Let $h_1 . h_2 \geq k_1 . k_2$ when $h_j \geq k_j$, for all such h_j, k_j in K ($j = 1, 2$). If $h(I, t) \geq k(I, t)$ for $(I, t) \in \mathscr{U}$ and some $\mathscr{U} \in A|E$, if $P(h)$ is a product of $h(I_j, t_j) \in K$ ($j = 1, \ldots, n$) with a given arrangement of brackets and multiplications, and if $P(k)$ is the corresponding product of $k(I_j, t_j) \in K$, with the same arrangement of brackets and multiplications, then an induction on the number and arrangement of multiplications shows that $P(h) \geq P(k)$. □

Theorem 2.3.4. *If $h_j: \mathscr{U}^1 \to K$ is integrable ($j = 1, 2$) with $h_1 \geq h_2$ in $\mathscr{U}^1$, then*

(2.3.9) $$\int_E dh_1 \geq \int_E dh_2.$$

If $h_j: \mathscr{U}^1 \to K$, integrable, $h_j \geq 0$ ($j = 1, 2, \ldots$), K an additive complete lattice with $x + y$ o-continuous in (x, y) and $h = \sum_{j=1}^{\infty} h_j$, an o-limit of a monotone increasing sequence, and if h is also integrable,

(2.3.10) $$\int_E dh \geq \sum_{j=1}^{\infty} \int_E dh_j.$$

Let there be scalar multiplication in K for real scalars. If $f = 0$ on $\bar{X} \setminus X$, f bounded on X with bounds $m = \inf f$, $M = \sup f$ there, and $h \geq 0$, and if $h(I, t; X)$ and fh are integrable, then

(2.3.11) $$m \int_E dh(I, t; X) \leq \int_E f\, dh(I, t, X) \leq M \int_E dh(I, t; X).$$

This is the mean value theorem for integrals. If in (2.3.11) we also have $K = \mathbb{R}$, $E = [a, b) \subset \mathbb{R}$, $X \subseteq E$, and $f: \bar{E} \to \mathbb{R}$ is continuous in $\bar{E}$, then for some $c \in \bar{E}$,

(2.3.12) $$\int_E f\, dh(I, t; X) = f(c) \int_E dh(I, t; X).$$

Proof. (2.3.9) is straightforward, and (2.3.10) follows from $h \geq \sum_{j=1}^{n} h_j$, the analogue of Theorem 2.3.3(2.3.7) using o-convergence, and (2.3.9). In (2.3.11) we use $mh \leq fh \leq Mh$ in X. Then as $h \geq 0$, if the integral of $h(I, t; X)$ is 0, so is the integral of $f(t)h(I, t; X)$, and (2.3.12) is true with 0 both sides. Otherwise

$$\int_E f\, dh \bigg/ \int_E dh$$

2.3 GENERALIZED RIEMANN AND VARIATIONAL INTEGRALS

lies between m and M, and (2.3.12) is true from the intermediate value theorem for continuous functions. □

Exercise 2.3.5 shows that there need not be equality in (2.3.10), and Exercise 2.3.6 shows that h need not be integrable. More can be proved for special kinds of h_j, see Chapter 3.

In this section all divisions have to be divisions of the elementary set E, but this fact does not affect the proofs of analogues of Theorems 2.1.1–13, so that the analogues of these theorems are true and need not be rewritten, except that Theorem 2.1.10 is partially rewritten here below.

Theorem 2.3.5.

(2.3.13) *When K is a Banach space with norm $p = \|\cdot\|$, and $h: \mathscr{U}^1 \to K$, $\overline{VS}(h; A; E)$ is contained in the closed sphere centre the origin and radius $\bar{V}(h; A; E)$.*

(2.3.14) *If also K is the real line ($q = 2$) or complex plane ($q = 4$), then some point $k \in \overline{VS}(h; A; E)$ has $|k| \geq \bar{V}(h; A; E)/q$ if the right side is finite; or if $\bar{V}(h; A; E) = +\infty$, then for all $\mathscr{U} \in A|E$, $\mathscr{G} \ VS(h; \mathscr{U}; E)$ is not compact.*

Proof. See that of Theorem 2.1.10. □

Theorem 2.3.6. *For K a Banach space with norm $\|\cdot\|$, and $h: U^1 \to K$, with h and $\|h\|$ integrable in the elementary set E,*

(2.3.15)
$$\left\| \int_E dh \right\| \leq \int_E d\|h\|.$$

Proof. For every division $\mathscr{E}$ of E,
$$\|(\mathscr{E}) \sum h\| \leq (\mathscr{E}) \sum \|h\|.$$
Now take limits as $\mathscr{U}$ shrinks. □

Exercise 2.3.1 For a fixed positive integer n let K be the set of vectors k with n real components k_j ($j = 1, 2, \ldots, n$) and $p(k) = \{k_1^2 + \cdots + k_n^2\}^{1/2}$. Show that there is an analogue of Theorem 2.3.3(2.3.8).

Example 2.3.2 Let K be a set of bounded functions $k: W \to R$, with $p(k) \equiv \sup_{w \in W} |k(w)|$, Then

$$\varepsilon > \sup_{w \in W} |(\mathscr{E}) \sum h(w; I, t) - H(w)| \geq |(\mathscr{E}) \sum h(w_0; I, t) - H(w_0)|$$

and $H(w_0)$ is the integral of $h(w_0; I, t)$ for each $w_0 \in W$, if H is the integral of h, another analogue of Theorem 2.3.3(2.3.8).

Example 2.3.3 If the group operation in $\mathbb{R}$ or $\mathbb{C}$ is multiplication, there is only one component, and here, Theorem 2.3.3(2.3.8) is trivial.

Exercise 2.3.4 For addition and scalar multiplication in K, and a scalar constant b, an integrable h, and continuity of bk in the $k \in K$, prove that

$$\int_E b\, dh = b \int_E dh.$$

Exercise 2.3.5 On the real line, for each fixed integer $j \geq 2$ let $h_j([u, v), t) = v - u$ if $(j + 1)u/j < v \leq ju/(j - 1)$ $(u > 0)$, and otherwise let $h_j = 0$. Then

$$h \equiv \sum_{j=2}^{\infty} h_j = v - u \quad (u < v \leq 2u) \text{ and otherwise } h = 0.$$

For **A** from Section 1.4 (the gauge integral) show that

$$\int_{[0,1)} h_j = 0 \quad (\text{all } j \geq 2), \quad \int_{[0,1)} h = 1$$

(Henstock (1963c) Ex. 20.1, p. 29). (If for $t > 0, 0 < \delta(t) \leq t/(n + 1)$ then $h_j = 0$ $(2 \leq j \leq n)$ for δ-fine interval-point pairs. But if $\delta(t) \leq \tfrac{1}{2}t$ $(t > 0), h([u, v), t) = v - u$ for δ-fine pairs.)

Exercise 2.3.6 In Exercise 2.3.5 prove that $h^* \equiv \sum_{j=1}^{\infty} h_{2j}$ is not integrable over any interval of $[0, 1)$. Henstock (1963c), Ex. 20.2 p. 29. (For $\mathcal{U} \in \mathbf{A}|[0, 1)$ defined by $\delta(t) > 0$, let X be the set of all x in $[0, 1]$ such that either $x = 1$, or the supremum of finite sums of h^* over divisions from $\mathcal{U}$ over $[x, 1)$, is $1 - x$. Let $s = \inf X$. Then for $x \in X$ tending to $s+$, the supremum for $[s, 1)$ is not less than $1 - x$, hence $s \in X$ and $0 \leq s \leq 1$. If $s > 0$ then $\delta(s) > 0$ and there is an integer n with $s/(2n) < \delta(s)$, $([(2n - 1)s/(2n), s), s)$ is δ-fine, and $h_{2n} = s/(2n)$. Hence $(2n - 1)s/(2n) \in X$ while $s = \inf X$. This contradiction shows that $s = 0$. Similarly $y - x$ is the supremum of finite sums of h^* over division from $\mathcal{U}$ of $[x, y)$, for $0 \leq x < y \leq 1$, while the infimum of finite sums of h^* over divisions from $\mathcal{U}$ of $[x, y)$, is 0, proved by replacing $2n$ by $2n + 1$. Hence the result.)

Example 2.3.7 Two disjoint partial intervals from different partitions of E cannot always be part of another partition of E. For let $T = [0, 1]$ and $\mathcal{T}$ the family of intervals $[u, v)$ with $v - u$ rational, or $u = 0$, or $v = 1$. Then the disjoint $[0, \tfrac{1}{2})$ and $[2^{-1/2}, 1)$ are partial intervals from divisions of $[0, 1)$. But a partition with partition point $\tfrac{1}{2}$ has only rational partition points and $2^{-1/2}$ cannot be one of them. Such an arrangement cannot therefore be part of a division space arrangement.

Example 2.3.8 If h and h^* are p-variationally equivalent and if $X \subseteq T$, then
$$\bar{V}_p(h; \mathbf{A}; \mathscr{E}; X) = \bar{V}_p(h^*; \mathbf{A}; \mathscr{E}; X).$$
(See Example 2.1.2.)

2.4 DECOMPOSABLE DIVISION SYSTEMS

In this section $(T, \mathscr{T}, \mathbf{A})$ is a decomposable division system for an elementary set E. Adding to the definitions at the beginning of Section 2.3, $(T, \mathscr{T}, \mathbf{A})$ is *fully decomposable* {respectively, *decomposable*, or *measurably decomposable relative to a measure*, see (2.1.12), or a *measure space* defined later} if to every family (respectively, countable family or countable family of measurable sets) $\mathscr{X}$ of mutually disjoint subsets $X \subseteq T$, and every function $\mathscr{U}(.): \mathscr{X} \to \mathbf{A}|E$, there is a $\mathscr{U} \in \mathbf{A}|E$ with
$$\mathscr{U}[X] \equiv \{(I, t): (I, t) \in \mathscr{U}, \ t \in X\} \subseteq \mathscr{U}(X) \quad (X \in \mathscr{X}).$$
There is no need for the union of the $X \in \mathscr{X}$ to be T or E^*. Usually we assume decomposability if any one of the three conditions is required.

Theorem 2.4.1. *Let* $h: \mathscr{U}^1 \to K$ *where* K *has a p-modulus, and let* (X_j) *be a sequence of sets* $X_j \subseteq T$ *with union* X. *Then*

(2.4.1) $\qquad \bar{V}_p(h; \mathbf{A}: E; X) \leqslant \sum_{j=1}^{\infty} \bar{V}_p(h; \mathbf{A}; E; X_j).$

If the X_j are mutually disjoint and p-h measurable, then X is p-h-measurable and, for each $Y \subseteq T$,

(2.4.2) $\bar{V}_p(h; \mathbf{A}; E; Y) = \sum_{j=1}^{\infty} \bar{V}_p(h; \mathbf{A}; E; Y \cap X_j) + \bar{V}_p(h; \mathbf{A}; E; Y \setminus X),$

(2.4.3) $\bar{V}_p(h; \mathbf{A}; E; X) = \sum_{j=1}^{\infty} \bar{V}_p(h; \mathbf{A}; E; X_j),$

$\bar{V}_p(h; \mathbf{A}; E; X)$ *being countably additive in X. Let S be a set of scalars with $S \times K \subseteq K$ and $p(sk) = |s|p(k)$ $(s \in S, k \in K)$.*

(2.4.4) *If $\bar{V}_p(h; \mathbf{A}; E; X) = 0$ with $f: T \to S$, then $\bar{V}_p(fh; \mathbf{A}; E; X) = 0$.*

(2.4.5) *Conversely, if $\bar{V}_p(fh; \mathbf{A}; E; X) = 0$ with $f \neq 0$ in $X_0 \subseteq X$, then $V_p(h; \mathbf{A}; E; X_0) = 0$.*

(2.4.6) *For $f: T \to S$, $h: \mathscr{U}^1 \to K$, $m: \mathscr{U}^1 \to K$, $\bar{V}_p(h - m; \mathbf{A}; E) = 0$, and fh integrable over E, then fm is also integrable over E and the two integrals are equal.*

Proof. The fact that in this section all divisions have to be divisions of the elementary set E does not affect the proof of the first five results, so that the proof of Theorem 2.2.1 can be used. Thus we prove (2.4.6), using (2.4.4) to have $\bar{V}_p(f(h-m); \mathbf{A}; E) = 0$. By Theorem 2.1.10(2.1.31) (analogue), $\overline{VS}(f(h-m); \mathbf{A}; E) = \text{sing}(z)$, z being the zero. By (2.3.3),

$$\int_E f\,d(h-m) = 0.$$

Thus Theorem 2.3.3(2.3.7), the finite additivity of the integral relative to the integrand, gives (2.4.6). □

It is the first part of the result on integration by substitution.

Example 2.4.1 If the division system $(T, \mathcal{T}, \mathbf{A})$ for E is not decomposable, we can construct a decomposable $(T, \mathcal{T}, \mathbf{A}^*)$ from it by decomposing the $\mathcal{U}$ in the following way. $\mathbf{A}^*$ is the family of all $\mathcal{U} \subseteq \mathcal{U}^1$ for which there are an elementary set E, a sequence $(\mathcal{U}_j) \subseteq \mathbf{A}|E$, and a sequence (X_j) of mutually disjoint subsets of T with union T, such that $\mathcal{U}[X_j] = \mathcal{U}_j[X_j]$ ($j = 1, 2, \ldots$). Thus $\mathbf{A} \subseteq \mathbf{A}^*$. We can say that the $\mathcal{U} \in \mathbf{A}|E$ have been decomposed to form the $\mathcal{U} \in \mathbf{A}^*$.

But there are $(T, \mathcal{T}, \mathbf{A})$ for which not every $\mathcal{U} \in \mathbf{A}^*$ divides E. For let T be the interval $[a, b]$ on the real axis, let $\mathcal{T}$ be the family of $[u, v) \subseteq T$, and let $\mathcal{U}^1$ be the family of $([u, v), t)$ with $u \leq t \leq v$ and $t < b$. Let $\mathbf{A}$ be the family of all $\mathcal{U} \subseteq \mathcal{U}^1$ for which there is a $\delta > 0$ such that if $v - u < \delta$, $([u, v), t) \in \mathcal{U}^1$, then $([u, v), t) \in \mathcal{U}$. Let $\mathcal{U}_j$ be defined by $\delta_j = (b-a)2^{-j}$ and take $X_1 = \text{sing}(b)$,

$$X_j = [b - (b-a)2^{-j+2}, b - (b-a)2^{-j+1}) \quad (j > 1).$$

Then the corresponding $\mathcal{U} \in \mathbf{A}^*$ does not divide $[a, b)$. For if $\mathscr{E}$ is a finite collection of interval–point pairs that comes from certain $\mathcal{U}_j$ let J be the greatest of the j. As

$$b - (b-a)2^{-J+1} + \delta_J = b - (b-a)2^{-J}, \quad [b - (b-a)2^{-J}, b)$$

is free from intervals from $\mathscr{E}$. Here the reason why $\mathcal{U}$ does not divide $[a, b)$, is that $\mathcal{U}[X_1]$ is empty.

Example 2.4.2 If in Example 2.4.1 every $\mathcal{U} \in \mathbf{A}^*$ divides E, then $(T, \mathcal{T}, \mathbf{A}^*)$ is a division system for E at least. For let $\mathcal{U}_j, \mathcal{U}_j^\circ \in \mathbf{A}|E; X_j, X_j^\circ \subseteq T$ ($j = 1, 2, \ldots$) with the (X_j) mutually disjoint with union T; and the (X_j°) mutually disjoint with union T. Let $\mathcal{U} \in \mathbf{A}^*$ be defined from $(\mathcal{U}_j)$ using (X_j), and $\mathcal{U}^\circ \in \mathbf{A}^*$ from $(\mathcal{U}_j^\circ)$ using (X_j°). Then we can put the non-empty sets $X_j \cap X_k^\circ$ ($j, k = 1, 2, \ldots$) in sequence say as (Y_r) and replace $(X_j), (X_j^\circ)$ in the constructions by (Y_j), repeating some $\mathcal{U}_j, \mathcal{U}_k^\circ$ where necessary to obtain new sequences to correspond to the (Y_r), again written $(\mathcal{U}_j), (\mathcal{U}_j^\circ)$. As $(T, \mathcal{T}, \mathbf{A})$ is a division system for E there is a $\mathcal{U}_j^+ \in \mathbf{A}|E$ with $\mathcal{U}_j^+ \subseteq \mathcal{U}_j \cap \mathcal{U}_j^\circ$, and from

$(\mathcal{U}_j^+)$, (Y_j) we can define $\mathcal{U}^+ \in \mathbf{A}^*$ with $\mathcal{U}^+ \subseteq \mathcal{U} \cap \mathcal{U}^\circ$, and by hypothesis $\mathcal{U}^+$ divides E. Thus $\mathbf{A}^*$ is directed for divisions of E.

Example 2.4.3 Clearly a fully decomposable (or decomposable) $(T, \mathcal{T}, \mathbf{A})$ is decomposable (or measurably decomposable). A decomposable $(T, \mathcal{T}, \mathbf{A}^*)$ in Example 2.4.2 is fully decomposable if the $\mathcal{U} \in \mathbf{A}^*$ have been formed by decomposing a single fixed sequence $(V_j) \subseteq \mathbf{A}|E$ and a single fixed sequence (X_j). For then, to each $t \in T$ corresponds a $\mathcal{U}(t) \in \mathbf{A}^*|E$, and to each $\mathcal{U}(t)$, constructed from $(V_j), (X_j)$, there corresponds an integer k such that $t \in X_k$. We can thus replace $\mathcal{U}(t)$ by $V_{k(t)}$ where $k(t) = k$, and so have full decomposability from decomposability. For example, let T be a metric space with the metric topology. If to each $\mathcal{U} \in \mathbf{A}|E$ corresponds a fixed $\delta > 0$ such that $(I, t) \in \mathcal{U}$ if $I \in \mathcal{T}, t \in \bar{I}$, and I lies in the sphere centre t, radius δ, then functions $\delta(t) > 0$ correspond to the $\mathcal{U} \in \mathbf{A}^*|E$, and $\mathbf{A}^*$ is automatically fully decomposable. In fact we can choose V_j using $\delta_j = 1/j$ $(j = 1, 2, \ldots)$. This is precisely the difference between the $(T, \mathcal{T}, \mathbf{A})$ giving Riemann integration and the $(T, \mathcal{T}, \mathbf{A}^*)$ giving gauge or generalized Riemann integration.

2.5 DIVISION SPACES

Throughout this section $(T, \mathcal{T}, \mathbf{A})$ is a division space. Thus we begin by reciting the conditions obeyed by the space. In the non-empty base space T of points we choose some non-empty subsets called *(generalized) intervals* I, their family being $\mathcal{T}$, while $\mathcal{U}^1$ is a fixed non-empty family of interval–point pairs (I, t) $(I \in \mathcal{T}, t \in T)$, t being called an *associated point of* I. $\mathbf{A}$ is a non-empty family of some non-empty subsets $\mathcal{U} \subseteq \mathcal{U}^1$. An *elementary set* E is an interval or a union of a finite number of mutually disjoint intervals. A subset $\mathcal{U} \subseteq \mathcal{U}^1$ *divides* E if, for a *finite* subset $\mathcal{E} \subseteq \mathcal{U}$, called a *division of* E *from* $\mathcal{U}$, the $(I, t) \in \mathcal{E}$ have mutually disjoint I with union E, and called *partial intervals of* E *from* $\mathcal{U}$. The collection of these I is called a *partition of* E. A non-empty subset $\mathcal{P}$ of $\mathcal{E}$, including $\mathcal{E}$ itself, is a *partial division of* E *from* $\mathcal{U}$, and the union of $I \in \mathcal{P}$ is a *partial set* P *of* E *that comes from* $\mathcal{E}$ *and* $\mathcal{U}$. P is *proper* if $P \neq E$. If there are partial sets $P_1, \ldots, P_n$ from the same $\mathcal{E}$, then P_j are called *copartitional*, their union being a partial set. For each elementary set E, and $\mathcal{U}.E$ the set of all $(I, t) \in \mathcal{U}$ with I a partial interval of E, let $\mathbf{A}|E$ be the set of all $\mathcal{U}.E$ dividing E with $\mathcal{U} \in \mathbf{A}$. We assume that $\mathbf{A}|E$ is not empty so that $\mathbf{A}$ *divides* E; and that $\mathbf{A}$ *is directed for divisions of* E, i.e. given $\mathcal{U}_j \in \mathbf{A}|E$ $(j = 1, 2)$, there is a $\mathcal{U} \in \mathbf{A}|E$ with $\mathcal{U} \subseteq \mathcal{U}_1 \cap \mathcal{U}_2$. This is the direction 'as $\mathcal{U}$ shrinks', giving a Moore–Smith type of limit. A *restriction of* $\mathcal{U}$ *to* P is defined to be a non-empty family $\mathcal{U}_1 \subseteq \mathcal{U}.P$. We assume that $\mathbf{A}$ *has the restriction property*, i.e. for each elementary set E, each partial set P, and each $\mathcal{U} \in \mathbf{A}|E$, there is in $\mathbf{A}|P$ a restriction of $\mathcal{U}$ to P. Example 2.3.7 necessitates the definition of 'co-partitional' and, to avoid such a pathology, we assume that if $P_1, \ldots, P_n$ are partial sets of E, then the P_j are co-partitional.

If $(T, \mathcal{T}, \mathbf{A})$ satisfies all these conditions for all elementary sets and all partial sets, it is called a *division space*.

Theorem 2.5.1.

(2.5.1) *For P a proper partial set of an elementary set E, and $\mathcal{U} \in \mathbf{A}|E$, then $\mathcal{U}$ divides P.*

(2.5.2) *If also $\mathcal{E}$ is a division of P from $\mathcal{U}$, with $\mathcal{U}_1 \in \mathbf{A}|E\backslash P$, there is a division $\mathcal{E}_1$ of $E\backslash P$ from $\mathcal{U}_1$, depending only on P and not otherwise on E, such that $\mathcal{E} \cup \mathcal{E}_1$ is a division of E from $\mathcal{U}$.*

(2.5.3) *If P_1, P_2 are partial sets of E, then $P_1 \cap P_2, P_1 \backslash P_2, P_1 \cup P_2$ are either empty or are partial sets of E, the first two also being either empty or partial sets of P_1.*

Proof. For (2.5.1), a restriction of $\mathcal{U}$ to P lies in $\mathbf{A}|P$ and so divides P. Hence $\mathcal{U}$ divides P. Now let $\mathcal{U}_2, \mathcal{U}_3 \in \mathbf{A}|E\backslash P$, $\mathcal{U}_2$ a restriction of $\mathcal{U}$ to $E\backslash P$ and $\mathcal{U}_3 \subseteq \mathcal{U}_1 \cap \mathcal{U}_2$. For $\mathcal{E}_1$ a division of $E\backslash P$ from $\mathcal{U}_3$, $\mathcal{E}_1$ is from $\mathcal{U}_1$ and $\mathcal{U}_2$ and so from $\mathcal{U}$, and $\mathcal{E} \cup \mathcal{E}_1$ is as given for (2.5.2). Then for (2.5.3), P_1 and P_2 are given co-partitional i.e. there is a division $\mathcal{E}_2$ of E from a $\mathcal{U} \in \mathbf{A}/E$ that gives a partition $\mathcal{P}$ of E, such that P_1 and P_2 are unions of intervals from P. Then $P_1 \cap P_2$ is the union of those intervals of $\mathcal{P}$ in P_1 that are also in P_2, and similarly for the other two sets. □

Theorem 2.5.2. *Let K have a multiplication $x \cdot y$ for which it is complete$(\mathbf{A}; P)$ for each elementary set P. Let $h: \mathcal{U}^1 \to K$, and E an elementary set.*

(2.5.4) *If $H(E) \equiv \int_E dh$ exists then $H(P) \equiv \int_P dh$ exists for all partial sets P of E.*

(2.5.5) *If also $x \cdot y$ is continuous in x and y separately, with $\mathcal{G}$ regular (i.e. for each $\mathcal{G}$-closed set F and each point $x \in K\backslash F$, there are disjoint neighbourhoods of F and x), then*

$$H(E) = H(P) \cdot H(E\backslash P) = H(E\backslash P) \cdot H(P).$$

(2.5.6) *More generally, if in (2.5.5) P_1, P_2 are disjoint (and so co-partitional) partial sets of E, $P_1 \cup P_2$ is a partial set and H is finitely multiplicative,*

$$H(P_1) \cdot H(P_2) = H(P_1 \cup P_2) = H(P_2) \cdot H(P_1).$$

(2.5.7) *If in (2.5.5) $H(P) = u$ for a unit $u \in K$ and every partial set P of E, h is of variation zero in E, and conversely.*

2.5 DIVISION SPACES

(2.5.8) *If in (2.5.5) K has a cancellation law, unit u, all G with $u \in G \in \mathscr{G}$, and*

$$S(h; \mathscr{U}; E) \subseteq H(E).G \quad (some\ \mathscr{U} \in \mathbf{A}|E)$$

then

$$S(h; \mathscr{U}; P) \subseteq H(P).\bar{G}$$

for every partial set P of E.

Note that for G small enough, $\bar{G}$ is contained in an arbitrarily small neighbourhood of u.

Proof. For (2.5.4) and P a proper partial set of E, and $\mathscr{U} \in \mathbf{A}|E$, then by Theorem 2.5.1(2.5.1), $\mathscr{U}$ divides P and $E \setminus P$, co-partitional partial sets of E. If G is any neighbourhood of $H(E)$,

(2.5.9) $$S(h; \mathscr{U}; E) \subseteq G \quad (some\ \mathscr{U} \in \mathbf{A}|E).$$

For q_1, q products over divisions from $\mathscr{U}$ of P and $E \setminus P$, respectively, $q.q_1$ is a product over a division of E from $\mathscr{U}$. By (2.5.9),

(2.5.10) $$q.S(h; \mathscr{U}; P) \subseteq G$$

and $S(h; \mathscr{U}; P)$ is fundamental $(\mathbf{A}; P)$ for the $\mathscr{U} \in \mathbf{A}|P$ in the semigroup sense of Section 0.5, and so is convergent $(\mathbf{A}; P)$ to a limit denoted by

$$H(P) \equiv \int_P dh.$$

Similarly $S(h; \mathscr{U}; E \setminus P)$ is convergent $(\mathbf{A}; E \setminus P)$ to $H(E \setminus P)$. In (2.5.5), $H(E \setminus P).q_1$ is the limit of some $q.q_1$ and so lies in $\bar{G}$. Hence

$$H(E \setminus P).H(P) \in \bar{G}; \quad H(E \setminus P).H(P) = H(E)$$

by regularity and Hausdorff. Similarly

$$H(P).H(E \setminus P) = H(E).$$

In (2.5.6), using the intervals that define P_1, P_2 as co-partitional partial sets of E, P_1 and P_2 are partial sets of $P_1 \cup P_2$. Replacing $P, E \setminus P, E$ by $P_1, P_2, P_1 \cup P_2$ in (2.5.5) gives (2.5.6). For (2.5.7) we have a uniformity result from (2.5.10) as $S(h; \mathscr{U}; P)$ is given to tend to u. Thus $q = q.u \in \bar{G}$ for every product q over every division from $\mathscr{U}$ of every partial set $E \setminus P$, and so every partial division of E from $\mathscr{U}$. Thus

$$VS(h; \mathscr{U}; E) \subseteq \bar{G}$$

and G is an arbitrary neighbourhood of $H(E) = u$. $\mathscr{G}$ being regular, h is of variation zero in E. The converse is trivial. For (2.5.8), as $x.y$ is continuous in y we can replace G in (2.5.9) by $H(E).G$ for arbitrary neighbourhoods G of u, and so

$$q.q_1 \in H(E).G; \quad H(E \setminus P).q_1 \in H(E).\bar{G}.$$

by the continuity of $x.y$ in x. Using (2.5.5) and the cancellation law, (2.5.8) follows

It is of help to translate this into a Banach space version, as follows.

Theorem 2.5.3. *Let K be a Banach space with norm $\|.\|$, let $h: \mathcal{U}^1 \to K$, and let $\mathcal{U} \in \mathbf{A}|E$ for an elementary set E. If, given $\varepsilon > 0$, we have for all divisions $\mathscr{E}$ of E from $\mathcal{U}$ depending on ε, and some value $H(E) \in K$,*

$$\|(\mathscr{E})\sum h - H(E)\| < \varepsilon$$

then for all partial sets P of E and all divisions $\mathscr{P}$ of P from $\mathcal{U}$, and some $H(P) \in K$,

$$\|(\mathscr{P})\sum h - H(P)\| \leq \varepsilon,$$

the H being finitely additive in P. Thus we can write the second inequality as

$$\|(\mathscr{P})\sum (h - H)\| \leq \varepsilon.$$

Proof. $H(E)$ is the integral of h over E. Following the proof of Theorem 2.5.2(2.5.4) $S(h; \mathcal{U}; P)$ is fundamental $(\mathbf{A}; P)$ for the $\mathcal{U} \in \mathbf{A}|P$. The Banach space is linear and complete, and so is a topological group. By Theorem 0.5.6 it is complete $(\mathbf{A}; P)$. Thus $S(h; \mathcal{U}; P)$ is convergent $(\mathbf{A}; P)$, say to $H(P)$. By the rest of Theorem 2.5.2, $H(P)$ is finitely additive with the inequality involving P and $\mathscr{P}$.

Theorem 2.5.4. *Let K be an additive group with a regular topology $\mathscr{G}$, $x + y$ continuous in x, y separately, and with K complete $(\mathbf{A}; P)$ for each elementary set P.*

(2.5.11) *If $h: \mathcal{U}^1 \to K$ is integrable to $H(P)$ over each partial set P of an elementary set E, then $h - H$ is of variation zero.*

(2.5.12) *Conversely, if $h - H$ is of variation zero with H finitely additive over co-partitional partial sets of E, then h is integrable to $H(P)$ over each elementary set P.*

(2.5.13) *If $h - H$ is of variation zero in E, then $h - H$ is uniformly continuous in E, in the sense that, given $G \in \mathscr{G}$ with the zero $z \in G$, there is a $\mathcal{U} \in \mathbf{A}|E$ with $h(I, t) - H(I) \in G$ when $(I, t) \in \mathcal{U}$. However, the $\mathcal{U}$ themselves usually shrink non-uniformly, so that the continuity is not uniform in the usual sense.*

Proof. Let the union of intervals from a partial division $\mathscr{P}$ of E be P. Then, given G in $z \in G \in \mathscr{G}$, there is a $\mathcal{U} \in \mathbf{A}|E$ satisfying (2.5.9), and

$$(\mathscr{P})\sum (h - H) = (\mathscr{P})\sum h - H(P) \in \bar{G}, \quad VS(h - H; \mathcal{U}; E) \subseteq \bar{G}$$

by the finite additivity of H, Theorem 2.5.2(2.5.6), (2.5.8). Regularity of $\mathscr{G}$ gives (2.5.11). For (2.5.12) the finite additivity of H with Theorem 2.5.1, a division $\mathscr{P}$ from $\mathcal{U}$ of a partial set P of E, is a partial division of E from $\mathcal{U}$, so that for

2.5 DIVISION SPACES

$G \in \mathcal{G}$ with $z \in G$, and for $\mathcal{U} \in \mathbf{A}|E$ and depending on G,

$$(\mathcal{P})\sum h - H(P) = (\mathcal{P})\sum(h - H) \in VS(h - H; \mathcal{U}; E) \subseteq G.$$

Hence (2.5.12). For (2.5.13) we take one (I, t) in $\mathcal{P}$. □

If K is an additive group with $H(P)$ finitely additive for co-partitional partial sets P, with $h(I, t) - H(I)$ of variation zero in E, we say that H is the *variational integral of h in E*. This corresponds to a free variational integral defined using the free variational sets and suggested by the end of Section 2.1. Theorem 2.5.4 shows that the generalized Riemann and variational integrals are equivalent if also $\mathcal{G}$ is regular with $x + y$ continuous in x and y separately, and so in this case the variational integral is uniquely defined. Similarly, as in Section 2.1, we can use $V_p(h - H; \mathcal{U}; E)$ instead of $VS(h - H; \mathcal{U}; E)$, for the *p-variational integral*. If p is the norm we call the integral the *strong variational integral*. It involves

(2.5.14) $\qquad (\mathcal{E})\sum \|h(I, t) - H(I)\| < \varepsilon$

and the norm variation. For finite-dimensional Banach spaces and possibly others there is often a link between variation zero and norm variation zero, giving the next theorem.

Theorem 2.5.5. *Let K be a Banach space with norm $\|\cdot\|$, addition, and satisfying Theorem 2.1.10(2.1.33) with q components. (It is likely that such a space is finite dimensional.) Let $h: \mathcal{U}^1 \to K$ be integrable to $H(P)$ over each partial set P of an elementary set E. Then*

(2.5.15) $\quad h - H$ *is of norm variation zero, and*

(2.5.16) $\qquad \bar{V}(h; \mathbf{A}; E) = \bar{V}(H; \mathbf{A}; E).$

Proof. For (2.5.15) combine Theorem 2.5.4(2.5.11) with Theorem 2.1.10(2.1.34), and then follow with Theorem 2.1.1(2.1.7) for (2.5.16). □

In this section we have to deal sometimes with more than one partial set at the same time, so we need to know what is the behaviour of the $p - h$-variation.

Theorem 2.5.6. *For K having a multiplication let P_1, P_2 be co-partitional partial sets of E with $\mathcal{U} \in \mathbf{A}|E$ and $\mathcal{U}_j \in \mathbf{A}|P_j$ a restriction of $\mathcal{U}$ to P_j $(j = 1, 2)$. For $h: \mathcal{U}^1 \to K$ and the p-modulus,*

(2.5.17) $\quad V_p(h; \mathcal{U}_1; P_1) + V_p(h; \mathcal{U}_2; P_2) \leq V_p(h; \mathcal{U}; P_1 \cup P_2),$

(2.5.18) $\quad \bar{V}_p(h; \mathbf{A}; P_1) + \bar{V}_p(h; \mathbf{A}; P_2) \leq \bar{V}_p(h; \mathbf{A}; P_1 \cup P_2),$

(2.5.19) $\quad VS(h; \mathcal{U}_1; P_1) \cdot VS(h; \mathcal{U}_2; P_2) \subseteq VS(h; \mathcal{U}; P_1 \cup P_2),$

and if multiplication is continuous in K,

(2.5.20) $\quad \overline{VS}(h; \mathbf{A}; P_1) . \overline{VS}(h; \mathbf{A}; P_2) \subseteq \overline{VS}(h; \mathbf{A}; P_1 \cup P_2)$.

Proof. For (2.5.17) let $\mathscr{P}_j$ be a division of P_j from $\mathscr{U}_j$ and so from $\mathscr{U}$ ($j = 1, 2$). Then $\mathscr{P}_1 \cup \mathscr{P}_2$ is a division of $P_1 \cup P_2$ from $\mathscr{U}$ and hence

$$(\mathscr{P}_1)\sum p(h) + (\mathscr{P}_2)\sum p(h) = (\mathscr{P}_1 \cup \mathscr{P}_2)\sum p(h) \leqslant V_p(h; \mathscr{U}; P_1 \cup P_2).$$

We have (2.5.17), taking suprema for varying $\mathscr{P}_j$. The left side of (2.5.18) is not greater than the left side of (2.5.17), so by taking the infimum of the right side of (2.5.17) for all $\mathscr{U} \in \mathbf{A}|E$, we have (2.5.18). If $u_j \in VS(h; \mathscr{U}_j; P_j)$ ($j = 1, 2$), $u_1 . u_2$ is a product for a partial division of $P_1 \cup P_2$ from $\mathscr{U}_1 \cup \mathscr{U}_2$ and so from $\mathscr{U}$. Hence (2.5.19). When multiplication is continuous, Theorem 0.4.8 in two dimensions shows that if X and Y are sets of K, then $\overline{X} . \overline{Y} \subseteq \overline{X . Y}$. This and (2.5.19) give

$$\overline{VS}(h; \mathbf{A}; P_1) . \overline{VS}(h; \mathbf{A}; P_2) \subseteq \mathscr{G}VS(h; \mathscr{U}_1; P_1) . \mathscr{G}VS(h; \mathscr{U}_2; P_2)$$
$$\subseteq \mathscr{G}\{VS(h; \mathscr{U}_1; P_1) . VS(h; \mathscr{U}_2; P_2)\}$$
$$\subseteq \mathscr{G}VS(h; \mathscr{U}; P_1 \cup P_2).$$

(2.5.20) follows on taking the intersection of the right side for all $\mathscr{U} \in \mathbf{A}|E$. □

Theorem 2.5.7.

(2.5.21) *For P a partial set of E, $\mathscr{U} \in \mathbf{A}|E$,*

$$V_p(h; \mathscr{U}; P) \leqslant V_p(h; \mathscr{U}; E), \quad VS(h; \mathscr{U}; P) \subseteq VS(h; \mathscr{U}; E).$$

(2.5.22) *An h of p-variation zero in E is the same in every partial set of E, and similarly for the variation set.*

(2.5.23) *If $K = \mathbb{R}$, $h(I, t) \geqslant 0$ (($I, t) \in \mathscr{U} \in \mathbf{A}|E$), P a partial set of E, and h integrable over E, then*

$$\int_P dh \leqslant \int_E dh.$$

(2.5.24) *If in (2.5.23) the integral on the right is 0, then h is of variation zero in E.*

Proof. As a partial division of P is a partial division of E, and as the integral over P exists in (2.5.23), these results are obvious. □

We now have a Jordan-type decomposition of a finitely additive real function of bounded variation into its positive and negative variations, remembering that the empty set is not an elementary set.

2.5 DIVISION SPACES

Theorem 2.5.8. *For $(T, \mathcal{T}, \mathbf{A})$ an infinitely divisible division space and W a finitely additive real function of bounded variation of elementary sets, if for all proper partial sets P of E,*

(2.5.25) $\qquad \bar{W}(E) = \sup W(P), \quad \underline{W}(E) = \inf W(P),$

then these are finite, finitely additive, and of bounded variation, with

(2.5.26) $\qquad \bar{W}(E) \geq 0 \geq \underline{W}(E), W(E) = \underline{W}(E) + \bar{W}(E), \bar{V}(W; \mathbf{A}; E)$
$\qquad\qquad\qquad = \bar{W}(E) + \underline{W}(E).$

Proof. As the division space is infinitely divisible, Example 2.5.1 shows that, given an integer $N > 1$, a proper partial set P of E, and $\mathscr{U} \in \mathbf{A}|E$, there are divisions $\mathscr{E}_1$ of P and $\mathscr{E}_2$ of $E \setminus P$ with M pairs (I, t) in $\mathscr{E}_1 \cup \mathscr{E}_2$ from $\mathscr{U}$, for some $M \geq N$, and as W is finitely additive,

(2.5.27) $\quad |W(P)| = |(\mathscr{E}_1) \sum W(I)| \leq (\mathscr{E}_1 \cup \mathscr{E}_2) \sum |W(I)| \leq V(W; \mathscr{U}; E),$
$\qquad \bar{W}(E) \leq V(W; \mathscr{U}; E),$
$\qquad |\underline{W}(E)| \leq V(W; \mathscr{U}; E), \quad W(E) = (\mathscr{E}_1 \cup \mathscr{E}_2) \sum W(I) \geq M \cdot \underline{W}(E),$
$\qquad \underline{W}(E) \leq W(E)/M, \quad \underline{W}(E) \leq 0, \quad \bar{W}(E) \geq W(E)/M, \bar{W}(E) \geq 0,$

as $N \to \infty$, and so $M \to \infty$. Given $\varepsilon > 0$, proper partial sets P, Q of E have

$W(P) > \bar{W}(E) - \varepsilon, \quad W(E) = W(P) + W(E \setminus P) > \bar{W}(E) - \varepsilon + \underline{W}(E),$
$W(Q) < \underline{W}(E) + \varepsilon, \quad W(E) = W(Q) + W(E \setminus Q) < \underline{W}(E) + \varepsilon + \bar{W}(E);$
$W(E) = \underline{W}(E) + \bar{W}(E) \leq \bar{W}(E), W(E) \geq \underline{W}(E),$

so that we can now drop the word 'proper' from the definitions of $\underline{W}, \bar{W}$. Let P_1, P_2 be co-partitional disjoint partial sets with the partial set $P \subseteq P_1 \cup P_2$. By Theorem 2.5.1(2.5.3), $P \cap P_j$ is a partial set, or is empty and we omit $W(P \cap P_j)$. As $P \cap P_j \subseteq P_j$ $(j = 1, 2)$,

(2.5.28) $\qquad W(P) = W(P \cap P_1) + W(P \cap P_2) \leq \bar{W}(P_1)$
$\qquad\qquad\qquad + \bar{W}(P_2), \bar{W}(P_1 \cup P_2) \leq \bar{W}(P_1) + \bar{W}(P_2).$

Further, given $\varepsilon > 0$, there are partial sets P_{j+2} of P_j such that

$$W(P_{j+2}) > \bar{W}(P_j) - \tfrac{1}{2}\varepsilon \quad (j = 1, 2).$$

P_1, P_2 being disjoint, so are P_3, P_4, and the finite additivity of $\bar{W}$ comes from (2.5.28) and

$$\bar{W}(P_1) + \bar{W}(P_2) - \varepsilon < W(P_3) + W(P_4) = W(P_3 \cup P_4) \leq \bar{W}(P_1 \cup P_2).$$

As also $\bar{W} \geq 0$, it is of bounded variation. Similarly for $\underline{W}$. For $\mathscr{E}$ a division of E,

$$|W(I)| = |\bar{W}(I) + \underline{W}(I)| \leq |\bar{W}(I)| + |\underline{W}(I)| = \bar{W}(I) - \underline{W}(I),$$

(2.5.29) $\quad (\mathscr{E})\sum |W(I)| \leq (\mathscr{E})\sum (\bar{W}(I) - \underline{W}(I))$,

$$V(W; \mathbf{A}; E) \leq \bar{W}(E) - \underline{W}(E) = \bar{W}(E) + |\underline{W}(E)|.$$

If $\bar{W}(E) = 0$ or $\underline{W}(E) = 0$, (2.5.27) and (2.5.29) give the last part of (2.5.26). If not, let a proper partial set P of E satisfy

$$W(P) > \bar{W}(E) - \varepsilon = W(E) - \underline{W}(E) - \varepsilon \quad (0 < \varepsilon \leq \tfrac{1}{2}|\underline{W}(E)|).$$

Then

$$W(E\setminus P) = W(E) - W(P) < \underline{W}(E) + \varepsilon < 0,$$

$$|W(E\setminus P)| > |\underline{W}(E)| - \varepsilon,$$

$$\bar{W}(E) + |\underline{W}(E)| - 2\varepsilon < W(P) + |W(E\setminus P)| = (\mathscr{E}_1)\sum W(I)$$
$$+ |(\mathscr{E}_2)\sum W(I)|$$
$$\leq (\mathscr{E}_1 \cup \mathscr{E}_2)\sum |W(I)| \leq V(W; \mathscr{U}; E),$$

for divisions $\mathscr{E}_1, \mathscr{E}_2$ of $P, E\setminus P$, respectively, from a $\mathscr{U} \in \mathbf{A}|E$, since W is finitely additive. This and (2.5.9) finish (2.5.26).

Theorem 2.5.9. *In Theorem 2.5.8, if a proper partial set P of E has $W(P) > \bar{W}(E) - \varepsilon$, then $\underline{W}(P) > -\varepsilon$ and $\bar{W}(E\setminus P) < \varepsilon$, for given $\varepsilon > 0$.*

Proof. By Theorem 2.5.8,

$$\bar{W}(P) + \underline{W}(P) = W(P) > \bar{W}(E) - \varepsilon \geq \bar{W}(P) - \varepsilon, \quad \underline{W}(P) > -\varepsilon,$$

$$\bar{W}(E\setminus P) + \underline{W}(E\setminus P) = W(E\setminus P) = W(E) - W(P) < W(E) - \bar{W}(E) + \varepsilon$$
$$= \underline{W}(E) + \varepsilon \leq \underline{W}(E\setminus P) + \varepsilon, \quad \bar{W}(E\setminus P) < \varepsilon. \quad \square$$

Thus, in P, W is within ε of being non-negative, and in $E\setminus P$, W is within ε of being non-positive.

Saks (1927) p. 214, gave a lemma in Burkill integration that the author translated into the language of generalized Riemann integration, first using gauges and then division spaces, and it became known as the Henstock lemma. A better title would be the Saks–Henstock lemma. Here it appears as Theorem 2.5.2(2.5.8) with its Banach space translation in Theorem 2.5.3.

Example 2.5.1 A subfamily $\mathscr{U} \subseteq \mathscr{U}^1$ is called *infinitely divisible* if each elementary set divided by $\mathscr{U}$ contains a proper partial set also divided by $\mathscr{U}$. If also $(T, \mathscr{T}, \mathbf{A})$ is a division space, $\mathscr{U}$ divides an elementary set E, and N is any integer, there is a division of E from $\mathscr{U}$ that contains at least N interval–point pairs (I, t). (For in Theorem 2.5.1(2.5.1) we are allowed to take P proper, so that $E\setminus P$ is a partial set and $\mathscr{U}$ divides P and $E\setminus P$. Thus there is a division of E from $\mathscr{U}$ having two or more (I, t). For each such I there is similarly a division from $\mathscr{U}$ containing two or more (J, x), and so on.)

2.6 DECOMPOSABLE DIVISION SPACES AND INTEGRATION BY SUBSTITUTION

In this section $(T, \mathcal{T}, \mathbf{A})$ satisfies the conditions at the beginning of Section 2.5, together with one of the conditions of full decomposability, decomposability, and measurable decomposability, as quoted at the beginning of Section 2.4. If not specially mentioned, we assume the second, i.e. decomposability.

We complete Theorem 2.4.1, particularly (2.4.6), and continue with results on integration by substitution.

Theorem 2.6.1.

(2.6.1) Let $h: \mathcal{U}^1 \to \mathbb{C}$ and $f: T \to \mathbb{C}$. Then $F(I) \equiv \int_I f \, dh = 0$ for all partial intervals I of an elementary set E, if and only if, for X the set where $f \neq 0$,

(2.6.2) $$\bar{V}(h; \mathbf{A}; E; X) = 0.$$

(2.6.3) If $f, g: T \to \mathbb{C}, h: \mathcal{U}^1 \to \mathbb{C}$, have $\int_I f \, dh = \int_I g \, dh$

for all partial intervals I of E, then $f = g$ except in a set X satisfying (2.6.2), and conversely.

(2.6.4) For $h, k: \mathcal{U}^1 \to \mathbb{C}$ and $f: T \to \mathbb{C}$, then $\int_I f \, dh = \int_I f \, dk$ for all partial intervals of E, if and only if, for X the set where $f \neq 0$,

(2.6.5) $$\bar{V}(h - k; \mathbf{A}; E; X) = 0$$

(2.6.6) If $h: T \to \mathbb{C}, k: \mathcal{U}^1 \to \mathbb{C}$, and $f, g: T \to \mathbb{C}$, have $h(I) = \int_I g \, dk$ for all partial intervals I of E, then

(2.6.7) $$\int_E f \, dh = \int_E f g \, dk$$

provided that one side exists, and then the other side exists, too.

Proof. If $F(I) = 0$ for all partial intervals I of E, then by Theorem 2.5.5(2.5.16),
$$\bar{V}(fh; \mathbf{A}; E) = \bar{V}(F; \mathbf{A}; E) = 0.$$

Now use Theorem 2.4.1(2.4.5) for (2.6.2). Conversely, if (2.6.2) holds, then $\bar{V}(fh; \mathbf{A}; E; X) = 0$ by Theorem 2.4.2(2.4.4), and so $\bar{V}(fh; \mathbf{A}; E) = 0$ since $f = 0$ in $\setminus X$. Hence the variational integral is 0, and $\int_I f \, dh$ exists with value 0, for every partial interval I of E. (2.6.3) and (2.6.4) follow from this since, respectively,

$$\int_I (f - g) \, dh = 0, \quad \int_I f \, d(h - k) = 0.$$

Then (2.6.6) follows from (2.6.4) since h and gk are variationally equivalent. □

This is a most important theorem, it is the modern version of the calculus integration by substitution. In particular, too, we can relax the definitions of the generalized Riemann integral in that f in the integral of fh can be replaced by any point function g with $f = g$ h-almost everywhere, and in that sense we can allow f to be infinite in a set of h-variation 0, replacing f by such a g that is finite everywhere, before integration.

2.7 ADDITIVE DIVISION SPACES

In this section we assume that $(T, \mathcal{T}, \mathbf{A})$ is an additive division space, and begin by reciting the definitions. Thus in the non-empty base space T of points we choose some non-empty subsets called (*generalized*) *intervals* I, with $\mathcal{T}$ their family, and $\mathcal{U}^1$ a fixed non-empty family of interval–point pairs (I, t) $(I \in \mathcal{T}, t \in T)$, t being an *associated point of* I. We use non-empty families $\mathbf{A}$ of some non-empty subsets $\mathcal{U} \subseteq \mathcal{U}^1$. An *elementary set* E is an interval or a union of a finite number of mutually disjoint intervals. A subset $\mathcal{U} \subseteq \mathcal{U}^1$ *divides* E if, for a *finite* subset $\mathcal{E} \subseteq \mathcal{U}$, called a *division of* E *from* $\mathcal{U}$, the $(I, t) \in \mathcal{E}$ have mutually disjoint I with union E. The collection of these I is called a *partition of* E, the I being called *partial intervals of* E *from* $\mathcal{U}$. A non-empty subset of $\mathcal{E}$, including $\mathcal{E}$ itself, is called a *partial division of* E *from* $\mathcal{U}$, and the union of I from a partial division is called a *partial set* P *of* E *that comes from* $\mathcal{E}$ *and* $\mathcal{U}$, and P is *proper* if $P \neq E$. Partial sets from the same $\mathcal{E}$ have been called *co-partitional*. But here, Theorem 2.7.1 shows that we do not need this definition in additive division spaces as we prove that partial sets are co-partitional.

For each elementary set E, and $\mathcal{U}.E$ the set of all $(I, t) \in \mathcal{U}$ with I a partial interval of E, let $\mathbf{A}|E$ be the set of all $\mathcal{U}.E$ dividing E with $\mathcal{U} \in \mathbf{A}$. Assuming in this section that the set $\mathbf{A}|E$ of all $\mathcal{U}.E$ dividing E with $\mathcal{U} \in \mathbf{A}$ is non-empty (we say that $\mathbf{A}$ *divides* E) and that $\mathbf{A}$ *is directed for divisions of* E (i.e. given $\mathcal{U}_j \in \mathbf{A}|E (j = 1, 2)$, there is a $\mathcal{U} \in \mathbf{A}|E$ with $\mathcal{U} \subseteq \mathcal{U}_1 \cap \mathcal{U}_2$), this direction being called the direction *as* $\mathcal{U}$ *shrinks*; it gives rise to a Moore–Smith limit. A *restriction of* $\mathcal{U}$ *to a partial set* P is a non-empty family $\mathcal{U}_1 \subseteq \mathcal{U}.P$. We assume that $\mathbf{A}$ has a property called the *restriction property*, i.e. for each elementary set E, each partial set P, and each $\mathcal{U} \in \mathbf{A}|E$, there is in $\mathbf{A}|P$ a restriction of $\mathcal{U}$ to P. Finally we assume that $\mathbf{A}$ *is additive*, the property that, given disjoint elementary sets E_j, and $\mathcal{U}_j \in \mathbf{A}|E_j$ $(j = 1, 2)$, there is a $\mathcal{U} \in \mathbf{A}|E_1 \cup E_2$ with $\mathcal{U} \subseteq \mathcal{U}_1 \cup \mathcal{U}_2$. The extra condition beyond Section 2.5 is the additivity of $\mathbf{A}$, which simplifies many results. Previously, when $\mathbf{A}$ is not additive, other definitions off the mainstream have been used. But later in this section we return to an older construction, a special case being in Henstock (1946, 1948), which gives an additive division space from a division space.

2.7 ADDITIVE DIVISION SPACES

Theorem 2.7.1.

(2.7.1) *If $\mathscr{E}$ is a division of an elementary set E from a $\mathscr{U} \in \mathbf{A}|E$, there is a $\mathscr{U}_1 \in \mathbf{A}|E$ such that every division $\mathscr{E}_1$ of E from $\mathscr{U}_1$ refines E, i.e. there is a division $\mathscr{E}_1 . I$ of each I with $(I, t) \in \mathscr{E}$, formed of those $(J, x) \in \mathscr{E}_1$ with $J \subseteq I$. We write $\mathscr{E}_1 \leqslant \mathscr{E}$.*

(2.7.2) *Each two partial sets of E are co-partitional, and Theorem 2.5.1(2.5.3) holds.*

Proof. For (2.7.1), as **A** has the restriction property there is a $\mathscr{U}(I) \in \mathbf{A}|I$ with $\mathscr{U}(I) \subseteq \mathscr{U}$, for each $(I, t) \in \mathscr{E}$. As **A** is additive there is a $\mathscr{U}_1 \in \mathbf{A}|E$ that lies in the union of the $\mathscr{U}(I)$. If $\mathscr{E}_0$ is a division of E from $\mathscr{U}_1$ then each $(J, x) \in \mathscr{E}_0$ has $(J, x) \in \mathscr{U}_1$ and so $(J, x) \in \mathscr{U}(I)$ for some $(I, t) \in \mathscr{E}$. Then $J \subseteq I$ and so J is disjoint from every other I_0 with $(I_0, t_0) \in \mathscr{E}$. As E is the union of all J with $(J, x) \in \mathscr{E}_0$, that union includes I, and I is the union of those $J \subseteq I$. Hence $\mathscr{E}_0 \leqslant \mathscr{E}$.

For (2.7.2) let the partial set P_j be the union of I from some $(I, t) \in \mathscr{E}_j$ where $\mathscr{E}_j$ is a division of E ($j = 1, 2$). By (2.7.1) there is a $\mathscr{U}_j \in \mathbf{A}|E$ such that every division of E from $\mathscr{U}_j$ refines $\mathscr{E}_j$ ($j = 1, 2$). By direction there is a $\mathscr{U} \in \mathbf{A}|E$ with $\mathscr{U} \subseteq \mathscr{U}_1 \cap \mathscr{U}_2$, and then $\mathscr{U}$ has the properties of $\mathscr{U}_1$ and $\mathscr{U}_2$, and every division $\mathscr{E}$ of E from $\mathscr{U}$ refines $\mathscr{E}_1$ and $\mathscr{E}_2$. Hence P_1 and P_2 are unions of some I with $(I, t) \in \mathscr{E}$ and so P_1, P_2 are co-partitional. Thus Theorem 2.5.1(2.5.3) follows. □

Let K have a multiplication $x.y$. An $h: T \to K$ is *finitely multiplicative* on $\mathcal{T}_0 \subseteq \mathcal{T}$ if for each $I \in \mathcal{T}_0$ and each partition $\mathscr{P}$ of I, consisting of intervals $J \in \mathcal{T}_0$,

(2.7.3) $$(\mathscr{P}) \prod h(J) = h(I),$$

independent of the order of the $J \in \mathscr{P}$. Then h can be defined unambiguously for finite unions E of disjoint intervals $I \in \mathcal{T}_0$, provided that if $\mathscr{P}, \mathscr{P}'$ are two partitions of E from $\mathcal{T}_0$, there is a partition $\mathscr{P}''$ of E from $\mathcal{T}_0$ that refines $\mathscr{P}$ and $\mathscr{P}'$. This follows from Theorem 2.7.1(2.7.1) if $(T, \mathcal{T}_0, \mathbf{A})$ is an additive division space, and in particular if $\mathcal{T}_0 = \mathcal{T}$.

If K has an order $\geqslant$ and if (2.7.3) is replaced by

(2.7.4) $$(\mathscr{P}) \prod h(J) \geqslant h(I)$$

we say that h is *finitely submultiplicative* on $\mathcal{T}_0$, while if (2.7.3) is replaced by

(2.7.5) $$h(I) \geqslant (\mathscr{P}) \prod h(J)$$

we say that h is *finitely supermultiplicative* on $\mathcal{T}_0$. If K is additive we replace 'multiplicative' by 'additive' in these definitions.

If $K = \mathbb{R}$ or $\mathbb{C}$ there are two semigroup operations. We can either use addition, possibly with scalar multiplication, or we can use ordinary multiplication and obtain product integrals.

Theorem 2.7.2. *Let $h: T \to \mathbb{R}$ be finitely subadditive on T. If, for a $\mathcal{U} \in \mathbf{A}|E$, $S(h; \mathcal{U}; E)$ is bounded with supremum s, then h is integrable in E with integral equal to s. However, if for some $\mathcal{U} \in \mathbf{A}|E$, $S(h; \mathcal{U}; E)$ is unbounded above, then for each integer n there is a $\mathcal{U}_n \in \mathbf{A}|E$ such that $S(h; \mathcal{U}_n; E) \subseteq (n, +\infty)$.*

Proof. We use Theorem 2.7.1(2.7.1) and the finite subadditivity to prove that if $k \in S(h; \mathcal{U}; E)$, there is a $\mathcal{U}_k \in \mathbf{A}|E$ such that $S(h; \mathcal{U}_k; E) \subseteq [k, +\infty)$. If s is finite, then for each $\varepsilon > 0$ we can take a $k > s - \varepsilon$, giving the integrability of h to s in this case. If $s = +\infty$ we can find a k greater than any given integer n, proving the other part. □

We now turn to various examples of use of the fundamental Theorem 2.7.2.

Theorem 2.7.3. *For K a Banach space with norm $\|\cdot\|$, let $h: \mathcal{U}^1 \to K$ be strong variationally integrable to $H(P)$ over each partial set P of an elementary set E. If h is of bounded variation, then $\|h\|$ and $\|H\|$ are integrable over P to*

$$V(P) \equiv \bar{V}(h; \mathbf{A}; P) \equiv \bar{V}(H; \mathbf{A}; P) \text{ with } \|H(P)\| \leq V(P), \quad \text{i.e.}$$

(2.7.6)
$$\left\| \int_P dh \right\| \leq \int_P d\|h\| = \bar{V}(h; \mathbf{A}; P).$$

If h is of unbounded variation, $\|h\|$ and $\|H\|$ cannot be integrable over P.

Proof. By definition H is finitely additive for partial sets, so that $\|H\|$ is finitely subadditive. By Theorem 2.7.2, $\|H\|$ is integrable if and only if $S(\|H\|; \mathcal{U}; E)$ is bounded above for some $\mathcal{U} \in \mathbf{A}|E$. As h is strong variationally integrable to H, by definition $h-H$ is of norm variation zero over E. By the analogue of Theorem 2.1.1(2.1.7), $\bar{V}(H; \mathbf{A}; E) = \bar{V}(h; \mathbf{A}; E)$, and $S(\|H\|; \mathcal{U}; E)$ is bounded above if and only if $\bar{V}(H; \mathbf{A}; E)$ and so $\bar{V}(h; \mathbf{A}; E)$ are finite. Further,

$$\big|\|h\| - \|H\|\big| \leq \|h - H\|, \quad \bar{V}(h - H; \mathbf{A}; E) = 0,$$

so that $\|H\|$ is integrable if and only if $\|h\|$ is, with equal integrals. The remaining result (2.7.6) is proved on using

$$\left\|(\mathscr{E})\sum h\right\| \leq (\mathscr{E})\sum \|h\|. \quad \square$$

It is worth while to write the corresponding results for $\mathbb{R}$ or $\mathbb{C}$.

Theorem 2.7.4.

(2.7.7) *For $K = \mathbb{R}$ or $\mathbb{C}$ with $x \cdot y$ meaning $x + y$, let $h: \mathcal{U}^1 \to K$ be generalized Riemann integrable over E. Then $|h|$ is also integrable over E if and only if h is of* *bounded variation in E, and*

$$\left| \int_E dh \right| \leq \int_E d|h| = \bar{V}(h; \mathbf{A}; E)$$

if the right side is finite.

(2.7.8) *For $K = \mathbb{R}$ or $\mathbb{C}$ with $x \cdot y$ the ordinary multiplication, let $h: \mathcal{U}^1 \to K$ be generalized Riemann integrable over E to $H(E)$, non-zero to avoid the anomaly (0.5.19). Then $|h|$ is generalized Riemann integrable over E to $|H(E)|$.*

Proof. For (2.7.7), by Theorems 2.5.2(2.5.6) and 2.5.5(2.5.15) h is (strong) variationally integrable, so that we need only apply Theorem 2.7.3. For (2.7.8), interpreting the neighbourhood of $H(E)$ as a circle centre $H(E)$ and radius $\varepsilon |H(E)|$, we have for some $\mathcal{U} \in \mathbf{A}|E$ and all divisions $\mathcal{E}$ of E from $\mathcal{U}$,

$$|(\mathcal{E})\sum |h| - |H(E)|| = ||(\mathcal{E})\sum h| - |H(E)|| \leq (\mathcal{E})\sum |h - H| \leq \varepsilon |H(E)|$$

In Lebesgue theory, to deal with the integrability of functions of interval–point functions of the type $f_j(t)\mu(I)$, a great use is made of the measurability of the point functions $f_j(t)$. Later we will show that if the interval–point functions are integrable then the point functions are measurable in some sense. Here we use a different method capable of dealing with the more general $h_j(I, t)$. We could use a continuous functional of h_j for the j in some set J, but for simplicity we use a function $r: K^n \to \mathbb{R}$, where K^n is the n-fold Cartesian product of K, here an additive normed group. We use two types of continuity condition; first,

(2.7.9) $|r(y_1, \ldots, y_n) - r(x_1, \ldots, x_n)|$
$$\leq B_1 \|y_1 - x_1\| + \cdots + B_n \|y_n - x_n\|,$$

for constants $B_1 > 0, \ldots, B_n > 0$, e.g. $K = \mathbb{R}$ and $|\partial r / \partial x_j| \leq B_j (j = 1, \ldots, n)$.

A more relaxed condition is the following. Given $\varepsilon > 0$, $x_1, \ldots, x_n$, we put $s(x_1, \ldots, x_n; \varepsilon) \equiv \sup |r(y_1, \ldots, y_n) - r(x_1, \cdots, x_n)|$ ($\|y_j - x_j\| \leq \varepsilon$, $j = 1, \ldots, n$). For arbitrarily large m, varying $\varepsilon_k > 0$ ($k = 1, \ldots, m$), and fixed

(2.7.10) $$\sum_{k=1}^{m} x_{jk} \quad (j = 1, \ldots, n),$$

(2.7.11) $$\sum_{k=1}^{m} s(x_{1k}, \ldots, x_{nk}; \varepsilon_k) \to 0 \text{ when } \sum_{k=1}^{m} \varepsilon_k \to 0.$$

Clearly (2.7.9) implies (2.7.11).

Theorem 2.7.5. *Let $r: K^n \to \mathbb{R}$ satisfy (2.7.9) or (2.7.11) with*

(2.7.12) $r(x_1 + y_1, \ldots, x_n + y_n) \leq r(x_1, \ldots, x_n) + r(y_1, \ldots, y_n)$

for all $(x_1, \ldots, x_n)$, $(y_1, \ldots, y_n)$ in K^n, and let H_j be the norm variational integral of $h_j(I, t)$ in an elementary set E. Then $r(h_1, \ldots, h_n)$ is integrable with integral equal to that of $r(H_1, \ldots, H_n)$, if and only if, for some $\mathcal{U} \in \mathbf{A}|E$ and some compact set C,

(2.7.13) $$S(r(h_1, \ldots, h_n); \mathcal{U}; E) \subseteq C.$$

Proof. Given $\varepsilon > 0$, there are $\mathcal{U}_j \in \mathbf{A}|E$ with

(2.7.14) $$V(h_j - H_j; \mathcal{U}_j; E) < \varepsilon \quad (j = 1, \ldots, n).$$

By direction in $\mathbf{A}|E$ we replace the $\mathcal{U}_j$ by a single $\mathcal{U} \in \mathbf{A}|E$ for which (2.7.14) holds for all the j. By (2.7.9) and each division $\mathcal{E}$ of E from $\mathcal{U}$, with

$$w \equiv r(h_1, \ldots, h_n) - r(H_1, \ldots, H_n),$$

$$(\mathcal{E})\sum |w| \leq \sum_{j=1}^{n} B_j(\mathcal{E}) \sum \|h_j - H_j\| < (B_1 + \cdots + B_n)\varepsilon,$$

and w has norm variation zero. If (2.7.11) holds, then for a division $\mathcal{E}$ of E from $\mathcal{U}$, consisting of (I_j, t_j) $(j = 1, \ldots, m)$, we take

$$y_{jk} = h_j(I_k, t_k), \quad x_{jk} = H_j(I_k), \quad \sum_{k=1}^{m} x_{jk} = H_j(E),$$

$$\varepsilon_k \equiv V(h_1 - H_1; \mathcal{U}; I_k) + \cdots + V(h_n - H_n; \mathcal{U}; I_k)$$

$$\sum_{k=1}^{m} |w(I_k, t_k)| \leq \sum_{k=1}^{m} S(H_1(I_k), \ldots, H_n(I_k); \varepsilon_k),$$

$$\sum_{k=1}^{m} \varepsilon_k \leq V(h_1 - H_1; \mathcal{U}; E) + \cdots + V(h_n - H_n; \mathcal{U}; E) < n\varepsilon,$$

using Theorem 2.5.6(2.5.17) and (2.7.14). Hence by (2.7.11), $V(w; \mathcal{U}; E) \to 0$ by choice of $\mathcal{U} \in \mathbf{A}|E$. Next, $c(P) \equiv r(H_1(P), \ldots, H_n(P))$ is finitely subadditive since

$$(\mathcal{E})\sum c(I) = (\mathcal{E})\sum r(H_1(I), \ldots, H_n(I)) \leq r((\mathcal{E})\sum H_1(I), \ldots, (\mathcal{E})\sum H_n(I))$$
$$= r(H_1(E), \ldots, H_n(E)) = c(E),$$

by (2.7.12), for a division $\mathcal{E}$ of E. Hence c is integrable if and only if $S(c; \mathcal{U}; E)$ is bounded for some $\mathcal{U} \in \mathbf{A}|E$. As $V(w; \mathcal{U}; E) \to 0$ by choice of $\mathcal{U} \in \mathbf{A}|E$, $r(h_1, \ldots, h_n)$ is integrable if and only if $S(c; \mathcal{U}; E)$, and so $S(r(h_1, \ldots, h_n); \mathcal{U}; E)$, are bounded, for some $\mathcal{U} \in \mathbf{A}|E$, giving (2.7.13) as a suitable condition. □

Theorem 2.7.6. *With $K = \mathbb{R}$, $r(x_1, \ldots, x_n) = \max(x_1, \ldots, x_n)$ and h_j integrable to H_j $(j = 1, \ldots, n)$, Theorem 2.7.5 shows that $\max(h_1, \ldots, h_n)$ and $\max(H_1, \ldots, H_n)$ are integrable in E (to the same integral) when for all choices of integers $j(I, t)$ in $1, \ldots, n$ and all divisions $\mathcal{E}$ of E from a $\mathcal{U} \in \mathbf{A}|E$, and a*

2.7 ADDITIVE DIVISION SPACES

compact set C,

(2.7.15) $\qquad (\mathscr{E}) \sum h_{j(I,t)}(I,t) \in C.$

Proof. (2.7.15) implies (2.7.13) for this r, for take $j(I,t) = J$ such that $h_J = \max(h_1, \ldots, h_n)$, while (2.7.13) for this r implies that the sum in (2.7.15) is bounded above, and bounded below since

$$r \geq h_j, \quad r - h_j \leq \sum_{k=1}^{n}(r - h_k), \quad h_j \geq \sum_{k=1}^{n} h_k - (n-1)r \quad (j = 1, \ldots, n),$$

and since the h_k are integrable. For (2.7.9) with $B_j = 1 (j = 1, \ldots, n)$,

$$x_j = (x_j - y_j) + y_j \leq |x_j - y_j| + y_j \leq \sum_{k=1}^{n} |x_k - y_k| + \max(y_1, \ldots, y_n),$$

$$\max(x_1, \ldots, x_n) - \max(y_1, \ldots, y_n) \leq \sum_{k=1}^{n} |x_k - y_k|,$$

giving (2.7.9) on interchanging the xs and ys. For (2.7.12),

$$x_j + y_j \leq \max(x_1, \ldots, x_n) + \max(y_1, \ldots, y_n).$$

Now take the maximum of the left side. Theorem 2.7.5 gives the results. □

Clearly there is a corresponding theorem with the minimum.

Theorem 2.7.7. *Writing $\mathbb{R}^+$ for the set of all non-negative real numbers, then for t fixed in $0 < t < 1$ let $r(x, y) = x^t y^{1-t}$ ($x, y \in \mathbb{R}^+$). If $h_1, h_2: \mathscr{U}^1 \to \mathbb{R}^+$ are integrable over E, so is $h_1^t h_2^{1-t}$.*

Proof. This r needs (2.7.11) and Hölder's inequality, in which we put

$$t = p^{-1}, \quad 1 - t = q^{-1}, \quad x_1 = u^p, \quad z_1 = s^p, \quad x_2 = v^q, \quad z_2 = w^q,$$

$$x_1^t x_2^{1-t} + z_1^t z_2^{1-t} \leq (x_1 + z_1)^t (x_2 + z_2)^{1-t} \quad (x_j \geq 0, z_j \geq 0, j = 1, 2).$$

Thus $-r$ satisfies (2.7.12), and $r \geq 0$. By the mean value theorem (2.7.9) is false. We prove (2.7.11), first noting that for the $f(x)$ below, $f(0) = 0$ and $f'(x) \geq 0$, so

(2.7.16) $\qquad f(x) \equiv x^t + z^t - (x + z)^t \geq 0 \quad (x \geq 0, z \geq 0).$

Using (2.7.16) and $x_j \geq 0$, $y_j = x_j + z_j \geq 0$ $(j = 1, 2)$,

$$y_1^t y_2^{1-t} \leq (x_1 + |z_1|)^t (x_2 + |z_2|)^{1-t} \leq (x_1^t + |z_1|^t)(x_2^{1-t} + |z_2|^{1-t})$$

$$= x_1^t x_2^{1-t} + x_1^t |z_2|^{1-t} + |z_1|^t x_2^{1-t} + |z_1|^t |z_2|^{1-t} = x_1^t x_2^{1-t} + E,$$

say, and $-D$, say, $= y_1^t y_2^{1-t} - x_1^t x_2^{1-t} \leq E$. Changing the xs to ys and vice versa, we have $D \leq E$ from

$$D \leq |z_1|^t (x_2 + z_2)^{1-t} + (x_1 + z_1)^t |z_2|^{1-t} + |z_1|^t |z_2|^{1-t},$$

when $z_1 \leq 0, z_2 \leq 0$. When $z_1 \geq 0, z_2 \geq 0, D \leq 0 \leq E$. The only remaining cases are when $z_1 < 0, z_2 > 0$, and when $z_1 > 0, z_2 < 0$, and clearly the second of these follows from the first. When $z_1 < 0, z_2 > 0$,

$$x_1^t = \{(x_1 + z_1) - z_1\}^t \leq (x_1 + z_1)^t + |z_1|^t,$$
$$x_1^t x_2^{1-t} \leq (x_1 + z_1)^t(x_2 + z_2)^{1-t} + |z_1|^t x_2^{1-t}, D \leq E.$$

Hence in all cases $|D| \leq E$, and (2.7.11) and the theorem follow on applying Hölder's inequality several times. □

As the division space is additive, the p-variation and variation set are finitely additive over elementary sets, strengthening Theorem 2.5.6(2.5.18), (2.5.20).

Theorem 2.7.8. *If E_1, E_2 are disjoint elementary sets, then*

(2.7.17) $\qquad \bar{V}_p(h; \mathbf{A}; E_1) + \bar{V}_p(h; \mathbf{A}; E_2) = \bar{V}_p(h; \mathbf{A}; E_1 \cup E_2).$

If multiplication is continuous in K with $\mathscr{G} VS(h; \mathscr{U}; E_1 \cup E_2)$ compact for some $\mathscr{U} \in \mathbf{A} | E_1 \cup E_2$, then

(2.7.18) $\qquad \overline{VS}(h; \mathbf{A}; E_1) . \overline{VS}(h; \mathbf{A}; E_2) = \overline{VS}(h; \mathbf{A}; E_1 \cup E_2).$

Proof. Theorem 2.5.6(2.5.18) shows that the left side of (2.7.17) is not greater than the right. Thus we can assume the left side finite. Given $\varepsilon > 0$ let $\mathscr{U}_j \in \mathbf{A} | E_j$ be such that

$$V_p(h; \mathscr{U}_j; E_j) < \bar{V}_p(h; \mathbf{A}; E_j) + \tfrac{1}{2}\varepsilon \quad (j = 1, 2).$$

By the additive property there is some $\mathscr{U} \in \mathbf{A} | E_1 \cup E_2$ with $\mathscr{U} \subseteq \mathscr{U}_1 \cup \mathscr{U}_2$. Hence

$$\bar{V}_p(h; \mathbf{A}; E_1) + \bar{V}_p(h; \mathbf{A}; E_2) > V_p(h; \mathscr{U}_1; E_1) + V_p(h; \mathscr{U}_2; E_2) - \varepsilon$$
$$= V_p(h; \mathscr{U}_1 \cup \mathscr{U}_2; E_1 \cup E_2) - \varepsilon$$
$$\geq V_p(h; \mathscr{U}; E_1 \cup E_2) - \varepsilon$$
$$\geq \bar{V}_p(h; \mathbf{A}; E_1 \cup E_2) - \varepsilon,$$

giving the opposite inequality to (2.5.18) as $\varepsilon \to 0+$, and so equality. Similarly

$$\overline{VS}(h; \mathbf{A}; E_1 \cup E_2) \subseteq \mathscr{G} VS(h; \mathscr{U}; E_1 \cup E_2) \subseteq \mathscr{G} VS(h; \mathscr{U}_1 \cup \mathscr{U}_2; E_1 \cup E_2)$$
$$= \mathscr{G}\{VS(h; \mathscr{U}_1; E_1) . VS(h; \mathscr{U}_2; E_2)\},$$

remembering the convention about empty products. By Theorem 0.4.10 we can take the closure inside the product, obtaining (2.7.18) as $\mathscr{U}_1, \mathscr{U}_2$ vary. □

2.7 ADDITIVE DIVISION SPACES

The error in approximation for partial sets is not more than the same error for the elementary set.

Theorem 2.7.9. *If $\bar{V}_p(h; A; E)$ is finite, and $\varepsilon > 0$ and $\mathcal{U} \in A | E$ satisfy*

(2.7.19) $\qquad V_p(h; \mathcal{U}; E) < \bar{V}_p(h; A; E) + \varepsilon,$

then for all proper partial sets P of E,

(2.7.20) $\qquad V_p(h; \mathcal{U}; P) < \bar{V}_p(h; A; P) + \varepsilon.$

(2.7.21) *In particular, $p(h(I, t)) < \bar{V}_p(h; A; I) + \varepsilon ((I, t) \in \mathcal{U})$.*

Proof. Using Theorem 2.5.6(2.5.17) and (2.7.19), (2.7.17),

$$V_p(h; \mathcal{U}; P) \leq V_p(h; \mathcal{U}; E) - V_p(h; \mathcal{U}; E \setminus P)$$
$$< \bar{V}_p(h; A; E) + \varepsilon - \bar{V}_p(h; A; E \setminus P) = \bar{V}_p(h; A; P) + \varepsilon. \quad \square$$

See also the similar Theorem 2.9.3. However, P. J. Muldowney has given an example to show that the analogue for variation sets is false. See Example 2.7.1.

One major use of additivity in division spaces is to ensure that the integral is additive when elementary sets are aggregated.

Theorem 2.7.10. *If the integral of h exists over disjoint elementary sets E_1, E_2, it exists over $E_1 \cup E_2$ and is finitely additive.*

Proof. Given $\varepsilon > 0$, let $\mathcal{U}_j \in A | E_j$ be such that each division $\mathscr{E}_j$ of E_j from $\mathcal{U}_j$ satisfies

$$|(\mathscr{E}_j) \sum h - H_j| < \tfrac{1}{2}\varepsilon,$$

H_j being the integral of h over E_j. Then a $\mathcal{U} \in A | E_1 \cup E_2$ exists with $\mathcal{U} \subseteq \mathcal{U}_1 \cap \mathcal{U}_2$, so that for $(I, t) \in \mathcal{U}$, either $I \subseteq E_1$ or $I \subseteq E_2$. Thus each division over $E_1 \cup E_2$ is $\mathscr{E}_1 \cup \mathscr{E}_2$ for $\mathscr{E}_j$ a division of E_j ($j = 1, 2$), and

$$|(\mathscr{E}) \sum h - (H_1 + H_2)| = |(\mathscr{E}_1) \sum h + (\mathscr{E}_2) \sum h - (H_1 + H_2)|$$
$$\leq |(\mathscr{E}_1) \sum h - H_1| + |(\mathscr{E}_2) \sum h - H_2| < \varepsilon. \quad \square$$

Having given theorems provable for an additive division space, we turn to an easy construction that produces an additive division space $(T, \mathcal{T}; A^+)$ from a division space $(T, \mathcal{T}, A)$, Henstock (1946, 1948) dealing with a special case.

We keep $T, \mathcal{T}, \mathcal{U}^1$ the same, while A^+ is the family of all $\mathcal{U}^+ \subseteq \mathcal{U}^1$ for which there are an elementary set E, a partition $\mathscr{P}$ of E, and a $\mathcal{U} \in A | E$ with restrictions $\mathcal{U}(I) \in A | I$ (all $I \in \mathscr{P}$), such that

$$\mathcal{U}^+ = \bigcup_{I \in \mathscr{P}} \mathcal{U}(I).$$

This $\mathbf{A}^+$ divides each $I \in \mathscr{P}$ by definition of $\mathbf{A}|I$, and so divides E, and E is an arbitrary elementary set, and $\mathscr{P}$ an arbitrary partition of E for $(T, \mathscr{T}, \mathbf{A})$.

To show that $\mathbf{A}^+$ is directed for divisions of E let $\mathscr{U}_j^+ \in \mathbf{A}^+|E$ be constructed using a partition $\mathscr{P}_j$ of E, $\mathscr{U}_j \in \mathbf{A}|E$, and restrictions $\mathscr{U}_j(I) \in \mathbf{A}|I$ (all $I \in \mathscr{P}_j$), for $j = 1, 2$. Let $\mathscr{U}_3 \in \mathbf{A}|E$ lie in $\mathscr{U}_1 \cap \mathscr{U}_2$, possible as $\mathbf{A}$ is directed for divisions of E. By the last requirement of a division space, the partial intervals $I \in \mathscr{P}_1$, $J \in \mathscr{P}_2$, of E are co-partitional, so that a partition $\mathscr{P}_3$ of E from $\mathscr{U}_3$ can be found to refine $\mathscr{P}_1$ and $\mathscr{P}_2$. Each $J \in \mathscr{P}_3$ is a partial interval of E and so there is a restriction $\mathscr{U}_3(J)$ of $\mathscr{U}_3$ to J and lying in $\mathbf{A}|J$. Also there are restrictions $\mathscr{U}_j(J) \in \mathbf{A}|J$ of the $\mathscr{U}_j(I)$ for $J \subseteq I$, $I \in \mathscr{P}_j$ ($j = 1, 2$) as J is then a partial interval of I, and there is a $\mathscr{U}_3^+(J) \in \mathbf{A}|J$ with

$$\mathscr{U}_3^+(J) \subseteq \mathscr{U}_1(J) \cap \mathscr{U}_2(J) \cap \mathscr{U}_3(J).$$

Then the union $\mathscr{U}_3^+$ of the $\mathscr{U}_3^+(J)$ ($J \in \mathscr{P}_3$) lies in $\mathbf{A}^+|E$ and by construction $\mathscr{U}_3^+ \subseteq \mathscr{U}_1^+ \cap \mathscr{U}_2^+$, and $\mathbf{A}^+|E$ is directed in the sense of divisions of E.

It can now be proved easily that $\mathbf{A}^+$ satisfies all requirements of an additive division space. Also, if $\mathscr{U}^+ \in \mathbf{A}^+|E$, with $\mathscr{P}$ the partition of E and $\mathscr{U} \in \mathbf{A}|E$ used to construct $\mathscr{U}^+$, then $\mathscr{U}^+$ divides each $I \in \mathscr{P}$, and the union of the separate divisions gives a division $\mathscr{E}$ of E from $\mathscr{U}^+$ and so from $\mathscr{U}$. If $h: \mathscr{U}^1 \to K$, with multiplication in K, the value of $(\mathscr{E}) \prod h(I, t)$ is a point in $S(h; \mathscr{U}; E)$, so that

$$S(h; \mathscr{U}^+; E) \subseteq S(h; \mathscr{U}; E).$$

This shows that in a sense the $\mathscr{U} \in \mathbf{A}|E$ are too big, and need to be slimmed down to the $\mathscr{U}^+ \in \mathbf{A}^+|E$.

The last in the list of requirements of a division space, that partial sets are co-partitional, can now be seen as necessary and sufficient that an additive division space can be constructed from the division space, see Theorem 2.7.1(2.7.2).

Theorem 2.7.2 first appeared as a theorem of J. C. Burkill (1924a, Section 5) for continuous finitely subadditive interval functions, and was then given in Henstock (1963c), Theorem 22.1, p. 34, for the gauge integral. Theorems 2.7.5, 2.7.6 are in Henstock (1964), Theorem 2.7.6 being already in Henstock (1963c) Theorem 25.1, p. 43.

Example 2.7.1 (P. J. Muldowney) The analogue for variation sets of Theorem 2.7.9 could be as follows. If $VS(h; \mathscr{U}; E)$ lies in a fixed compact set for each $\mathscr{U} \in \mathbf{A}|E$, if G is a neighbourhood of the origin, and if for some $\mathscr{U}_1 \in \mathbf{A}|E$,

$$VS(h; \mathscr{U}_1; E) \subseteq \overline{VS}(h; \mathbf{A}; E) + G$$

then

$$VS(h; \mathscr{U}_1; P) \subseteq \overline{VS}(h; \mathbf{A}; P) + G$$

for all partial sets P of E.

2.8 DECOMPOSABLE ADDITIVE DIVISION SPACES

However, the following example using the gauge integral shows that the result is false.

$$h([u, v); x) \equiv \begin{cases} v - u & ([u, v) \subseteq [1, 2)), \\ u & (0 < u < 1, v = 1), \\ 0 & \text{(otherwise)}, \end{cases}$$

$$0 < \delta(x) < |x - 1| \quad (x \neq 1)$$

Then

$$VS(h; \delta; [0, 2)) = [0, 2), \overline{VS}(h; \mathbf{A}; [0, 2)) = [0, 2],$$

$$VS(h; \delta; [0, 1)) = (1 - \delta, 1) \cup \text{sing}(0),$$

$$\overline{VS}(h; \mathbf{A}; [0, 1)) = \text{pair}(0, 1).$$

Taking $G = (-\varepsilon, \varepsilon)$ with $0 < \varepsilon < \delta$, we contradict the supposed analogue.

Exercise 2.7.2 Let K_j ($j = 1, 2, 3$) be Banach spaces such that if $x_j \in K_j$ ($j = 1, 2$) then $x_1 x_2 \in K_3$. For simplicity let the norms in K_j ($j = 1, 2, 3$) all be written $\|.\|$ with $\|x_1 x_2\| = \|x_1\| \cdot \|x_2\|$. Let $f: T \to K_1$, $h: \mathcal{U}^1 \to K_2$, be such that h, $\|h\|$, fh, $\|fh\|$ are all strong variationally integrable over E. Given that k is a fixed value in K_1, prove that $\|(f - k)h\|$ is integrable over E.

(*Hint:* For H, F the integrals of h and fh, respectively, $\|(f - k)h - (F - kH)\|$ has variation zero, $F - kH$ is finitely additive and so $\|F - kH\|$ is finitely subadditive, while $\|(f - k)h\| \leq \|fh\| + \|k\| \|h\|$.)

2.8 DECOMPOSABLE ADDITIVE DIVISION SPACES

This section deals with those division spaces that are additive and decomposable. Thus to the definitions of Section 2.7 we add the following. $(T, \mathcal{T}, \mathbf{A})$ is *fully decomposable* (respectively, *decomposable*, or *measurably decomposable* relative to a measure or measure space, defined later) if to every family (respectively, countable family or countable family of measurable sets) $\mathcal{X}$ of mutually disjoint subsets $X \subseteq T$, and every function $\mathcal{U}(.): \mathcal{X} \to \mathbf{A}|E$, there is a $\mathcal{U} \in \mathbf{A}|E$ with

$$\mathcal{U}[X] \equiv \{(I, t): (I, t) \in \mathcal{U}, t \in X\} \subseteq \mathcal{U}(X) \quad (X \in \mathcal{X}).$$

The union of the $X \in \mathcal{X}$ need not be T nor E^*. If, for the given $\mathcal{X}$,

$$\mathcal{U}[X] = \mathcal{U}(X)[X] \quad (X \in \mathcal{X})$$

we call $\mathcal{U}$ the *diagonal* of the $(\mathcal{U}(X), \mathcal{X})$.

In a decomposable additive division space the variation of each member of a monotone increasing sequence of sets is monotone increasing to the variation of the union of the sets, even if non-measurable sets occur in the sequence. The proof normally given for Lebesgue measure has to assume that it is *regular*, i.e. given a set $X \subseteq T$, there is an h-measurable set M containing X with $V(h; \mathbf{A}; E; M) = V(h; \mathbf{A}; E; X)$. We have no need of any such assumption.

Theorem 2.8.1. *Let $(T, \mathscr{T}, \mathbf{A})$ be a decomposable additive division space and let (X_j) be a monotone increasing sequence of sets in T with union X. Then for h: $\mathscr{U}^1 \to K$, K being a Banach space,*

(2.8.1) $$\lim_{j \to \infty} V(h; \mathbf{A}; E; X_j) = V(h; \mathbf{A}; E; X).$$

More generally, for an arbitrary sequence (X_j) of sets in T,

(2.8.2) $$V\left(h; \mathbf{A}; E; \liminf_{j \to \infty} X_j\right) \leqslant \liminf_{j \to \infty} V(h; \mathbf{A}; E; X_j),$$

where $\liminf_{j \to \infty} X_j$ denotes $\bigcup_{m=1}^{\infty} \bigcap_{j=m}^{\infty} X_j$.

Proof. From $X \supseteq X_j$ in (2.8.1),

(2.8.3)
$$V(h; \mathbf{A}; E; X_j) \leqslant V(h; \mathbf{A}; E; X), \quad \lim_{j \to \infty} V(h; \mathbf{A}; E; X_j) \leqslant V(h; \mathbf{A}; E; X).$$

Thus if the limit is infinite there is nothing to prove. For a finite limit let $\mathscr{U}_j \in \mathbf{A}|E$ satisfy, for given $\varepsilon > 0$,

(2.8.4) $\quad V(h; \mathscr{U}_j; E; X_j) < V(h; \mathbf{A}; E; X_j) + \varepsilon . 2^{-j} \quad (j = 1, 2, \dots).$

By decomposability there is a $\mathscr{U} \in \mathbf{A}|E$ with

(2.8.5) $\qquad \mathscr{U}[X_j \setminus X_{j-1}] \subseteq \mathscr{U}_j \quad (X_0 \text{ empty}, j = 1, 2, \dots).$

If $\mathscr{E}$ is a division of E from $\mathscr{U}$, and if $\mathscr{Q}, \mathscr{Q}_j$ are the partial divisions of E with associated points t in X and $X_j \setminus X_{j-1}$, respectively ($j = 1, 2, \dots$), there is a greatest integer m (depending on $\mathscr{E}$) such that $\mathscr{Q}_m$ is not empty (since $\mathscr{E}$ contains only a finite number of (I, t)). Let E_j be the union of the I from the $(I, t) \in \mathscr{Q}_j$ ($j = 1, \dots, m$). Then from Theorems 2.7.8, 2.7.9, and (2.8.a) ($a = 3, 4, 5$),

$$(\mathscr{E}) \sum \|h\| \chi(X; t) = (\mathscr{Q}) \sum \|h\| = \sum_{j=1}^{m} (\mathscr{Q}_j) \sum \|h\| \leqslant \sum_{j=1}^{m} V(h; \mathscr{U}_j; E_j; X_j)$$

$$< \sum_{j=1}^{m} \{V(h; \mathbf{A}; E_j; X_j) + \varepsilon . 2^{-j}\} < \sum_{j=1}^{m} V(h; \mathbf{A}; E_j; X_m) + \varepsilon$$

$$\leqslant V(h; \mathbf{A}; E; X_m) + \varepsilon \leqslant \lim_{j \to \infty} V(h; \mathbf{A}; E; X_j) + \varepsilon,$$

$$V(h; \mathbf{A}; E; X) \leqslant V(h; \mathscr{U}; E; X) \leqslant \lim_{j \to \infty} V(h; \mathbf{A}; E; X_j) + \varepsilon,$$

giving the opposite inequality to (2.8.3) and so (2.8.1).

2.9 STAR-SETS AND THE INTRINSIC TOPOLOGY

For an arbitrary sequence (X_j), $\bigcap_{j=m}^{\infty} X_j$ is monotone increasing in m. Hence

$$V(h; \mathbf{A}; E; \liminf_{j \to \infty} X_j) = V\left(h; \mathbf{A}; E; \bigcup_{m=1}^{\infty} \bigcap_{j=m}^{\infty} X_j\right) = \lim_{m \to \infty} V\left(h; \mathbf{A}; E; \bigcap_{j=m}^{\infty} X_j\right)$$

$$\leq \liminf_{m \to \infty} V(h; \mathbf{A}; E; X_m)$$

giving (2.8.2). Note that if (X_j) is monotone increasing to X, (2.6.2) gives

$$V(h; \mathbf{A}; E; X) \leq \lim_{j \to \infty} V(h; \mathbf{A}; E; X_j),$$

which with (2.8.3) gives (2.8.1). Thus (2.8.1) and (2.8.2) are equivalent for monotone increasing sequences. When the sequence is monotone decreasing the result is trivial and obtainable from reversing the inequalities in (2.8.3).

The result and proof can be found in Henstock (1968b) pp. 231–2, Theorem 44.9, and the result, with a different proof, is in Henstock (1963c), pp. 54–5, Theorem 28.6. In both, (2.8.2) is missing.

As the analogue of Theorem 2.7.9 is false for variation sets it appears that an analogous proof for variation sets cannot go through, and perhaps the analogue of Theorem 2.8.1 for variation sets is false.

2.9 STAR-SETS AND THE INTRINSIC TOPOLOGY

For $\mathcal{U} \subseteq \mathcal{U}^1$ and $\mathcal{U}.E$ the set of all $(I, t) \in \mathcal{U}$ with I a partial interval of E, in order that $\mathcal{U}.E$ is not empty, then at least E must have partial intervals and so divisions, and so must be an elementary set, which we assume. Next, let $E^*.\mathcal{U}$ be the set of all t with some I and $(I, t) \in \mathcal{U}.E$. Then the *star-set* E^* is the intersection of $E^*.\mathcal{U}$ for all $\mathcal{U} \in \mathbf{A}|E$, and here we need a division system for E.

If there is a $\mathcal{U}(E) \in \mathbf{A}|E$ for which every $\mathcal{U}_1 \in \mathbf{A}|E$ with $\mathcal{U}_1 \in \mathcal{U}(E)$, has $E^*.\mathcal{U}_1 = E^*.\mathcal{U}(E)$, then this is E^* and we say that $(T, \mathcal{T}, \mathbf{A})$ *is stable for E*. If true for all elementary sets E we say that $(T, \mathcal{T}, \mathbf{A})$ is *stable*. For the non-empty $\mathcal{T}_0 \subseteq \mathcal{T}$, and P a proper partial set of E that is from $\mathcal{T}_0$, i.e. the union of a finite number of disjoint partial intervals of E from $\mathcal{T}_0$, we say that a division space $(T, \mathcal{T}, \mathbf{A})$ *is weakly $\mathcal{T}_0$-compatible with P in E* when there is a $\mathcal{U}_P \in \mathbf{A}|E$ such that $(I, t) \in \mathcal{U}_P$ and $t \notin P^*$ imply $I \subseteq E \setminus P$, equivalent to $I \subseteq E$ and $I \cap P$ empty, and so with I, P co-partitional. If, for some such $\mathcal{U}_P$, we also have $I^* \subseteq E^* \setminus P^*$, equivalent to $I^* \subseteq E^*$ and $I^* \cap P^*$ empty, we say that $(T, \mathcal{T}, \mathbf{A})$ *is strongly $\mathcal{T}_0$-compatible with P in E*. We say that $(T, \mathcal{T}, \mathbf{A})$ is *infinitely divisible* if each elementary set contains a proper partial set.

The intrinsic topology $\mathcal{G}_I$ over E of the section's title has by definition a sub-base B consisting of the empty set, E^*, and complements $E^* \setminus P^*$ for the various partial sets P of E.

Theorem 2.9.1.

(2.9.1) B is a base for $\mathcal{G}_I$ if for all partial sets P_1, P_2 of E, $P_1 \cup P_2$ is also a partial set, with

(2.9.2) $$(P_1 \cup P_2)^* = P_1^* \cup P_2^*.$$

(2.9.3) If the division system $(T, \mathcal{T}, \mathbf{A})$ is weakly $\mathcal{T}_0$-compatible with co-partitional partial sets P_1, P_2 from $\mathcal{T}_0$, then (2.9.2) holds.

(2.9.4) A fully decomposable division system for E is stable for E.

(2.9.5) If the division space $(T, \mathcal{T}, \mathbf{A})$ is weakly $\mathcal{T}_0$-compatible with a partial set P of E from $\mathcal{T}_0$, then $(T, \mathcal{T}, \mathbf{A})$ is stable for P.

(2.9.6) A stable additive division space is weakly $\mathcal{T}$-compatible with all partial sets.

This theorem gives many connections between various definitions.

Proof. In (2.9.1), as intersections of complements of sets are complements of the unions of the sets, by (2.9.2) finite intersections of sets of B belong to B and B is a base. For (2.9.3), $P \equiv P_1 \cup P_2 \supseteq P_1$. As co-partitional, for

$$\mathcal{U} \in \mathbf{A}|E, \quad \mathcal{U}.P \supseteq \mathcal{U}.P_1, \quad P^*.\mathcal{U} \supseteq P_1^*.\mathcal{U} \supseteq P_1^*, \quad P^* \supseteq P_1^*, \quad P^* \supseteq P_1^* \cup P_2^*.$$

Again, by weak $\mathcal{T}_0$-compatibility, if

$$(I, t) \in \mathcal{U} = \mathcal{U}_{P_1} \cap \mathcal{U}_{P_2}, \quad t \in P^* \setminus (P_1^* \cup P_2^*),$$

then $t \in \setminus P_j^*$, $I \subseteq E \setminus P_j$ ($j = 1, 2$), $I \subseteq E \setminus P$. But $t \in P^* \subseteq P^*.\mathcal{U}$, so that there is an $I \subseteq P$, giving a contradiction. Thus no such t exists and (2.9.3) and (2.9.2) follow. For (2.9.4) see Chapter 1(1.1.2). For (2.9.5) and the $\mathcal{U}_P$ of weak $\mathcal{T}_0$-compatibility, $P^*.\mathcal{U}_P$ is the set of all t with $(I, t) \in \mathcal{U}_P.P$. Thus $I \subseteq P$ and so we cannot have $I \subseteq E \setminus P$. Hence by weak $\mathcal{T}_0$-compatibility,

$$t \in P^*, \quad P^*.\mathcal{U}_P \subseteq P^* \subseteq P^*.\mathcal{U}_P,$$

and there is equality and so stability. For (2.9.6) let P be the union of some I with (I, t) in a division $\mathcal{E}$ of E. For $\mathcal{U}(P)$ in the definition of stability let $\mathcal{U}_1 \in \mathbf{A}|E$ be as in Theorem 2.7.1(2.7.1) with $\mathcal{U}_1.P \subseteq \mathcal{U}(P)$. Then $P^*.\mathcal{U}_1 = P^*$. If $(I, t) \in \mathcal{U}_1$, $t \notin P^*$ then $I \not\subseteq P$ and so $I \subseteq E \setminus P$ and we have weak $\mathcal{T}$-compatibility.

Theorem 2.9.2.

(2.9.7) Let B be a Banach space with zero z, $h: \mathcal{U}^1 \to B$, and $V(X) \equiv V(h; \mathbf{A}; E; X)$, finite. Let $h(I, t) = z$ if $I \in \mathcal{T} \setminus \mathcal{T}_0$. Given $\varepsilon > 0$ and $(T, \mathcal{T}, \mathbf{A})$ weak $\mathcal{T}_0$-compatible, there is a proper partial set P_1 of E from $\mathcal{T}_0$ with

2.9 STAR-SETS AND THE INTRINSIC TOPOLOGY

(2.9.8) $V(X\backslash P_1^*) < \varepsilon$, $V(X \cap P_1^*) > V(X) - \varepsilon$, if either $(T, \mathcal{T}, \mathbf{A})$ is infinitely divisible, or if, for a partial set P of E from $\mathcal{T}_0$, $X = Y\backslash P^*$. Then $P \cap P_1 = \emptyset$. In either case $V(X)$ is the supremum of $V(X \cap P_1^*)$.

(2.9.9) In (2.9.7) with $Y = T$, if $V(\backslash P^*)$ is finite it is the supremum of $V(P_1^*)$ for all $P_1 \subseteq \backslash P$.

(2.9.10) Let $h(I, t) = 0$ $(I \in \mathcal{T}\backslash \mathcal{T}_0)$, P a partial set of E from $\mathcal{T}_0$, and $(T, \mathcal{T}, \mathbf{A})$ strongly $\mathcal{T}_0$-compatible with P in E, or $V(P_1^* \cap P_2^*) = 0$ if P_1, P_2 are disjoint partial sets. Then P^* is Carathéodory h-measurable, i.e.

(2.9.11) $\qquad V(X \cap P^*) + V(X\backslash P^*) = V(X)$ (all $X \subseteq T$).

(2.9.12) In (2.9.10), if P_1^*, P_2^* are disjoint star-sets, then

(2.9.13) $\qquad\qquad V(P_1^*) + V(P_2^*) = V(P_1^* \cup P_2^*)$.

(2.9.14) In (2.9.10) with full decomposability, arbitrary unions of various $\backslash P^*$ are Carathéodory h-measurable.

Proof. For (2.9.7), as $V(X)$ is finite let $\mathcal{U} \in \mathbf{A}|E$ have $V(h; \mathcal{U}; E; X)$ finite. For a division $\mathscr{E}$ of E let $t \in X$ when $(I, t) \in \mathscr{E}_1 \subseteq \mathscr{E}$, and $t \notin X$ or $I \in \mathcal{T}\backslash \mathcal{T}_0$ when $(I, t) \in \mathscr{E}\backslash \mathscr{E}_1$, in which case $h(I, t)\chi(X; t) = 0$. We can have $\mathscr{E}_1 = \mathscr{E}$. We can take E so that

(2.9.15) $\qquad V(h; \mathcal{U}; E; X) - \varepsilon < (\mathscr{E}_1)\sum \|h(I, t)\| \leq V(h; \mathcal{U}; E; X)$.

For $(T, \mathcal{T}, \mathbf{A})$ infinitely divisible, given an integer $N > 0$, there is a division of E from $\mathcal{U}$ with at least N pairs (I, t). In such a division $\mathscr{E}$, for N large enough the least $|h(I, t)| > 0$ can be as small as we please, and can be omitted from the sum in (2.9.15) without altering the inequalities, thus replacing $\mathscr{E}_1$ by a new $\mathscr{E}_1$. Then the union P_1 of I from the $(I, t) \in \mathscr{E}_1$, is proper. Alternatively, for $X = Y\backslash P^*$ let $(I, t) \in \mathcal{U} \subseteq \mathcal{U}_P$. For $t \in X$ then $t \in \backslash P^*$, so that $I \subseteq E\backslash P$, $P_1 \subseteq E\backslash P$, and P_1 is proper with $P \cap P_1 = \emptyset$, giving the stronger (2.9.15) again, and P, P_1 are co-partitional. Let $\mathcal{U}_1 \in \mathbf{A}|E$ have $\mathcal{U}_1 \subseteq \mathcal{U} \cap \mathcal{U}_{P_1}$. In a division of E to give $V(h; \mathcal{U}_1; E; X\backslash P_1^*)$ we omit those (I, t) with $t \notin X\backslash P_1^*$, the rest forming a partial division $\mathscr{E}_2$ with $I \subseteq E\backslash P_1$, disjoint from P_1, while $t \in X$. As $\mathscr{E}_1$ is a division of P_1, $\mathscr{E}_1 \cup \mathscr{E}_2$ is a partial division of E, and by varying $\mathscr{E}_2$,

(2.9.16) $\qquad (\mathscr{E}_1)\sum \|h(I, t)\| + V(h; \mathcal{U}_1; E; X\backslash P_1^*) \leq V(h; \mathcal{U}; E; X)$.

Using the first inequality in (2.9.15) and removing the sum over $\mathscr{E}_1$, we have

$$V(X\backslash P_1^*) \leq V(h; \mathcal{U}_1; E; X\backslash P_1^*) < \varepsilon,$$

the first result in (2.9.8). Finite subadditivity of $V(X)$ in X gives the second result. Hence (2.9.9). For (2.9.10), if $P = E$ or $V(X) = +\infty$, (2.9.11) is trivially true. If $P \neq E$, $V(X) < +\infty$, $\varepsilon > 0$, let the $\mathcal{U}$ for (2.9.15) with $X, \mathscr{E}_1$ replaced by $X \backslash P^*$ and $\mathscr{E}_3$, satisfy

(2.9.17) $$V(h; \mathscr{U}; E; X) < V(X) + \varepsilon.$$

From (2.9.15) with $X \backslash P^*$, (2.9.16), (2.9.17),

(2.9.18) $$V(X \backslash P^*) - \varepsilon + V(X \backslash P_1^*) < (\mathscr{E}_3) \sum \|h(I, t)\|$$
$$+ V(h; \mathscr{U}_1; E; X \backslash P_1^*)$$
$$\leqslant V(h; \mathscr{U}; E; X) < V(X) + \varepsilon.$$

Also by strong $\mathscr{T}_0$-compatibility, $I^* \cap P_1^* = \varnothing$, or by the other assumption its variation is zero. Using (2.9.18) with

$$X \cap P^* = (X \cap P^* \cap P_1^*) \cup (X \cap P^* \backslash P_1^*) \subseteq (P^* \cap P_1^*) \cup (X \backslash P_1^*),$$
$$V(X \backslash P^*) + V(X \cap P^*) < V(X) + 2\varepsilon,$$

and (2.9.11) follows. For $X = P_1^* \cup P_2^*$ and $P = P_1$, we prove (2.9.13). Finally, for (2.9.14) let Y be an arbitrary union of sets $\backslash P^*$. Then each $t \in Y$ lies in a $\backslash P(t)^*$ from Y and depending on t. By full decomposability we choose $\mathscr{U} \in \mathbf{A}|E$ so that

(2.9.19) $$\mathscr{U}[\text{sing}(t)] \subseteq \mathscr{U}_{P(t)} \quad (t \in Y).$$

By this the $(I, t) \in \mathscr{U}$ with $t \in Y$ have $t \in \backslash P(t)^*$ and so $I^* \subseteq \backslash P(t)^* \subseteq Y$, by strong T_0-compatibility. If also $\mathscr{U}$ satisfies the conditions in (2.9.7) that give (2.9.8), so that P_1 is the finite union of disjoint I with $I^* \subseteq Y$, then from (2.9.2), (2.9.11),

$$P_1^* \subseteq Y, \quad \backslash P_1^* \supseteq \backslash Y, \quad V(\backslash Y) + V(Y) < V(\backslash Y) + V(P_1^*) + \varepsilon$$
$$= V(\backslash Y) + V(E^*) - V(\backslash P_1^*) + \varepsilon \leqslant V(E^*) + \varepsilon.$$

Replacing h by $h. \chi(X; \cdot)$ and letting $\varepsilon \to 0+$, we have (2.9.11) with Y instead of P^*.

This theorem gives some approximation and measurability results, and in particular shows that the open sets of the intrinsic topology are measurable.

Thomson (1972a) pp. 504–5, Lemma 1, pointed out that in his division system we can replace $V(h; \mathbf{A}; P)$ by $V(P^*)$ and obtain (2.9.23) of the next theorem from the assumption (2.9.22).

Theorem 2.9.3. *Let $(T, \mathscr{T}, \mathbf{A})$ be a stable division space with weak $\mathscr{T}$-compatibility. Let B be a Banach space with $h: \mathscr{U}^1 \to B$. If P is a partial set of E then*

(2.9.20) $$V(h; \mathbf{A}; P) \leqslant V(h; \mathbf{A}; E; P^*) \equiv V(P^*).$$

(2.9.21) *There is equality when $\mathscr{F}(E; P)$ is of h-variation zero and also when the division space is additive.*

2.9 STAR-SETS AND THE INTRINSIC TOPOLOGY

If $(T, \mathcal{T}, \mathbf{A})$ is also strongly $\mathcal{T}$-compatible, given $\varepsilon > 0$ and P a partial set of E from $\mathcal{T}_0$, if

(2.9.22) $V(h; \mathcal{U}; E) < V(h; \mathbf{A}; E) + \varepsilon = V(h; \mathbf{A}; E; E^*) + \varepsilon \equiv V(E^*) + \varepsilon,$

then

(2.9.23) $$V(h; \mathcal{U}; P) < V(P^*) + \varepsilon.$$

Proof. For (2.9.20), if P is a partial set of E, $\mathcal{U} \subseteq \mathcal{U}_P$, and $(I, t) \in \mathcal{U}.P$, then $(I, t) \in \mathcal{U}$ and I is a *partial interval* of P. As $t \in \backslash P^*$ implies $I \subseteq E \backslash P$, whereas $I \subseteq P$, we must have $t \in P^*$ and

$$\|h(I, t)\chi(P^*, t)\| = \|h(I, t)\|.$$

Thus (I, t) can be used to calculate $V(h; \mathcal{U}; E; P^*)$, and (2.9.20) follows by

$$\bar{V}(h; \mathbf{A}; P) \leq V(h; \mathcal{U}; P) \leq V(h; \mathcal{U}; E; P^*).$$

Equality does not always occur in (2.9.20), for the calculation of $V(P^*)$ sometimes uses (I, t) with $t \in \mathcal{F}(E; P)$ and neither $I \cap P$ nor $I \backslash P$ empty. As $I \nsubseteq P$, $h(I, t)$ cannot be used for the calculation of $V(h; \mathbf{A}; P)$, nor for $V(h; \mathbf{A}; E \backslash P)$ since $I \nsubseteq E \backslash P$, so that $V(h; \mathbf{A}; P)$ need not be finitely additive in P. These difficulties disappear in an additive division space since eventually every I lies in P entirely or in $E \backslash P$ entirely; and they disappear too when $\mathcal{F}(E; P)$ has h-variation zero, as in (2.9.21). To prove (2.9.23) from (2.9.22) we need the finite additivity of $V(P^*)$ in P from Theorem 2.9.1(2.9.3) and Theorem 2.9.2(2.9.12). Let $\mathcal{U}_1 \in \mathbf{A} | E$ satisfy $\mathcal{U}_1 \subseteq \mathcal{U} \cap \mathcal{U}_P$. If $t \notin P^*$, $(I, t) \in \mathcal{U}_1$, then $I \subseteq E \backslash P$ and $I \cap J$ is empty if $J \subseteq P$. Thus

$$V(h; \mathcal{U}; P) + V(\backslash P^*) \leq V(h; \mathcal{U}; P) + V(h; \mathcal{U}_1; E; \backslash P^*)$$
$$\leq V(h; \mathcal{U}; E) < V(E^*) + \varepsilon.$$

Subtracting $V(\backslash P^*)$ from both sides, we have (2.9.23). □

There now follows the principal use of this section's theory.

Theorem 2.9.4. *If $(T, \mathcal{T}, \mathbf{A})$ is a fully decomposable division space strongly $\mathcal{T}$-compatible with every partial set of E, then E^* is compact in the intrinsic topology G_I.*

Proof. By Theorem 2.9.1(2.9.3), making B a base, we need only use covers of E^* that are families of $\backslash P^*$. Each $t \in E^*$ lies in one of the $\backslash P^*$, say, $\backslash P(t)^*$. For

$$\mathcal{U}[\text{sing}(t)] \subseteq \mathcal{U}_{P(t)} \quad (t \in E^*),$$

a division $\mathscr{E}$ of E from $\mathcal{U}$, and $(I, t) \in \mathscr{E}$, we have

$$(I, t) \in \mathcal{U}_{P(t)}, \quad t \notin P(t)^*, \quad I^* \subseteq E^* \backslash P(t)^*, \quad E^* = \bigcup_{\mathscr{E}} I^* \subseteq \bigcup_{\mathscr{E}} E^* \backslash P(t)^*,$$

i.e. a finite cover of E^* from $\mathscr{G}_I$, and E^* is compact in the intrinsic topology.

For Z_1, Z_2 disjoint sets in E^*, if Z_1 is Carathéodory h-measurable and $X = Z_1 \cup Z_2$,
$$V(X \cap Z_1) + V(X \setminus Z_1) + V(X),$$
(2.9.24) $$V(Z_1) + V(Z_2) = V(Z_1 \cup Z_2).$$

Thomson (1982/1983, p. 107, Def. 4.5) gives another condition for (2.9.24). $\mathbf{A}|E$ separates disjoint sets Z_1, Z_2 if there is a $\mathcal{U} \in \mathbf{A}|E$ with $I \cap J$ empty when $(I, t) \in \mathcal{U}[Z_1]$ and $(J, u) \in \mathcal{U}[Z_2]$. Then (2.9.24) is clearly true. □

Theorem 2.9.5. *Let $(T, \mathcal{T}, \mathbf{A})$ be a division space weakly $\mathcal{T}$-compatible with every partial set of E. If $P_2 = E \setminus P_1$ where P_1, P_2 are proper partial sets of E, and $Z_j \subseteq \setminus P_j^*$ ($j = 1, 2$), then $\mathbf{A}|E$ separates Z_1, Z_2 and (2.9.24) is true.*

Proof. Let $\mathcal{U} \in \mathbf{A}|E$ with $\mathcal{U} \subseteq \mathcal{U}_{P_1} \cap \mathcal{U}_{P_2}$. If $(I, t) \in \mathcal{U}$, $t \in Z_1$, then $(I, t) \in \mathcal{U}_{P_1}$, $t \in \setminus P_1^*$, $I \subseteq E \setminus P_1 = P_2$. Similarly, if $(J, u) \in \mathcal{U}$, $u \in Z_2$, then $J \subseteq P_1$ and I, J are disjoint. □

We now have a long proof of the Radon–Nikodym theorem for $K = \mathbb{C}$ or $\mathbb{R}$. The first part of the proof is written for general Banach spaces as it only involves the norm, whereas the second part is true only for $\mathbb{C}$ or $\mathbb{R}$.

Theorem 2.9.6. *Let $(T, \mathcal{T}, \mathbf{A})$ be a decomposable additive division space, let B be a Banach space and let $h: \mathcal{U}^1 \to B$ be strong variationally integrable and of bounded variation over an elementary set E, with variation*
$$V(P) \equiv \bar{V}(h; \mathbf{A}: E: P^*) = \bar{V}(h; \mathbf{A}; P),$$
such that

(2.9.25) $$V(\mathcal{F}(E; P)) = 0 \text{ for every partial set } P \text{ of } E.$$

Then there is an $f: E^* \to B$ with $\|f(t)\| = 1$ everywhere, and for all partial sets P of E,

(2.9.26) $$H(P) \equiv \int_P dh = \int_P f \, dV.$$

Proof. First, $V(P) = \bar{V}(h; \mathbf{A}; P)$ by Theorem 2.9.3(2.9.21), so that by Theorem 2.7.3, $\|h\|$ and $\|H\|$ are integrable to V. Using an extended definition of the signum function

(2.9.27) $$\operatorname{sgn}(w) = \begin{cases} w/\|w\| & (w \neq z) \\ 1 & (w = z) \end{cases}$$

so that $\|w\| \operatorname{sgn}(w) = w$, $\|\operatorname{sgn}(w)\| = 1$ $(w \in B)$;

(2.9.28) for $g(I) \equiv \operatorname{sgn}(H(I))$, $\|g(I)\| = 1$,
$$(\mathscr{E}) \sum \|H - gV\| = (\mathscr{E}) \sum |\|H\| - V| > 0$$

2.9 STAR-SETS AND THE INTRINSIC TOPOLOGY

as $\mathscr{U}$ shrinks, again by Theorem 2.7.3, and for all partial sets P of E,

(2.9.29) $$H(P) = \int_P d(gV).$$

The whole difficulty of the proof lies in replacing $g(I)$ by a suitable $f(t)$ independent of P.

By Theorem 2.7.3 again and (2.9.28) there are a monotone decreasing sequence $(\mathscr{U}_n) \subseteq \mathbf{A}|E$ and a sequence $(\mathscr{P}_n)$ of partitions of E with $\mathscr{P}_n$ from $\mathscr{U}_n$, $\mathscr{P}_{n+1}$ a refinement of $\mathscr{P}_n$, and

(2.9.30) $$(\mathscr{P}_n) \sum \|H\| > V(E) - 2^{-4n};$$

(2.9.31) $$(\mathscr{P}'_n) \sum \|H - gV\| < 2^{-n}$$

for all partitions $\mathscr{P}'_n$ of E from $\mathscr{U}_n$ $(n = 1, 2, \ldots)$. Let $\mathscr{Q}_n$ be the family and P_n the partial set of those $I \in \mathscr{P}_n$ with

(2.9.32) $$\|H(I)\| < (1 - 2^{-3n})V(I).$$

By (2.9.30), Theorem 2.7.3(2.9.32), the finite additivity of V, Theorem 2.9.3(2.9.21), and Theorem 2.4.1(2.4.1),

$$V(E) - 2^{-4n} < (\mathscr{P}_n)\sum \|H\| = (\mathscr{P}_n \setminus \mathscr{Q}_n)\sum \|H\| + (\mathscr{Q}_n)\sum \|H\| < (\mathscr{P}_n \setminus \mathscr{Q}_n)\sum V$$
$$+ (1 - 2^{-3n})(\mathscr{Q}_n)\sum V = V(E) - 2^{-3n}V(P_n),$$

(2.9.33) $$V(P_n) \leq 2^{-n}, \quad \bar{V}\left(h; \mathbf{A}; E; \bigcup_{m \geq n} P_m^*\right) \leq 2^{1-n},$$

$$X = \bigcap_{n=1}^{\infty} \bigcup_{m \geq n} P_m^*, \quad \bar{V}(h; \mathbf{A}; E; X) = 0.$$

In $\setminus X$, (2.9.32) is eventually false, so that for some integer $N = N(t)$ depending on the fixed point $t \in I^*$, and all $n \geq N$,

(2.9.34) $$\|H(I)\| \geq (1 - 2^{-3n})V(I) \quad (I \in \mathscr{P}_n).$$

We now take $B = \mathbb{C}$. If a partition $\mathscr{P}'$ of E refines $\mathscr{P}_n$, let $\mathscr{R}$ be the family of $J \in \mathscr{P}'$ with $J \not\subseteq P_n$ and

(2.9.35) $$\text{Real}\{H(J)/g(I)\} \leq (1 - 2^{-2n})|H(J)| \ (J \subseteq I \in \mathscr{P}_n), \ W = \bigcup_\mathscr{R} J.$$

As (2.9.34) is true, and by definition of g, (2.9.35), and Theorem 2.7.3,

$$(1 - 2^{-3n})V(I) \leq |H(I)| = \text{Real}\{(\mathscr{P}' \cap I)\sum H(J)/g(I)\}$$
$$\leq (\mathscr{R} \cap I)\sum |H(J)|(1 - 2^{-2n})$$
$$+ (\mathscr{P}' \cap I \setminus \mathscr{R})\sum |H(J)|$$
$$\leq (1 - 2^{-2n})(\mathscr{R} \cap I)\sum V(I)$$
$$+ (\mathscr{P}' \cap I \setminus \mathscr{R})\sum V(J)$$
$$= V(I) - 2^{-2n}(\mathscr{R} \cap I)\sum V(J),$$

(2.9.36) $\quad (\mathcal{R} \cap I) \sum V(J) \leq 2^{-n} V(I), (\mathcal{R}) \sum V(J) \leq 2^{-n} V(E).$

When $\mathcal{P}' = \mathcal{P}_{n+1}$ let W be W_n. By (2.9.36) and Theorem 2.4.1(2.4.1),

(2.3.37) $V(W_n) = (\mathcal{R}) \sum V(J) \leq 2^{-n} V(E), \bar{V}\left(h; \mathbf{A}; E; \bigcup_{m \geq n} W_m^*\right) < 2^{1-n} V(E).$

When $J \not\subseteq P_n \cup W_n$, (2.9.35) is false, so that if $r_n(t)$ is the $r(t)$ of Exercise 2.9.1 with $\mathscr{E}$ from $\mathcal{P}_n$ and $g(I) = \text{sgn}(H(I))$ as before, then, h-almost everywhere,

$$\text{Real}\{r_{n+1}(t)/r_n(t)\} = \text{Real}\{g(J)/g(I)\}$$
$$> 1 - 2^{-2n} \quad (t \in J \subseteq I, J \in \mathcal{P}_{n+1}, I \in \mathcal{P}_n),$$
$$|\arg(r_{n+1}(t)/r_n(t))| < \theta_n < \pi \sin(\tfrac{1}{2}\theta_n) = \pi \cdot 2^{-(2n+1)/2} < 2^{2-n}.$$

Hence by (2.9.33) and (2.9.37) there is a set X_n with

(2.9.38) $\quad\quad\quad \bar{V}(h; \mathbf{A}; E; X_n) < 2^{1-n}\{1 + V(E)\}$

and outside X_n,

$$|\arg(r_{m+1}(t)) - \arg(r_m(t))| < 2^{2-m},$$
$$|\arg(r_m(t)) - \arg(r_n(t))| < 2^{3-n} \quad (m > n),$$

and $\arg(r_n(t))$ tends to a limit, say $\theta(t)$. For

$$r(t) = \exp(i\theta(t)), \ |r(t)| = 1, \ \arg r(t) = \theta(t), \ |\arg r(t) - \arg r_n(t)| \leq 2^{3-n},$$
$$|r - r_n|^2 = |r/r_n - 1|^2 = 4\sin^2\{\tfrac{1}{2}\arg(r/r_n)\} \leq \{\arg(r/r_n)\}^2,$$

(2.9.39) $\quad\quad\quad |r(t) - r_n(t)| \leq 2^{3-n} \quad (t \notin X_n).$

From $|r_n(t)| = 1$, and so r_n is bounded, h-almost everywhere, with (2.9.38), (2.9.39), and integrable over E; it converges h-almost everywhere to $r(t)$ and so $r(t)V(I)$ is integrable and

$$\left|\int_E r(t)\,dV - H(E)\right| \leq 2^{-n} + 2^{3-n} V(E) + 2^{1-n}\{1 + V(E)\},$$

giving (2.9.26) for $P = E$. Finally, let P be a proper partial set from a partition $\mathcal{P}$ of E. By Theorem 2.7.1(2.7.1) there is a $\mathcal{U}_n^* \in \mathbf{A}|E$ with $\mathcal{U}_n^* \subseteq \mathcal{U}_n$, such that all divisions of E from $\mathcal{U}_n^*$ refine $\mathcal{P}_n$ and $\mathcal{P}$. Then, except in a set satisfying (2.9.37), the corresponding r_n^* satisfies $|r_n^*(t) - r_n(t)| < 2^{2-n}$, $\lim_{n \to \infty} r_n^*(t) = r(t)$, and $r_n^*(t)$ is an $r_n(t)$ for Exercise 2.9.1 with P for E. Hence (2.9.26) holds for all partial sets of E. □

The star-sets P^* were originally defined as a means of location of properties, so that for the (I, t) in divisions, the t lay in well-defined sets. The definition of P^* has gradually been refined, see Henstock (1961a, p. 118, axiom (X1); 1961b, p. 415, axiom (T1); 1968b, p. 218; 1973b, p. 320; 1978, p. 71;

2.9 STAR-SETS AND THE INTRINSIC TOPOLOGY

1980a, p. 397). The first clues that it might be useful to regard the P^* as closed sets came since in many division spaces the P^* are closed for some reasonable topology. For example, see McGill (1977) and Henstock (1973b, p. 335, elementary *-sets and the Tychonoff analogue, Theorem 8). From these considerations, the intrinsic topology was developed in Henstock (1980a, pp. 411–12, Theorems 15, 16). McGill (1973; 1974/1975; 1975a, b, c; 1976; 1977; 1980; 1981) gives an alternative theory.

Exercise 2.9.1 The following result connects the point structure of T with the structure of intervals and divisions. For $(T, \mathcal{T}, \mathbf{A})$ a stable division space let $\mathscr{E}$ be a division of an elementary set E from a $\mathscr{U} \in \mathbf{A}|E$. Let $r: T \to \mathbb{C}$ be such that for each $(I, t) \in \mathscr{E}$ and some numbers $g(I)$,

$$r(u) = g(I)(u \in I^* \backslash \mathscr{F}(E; I) = I^* \backslash (E \backslash I)^*), \quad \mathscr{F}(E; P) \equiv P^* \cap (E \backslash P)^*,$$

the frontier star-set of a proper partial set P of E. Otherwise let r be arbitrary. Then we say that *r is a step-function based on $\mathscr{E}$*; it is defined except on some frontier star-sets.

To make the definition useful when integrating rh for h; $\mathscr{U}^1 \to \mathbb{C}$ we assume that $\mathscr{F}(E; P)$ has h-variation zero, so that the arbitrariness of r can be ignored.

Let $f: T \to \mathbb{C}$ be bounded in E^* with fh integrable in E, and let r be bounded in E^*. Prove that there exists

(2.9.40) $$\int_E fr\,dh = (\mathscr{E}) \sum g(I) \int_I f\,dh.$$

Given that $(T, \mathcal{T}, \mathbf{A})$ is also decomposable show that f and r need only be finite.

(*Hint*: show that $\{r(x) - g(I)\} f(x) h(J, x)$ has variation zero over I, and so has zero integral.)

Example 2.9.2 The zero h-variation of $\mathscr{F}(E; P)$ is a kind of continuity condition on h. Some such condition seems necessary for the truth of Theorem 2.9.6(2.9.26); for $[u, v) \subseteq E = [-1, 1)$ let $h([u, v), t) = h(u, v)$ with $h(u, 0) = 1\ (u < 0)$, $h(0, v) = -1\ (v > 0)$, $h(u, v) = 0$ otherwise. Then for the gauge integral the neighbourhoods of non-zero points can be chosen so that they do not include 0, and then divisions of $[u, 0)$ and $[0, v)$ must necessarily include 0 as a division-point. Hence for

$$V(u, v) = \bar{V}(h; \mathbf{A}; [u, v)),$$

$$V(u, 0) = 1 = V(0, v),\ V(u, v) = 2 \quad (u < 0 < v),$$

$$\int_u^0 f\,dV = f(0) = \int_0^v f\,dV \quad (u < 0 < v),$$

and (2.9.26) is false since either

$$h(u, 0) \neq \int_u^0 f\,dV \text{ or } h(0, v) \neq \int_0^v f\,dV, \text{ or both.}$$

2.10 SPECIAL RESULTS

We begin with integration by parts, normally considered in $\mathbb{R}$ but sometimes in higher dimensions.

Theorem 2.10.1. *If $(\mathbb{R}, \mathcal{T}, \mathbf{A})$ is a division space, if $a < b$ are finite, if the integrals below exist with $f, g: \mathbb{R} \to \mathbb{C}$, and dg standing for dh with $h([p, q); t) = g(q) - g(p)$, and with $t = p$ or q,*

(2.10.1) $$\int_u^v f\,dg + \int_u^v g\,df = f(v)g(v) - f(u)g(u) \equiv \Delta(fg; [u, v))$$

holds for all $[u, v) \subseteq [a, b)$, only if

(2.10.2) $$\bar{V}((\Delta f)(\Delta g); \mathbf{A}; [a, b)) = 0.$$

Conversely, if (2.10.2) is true with one integral existing in (2.10.1), the other exists and (2.10.1) holds for all $[u, v) \subseteq [a, b)$.

Proof. We use Theorem 2.5.5(2.5.15) with the identity

(2.10.3) $$\{f(x) - f(t)\}\{g(x) - g(t)\} = f(x)\{g(x) - g(t)\} + g(x)\{f(x) - f(t)\}$$
$$- f(x)g(x) + f(t)g(t)$$

$$\bar{V}((\Delta f)(\Delta g); \mathbf{A}; [a, b)) = \bar{V}(f\Delta g + g\Delta f - \Delta(fg); \mathbf{A}; [a, b))$$
$$= \bar{V}\left(\int f\,dg + \int g\,df - \Delta(fg); \mathbf{A}; [a, b)\right) = 0.$$

Conversely, if the first integral in (2.10.1) exists, with (2.10.2), (2.10.3) gives

$$\bar{V}(g\Delta f + \int f\,dg - \Delta(fg); \mathbf{A}; [a, b)) = 0$$

with $\Delta(fg) - \int f\,dg$ finitely additive, which last difference is therefore the norm variational integral of $g\Delta f$, which is also the generalized Riemann integral of $g\Delta f$. □

We can rearrange (2.10.3) slightly to give

(2.10.4) $$f(x)\{g(x) - g(t)\} = f(x)g(x) - f(t)g(t) - g(t)\{f(x) - f(t)\},$$

2.10 SPECIAL RESULTS

leading to the definition of the *integral by parts*, **IP**,

(2.10.5) $\quad (\text{IP}) \int_{[a,b)} f \, dg = \Delta(fg; [a, b)) - \int_{[a,b)} g \, df$

on taking the general interval from t to x, with t as the associated point and x the opposite point. Thus from the family $\mathscr{U} \in \mathbf{A}|E$ of interval–point pairs $([p, q), t)$, where t is one of p, q and x is the other, we can construct a corresponding 'mirror image' family $\mathscr{U}^*$ of $([p, q), x)$, and then construct the corresponding $\mathbf{A}^*$ using $\mathscr{U}^*$, just as $\mathbf{A}$ was constructed using $\mathscr{U}$. Then all properties that do not involve the associated points pass from $(\mathbb{R}, \mathscr{T}, \mathbf{A})$ to $(\mathbb{R}, \mathscr{T}, \mathbf{A}^*)$. But the various decomposabilities, the star sets, the weak and strong compatibilities, the compatibility with a topology, and product space definitions, need careful examination. For an example of a decomposable $(\mathbb{R}, \mathscr{T}, \mathbf{A})$ that has $(\mathbb{R}, \mathscr{T}, \mathbf{A}^*)$ not decomposable we need Theorem 7.2.3 to show that if f is the indicator of the rationals and $g(t) = t$, and using the gauge integral, the right side of (2.10.5) does not exist. It exists for the first n rationals, so that the monotone convergence theorem is false for this $(T, \mathscr{T}, \mathbf{A}^*)$, and by Theorem 3.2.1 it cannot be decomposable.

The conditions in Theorem 2.10.1 are independent, in the sense that the integrals in (2.10.1) can exist without (2.10.2) being true, and, in another example, (2.10.2) can be true when the two integrals do not exist. See Example 2.10.1 and Exercise 2.10.2.

Theorem 2.10.1 can be written in the following form.

Theorem 2.10.2. *If the first integral exists, with* $(\mathbb{R}, \mathscr{T}, \mathbf{A})$ *a division space and* $a < b$, *finite,*

$$\int_{[a,b)} \{f(t) - f(a)\} \, dg = (\text{IP}) \int_{[a,b)} \{g(b) - g(x)\} \, df.$$

This enables us to give a simplified formula for integration by parts in higher dimensions. For real coordinates let $\mathbf{a} = (a_1, \ldots, a_n)$, $\mathbf{b} = (b_1, \ldots, b_n)$, with $[\mathbf{a}, \mathbf{b})$ the n-dimensional brick of all points $\mathbf{x} = (x_1, \ldots, x_n)$ satisfying

$$\min(a_j, b_j) \leq x_j \leq \max(a_j, b_j) \quad \text{with } a_j \neq b_j \quad (1 \leq j \leq n).$$

For $f(\mathbf{x}) = f(x_1, \ldots, x_n) : \mathbb{R}^n \to K$ we define

$$\Delta_k f = \Delta_k f(y_1, \ldots, y_n) = f(y_1, \ldots, y_{k-1}, \max(a_k, b_k), y_{k+1}, \ldots, y_n)$$
$$- f(y_1, \ldots, y_{k-1}, \min(a_k, b_k), y_{k+1}, \ldots, y_n),$$
$$\Delta(f(y); \mathbf{a}, \mathbf{b}) = \Delta_1 \cdots \Delta_n f.$$

In the last multiple difference all ys disappear. (2.10.4) can be extended to n dimensions.

(2.10.6) $\quad f(\mathbf{x})\Delta(g; \mathbf{x}, \mathbf{t}) = \Delta(fg; \mathbf{x}, \mathbf{t}) - \sum_1 + \sum_2 - \sum_3 + \cdots$

where $\sum_k$ is the sum of $\Delta(f(y)g(t_1, \ldots, t_k, y_{k+1}, \ldots, y_n); \mathbf{x}, \mathbf{t})$ and similar terms in which the k letters t are permuted with the $n - k$ letters y. Integrating (2.10.6) with the $\mathbf{x}$ replaced by $\mathbf{t}$, which can be done if g is sufficiently smooth, we have a formula of W. H. Young (1918). 2^n integrals occur, and if in the form using x, all but one (i.e. $2^n - 1 \geq 3$) integrals exist, we can prove the existence of the remaining integral.

On using the theory of Chapter 5, another formula needs only the existence of two integrals. Strictly, Theorems 2.10.3, 2.10.5 should really appear in Chapter 5, but for convenience they appear here.

Theorem 2.10.3. *Using a Fubini division space* $(\mathbb{R}^n, \mathcal{T}, \mathbf{A})$ *whose intervals are finite n-dimensional bricks* $[\mathbf{a}, \mathbf{b}]$, *if the two integrals below exist, then*

(2.10.7) $\quad \displaystyle\int_{[\mathbf{a},\mathbf{b}]} \Delta(f; \mathbf{a}, \mathbf{t})\,dg = (\mathbf{IP})\int_{[\mathbf{a},\mathbf{b}]} \Delta(g; \mathbf{x}, \mathbf{b})\,df,$

the second integral using $\mathbf{A}^*$; *it is equal to*

$$\int_{[\mathbf{a},\mathbf{b})} \Delta(g; \mathbf{t}, \mathbf{b})\,df$$

(2.10.8) $\quad$ *if* $\bar{V}(|\Delta(g; \mathbf{t}, \mathbf{b}) - \Delta(g; \mathbf{x}, \mathbf{b})|\Delta(f; \mathbf{t}, \mathbf{x}); \mathbf{A}; [\mathbf{a}, \mathbf{b})) = 0$.

Proof. As both integrals exist, Theorem 5.1.1 shows that the integrals are equal to n repeated integrals with respect to each variable in turn, so that Theorem 2.10.2 gives the results. □

The existence of one integral might be deduced from that of the other, by showing that the existence of the repeated integral, with measurability conditions, implies the existence of the integral in n dimensions. If no measurability conditions are given, Sierpinski (1920) stops this argument. Also a converse similar to that in Theorem 2.10.1 need not hold since the integrand on the right of (2.10.7) is a function of $\mathbf{b}$.

In $n = 1$ dimension we have the second mean value theorem for integrals.

Theorem 2.10.4. *If f is monotone in $[a, b]$, g bounded and attaining every value between its bounds, and $f\Delta g$ integrable in $[a, b)$, then for a ξ in $[a, b]$,*

(2.10.9) $\quad \displaystyle\int_{[a,b)} f\,dg = f(b)\{g(b) - g(\xi)\} + f(a)\{g(\xi) - g(a)\}.$

2.10 SPECIAL RESULTS 115

Proof. For f monotone increasing, Theorem 2.10.2 gives

$$\int_{[a,b)} f\,dg = \Delta(fg;[a,b)) - (\mathbf{IP})\int_{[a,b)} g(x)\,df,$$

$$\{f(b)-f(a)\}\inf g \leqslant (\mathbf{IP})\int_{[a,b)} g\,df \leqslant \{f(b)-f(a)\}\cdot\sup g,$$

$$(\mathbf{IP})\int_{[a,b)} g\,df = \{f(b)-f(a)\}g(\xi)$$

for suitable ξ. Hence (2.10.9). For f monotone decreasing take $-f$. □

Often g is an indefinite integral, and the gauge and general Denjoy integrals and all integrals that are included in them are continuous if the integrator is continuous, and so satisfy the Darboux condition on g. So does the not necessarily continuous Young's integral (Hobson (1926, §480, p. 720). (Henstock (1973a) gives many historical details on integration by parts.) These can therefore be used in the theorem. Note that (2.10.9) can be written

$$\int_{[a,b)} \{f(t)-f(a)\}\,dg = \{g(b)-g(\xi)\}\{f(b)-f(a)\}.$$

This leads to a similar result in $\mathbb{R}^n$.

Theorem 2.10.5. *For* $(\mathbb{R}^n, \mathcal{T}, \mathbf{A})$ *and* $f; \mathbb{R}^n \to \mathbb{R}$ *with* $\Delta(f;\mathbf{t},\mathbf{x})$ *of constant sign, and if* $g: \mathbb{R}^n \to \mathbb{R}$ *is such that* $\Delta(g;\mathbf{t},\mathbf{b})$ *is bounded in* $\mathbf{t}$ *and attains every value between its bounds, with* $\Delta(f;\mathbf{a},\mathbf{t})$ *integrable with respect to* g *on* $[\mathbf{a},\mathbf{b})$, *then*

$$\int_{[\mathbf{a},\mathbf{b}]} \Delta(f;\mathbf{a},\mathbf{t})\,dg = \Delta(f;\mathbf{a},\mathbf{b})\Delta(g;\mathbf{c},\mathbf{b}) \quad (\text{some } \mathbf{c}\in[\mathbf{a},\mathbf{b}]).$$

Proof. In the proof of Theorem 2.10.3, the right side of (2.10.7) exists as a repeated integral if the left side exists. Hence we can use a proof similar to that of Theorem 2.10.4. □

As has already been pointed out, these results are usually applied when g is an indefinite integral, and we use Theorem 2.6.1(2.6.6). Corresponding theorems hold for the N-variational integral.

Next we look at the Cauchy and Harnack extensions of the generalized Riemann integral. The Cauchy extension was first used on the Riemann integral to integrate some functions unbounded in the neighbourhood of certain points, and de la Vallée Poussin (1892a, b) tackled some non-absolute integrals before Lebesgue's famous paper in 1902. Denjoy (1912a, b) applied both extensions repeatedly to the Lebesgue integral to integrate some functions that are not Lebesgue integrable in the neighbourhood of certain points, and in a transfinite inductive process to integrate all derivatives. Perron (1914) defined a much simpler integral, see Section 7.1, that is equivalent to Denjoy's special integral, and equivalent also to the gauge integral of '$f(t)\,dt$'. Thus it is not surprising that the Cauchy and Harnack extensions are included in that particular case of the generalized Riemann integral. The extensions in higher dimensions are not so clear as the extension in $T = \mathbb{R}$, which we will now give.

Theorem 2.10.6. *Let $(\mathbb{R}, \mathcal{T}, \mathbf{A})$ be a decomposable additive division space, let $h: \mathcal{U}^1 \to \mathbb{C}$ be integrable to $H(u, v)$ over every partial set $[u, v)$ of $[a, b)$ that has $v < b$, and let $[u, v)^* \subseteq [u, v]$. If there exists*

$$H_1(u, b) \equiv \lim\{H(u, v) + h([v, b), b)\}$$

in the sense that, given $\varepsilon > 0$, there is a $\mathcal{U}_0 \in \mathbf{A}|[a, b)$ such that

(2.10.10) $\qquad |H(u, v) + h([v, b), b) - H_1(u, b)| < \varepsilon$

for all $([v, b), b) \in \mathcal{U}_0$, then $H(a, b)$ exists equal to $H_1(a, b)$.

Proof. Let (v_n) be a strictly increasing sequence in (a, b) that tends to b, with each $[v_{n-1}, v_n)$ a partial set of $[a, b)$, and with $v_0 = a$ $(n = 1, 2, \ldots)$. Then there is a $\mathcal{U}_n \in \mathbf{A}|[a, b)$ such that if $\mathcal{E}_n$ is a division of $[v_{n-1}, v_n)$ from $\mathcal{U}_n$, then

(2.10.11) $\qquad |H(v_{n-1}, v_n) - (\mathcal{E}_n)\sum h| < \varepsilon \cdot 2^{-n}$,

and if $(I, t) \in \mathcal{U}_n$ then $I \subseteq [v_{n-1}, v_n)$ or $I \cap [v_{n-1}, v_n)$ is empty. If also $v_{n-1} < v < v_n$ with $[v_{n-1}, v)$ a partial set of $[a, b)$ and if $\mathcal{E}'_n$ is a division of $[v_{n-1}, v)$ from $\mathcal{U}_n$, then

(2.10.12) $\qquad |H(v_{n-1}, v) - (\mathcal{E}'_n)\sum h| \leq \varepsilon \cdot 2^{-n}$.

Also by finite additivity of H, for $[a, u)$ a partial set of $[a, b)$,

$$H_1(a, b) = H(a, u) + H_1(u, b) \quad (a < u < b)$$

and we can replace (2.10.10) by

(2.10.13) $\qquad |h([v, b), b) - H_1(v, b)| < \varepsilon$

As $[u, v)^* \subseteq [u, v]$, $[v_{n-1}, v_n)^*$ and $[v_{m-1}, v_m)^*$ ($m \neq n$) have at most one point in common, and it does not lie in any other $[u_{p-1}, u_p)^*$. By direction there is a $\mathcal{U}_+ \in \mathbf{A}|[a, b)$ that lies in $\mathcal{U}_n \cap \mathcal{U}_m$. Using this for the common points, and decomposability for the points $[v_{n-1}, v_n)^* \cap (v_{n-1}, v_n)$, we can find a $\mathcal{U} \in \mathbf{A}|[a, b)$ that lies in $\mathcal{U}_0$ at b, and in $\mathcal{U}_n$ at $[v_{n-1}, v_n)^*$. If $\mathcal{E}$ is a division of $[a, b)$ from $\mathcal{U}$, and $(I, t) \in \mathcal{E}$, the only I with $b \in \bar{I}$ must have $t = b$ (again as $[u, v)^* \subseteq [u, v]$) and $I = [v, b)$ with $v_{m-1} < v \leq v_m$ for some integer

2.10 SPECIAL RESULTS

m. Thus by (2.10.n) ($n = 11, 12, 13$),

$$|(\mathscr{E})\sum h - H_1(a,b)| \leq \sum_{n=1}^{m-1} |(\mathscr{E} \cap [v_{n-1}, v_n))\sum h - H(v_{n-1}, v_n)|$$
$$+ |(\mathscr{E} \cap [v_{m-1}, v))\sum h - H(v_{m-1}, v)| + |h([v,b), b) - H_1(v, b)| < 2\varepsilon,$$

which proves the result.

Similarly we can deal with intervals $[a, b)$ with trouble at a, and can replace b by $+\infty$, or a by $-\infty$, or both. The conventional $([v, +\infty), +\infty)$ and $([-\infty, u), -\infty)$, have $h = 0$. $[a, +\infty)$ uses divisions, for example, $([v, +\infty)$ with $h = 0$, then a division of $[a, v)$, will give a division of $[a, +\infty)$. As $\mathscr{U}$ shrinks we let $v \to \infty$. Theorem 2.10.6 then shows that integrals defined by these divisions are the same as those defined over $[a, v)$ for suitable vs tending to $+\infty$. The importance of these new divisions is that proofs of results, such as for limits under the integral sign, go through for $[a, +\infty)$ just as they do for $[a, b)$, without using the extra limit as the $v \to \infty$. Similarly for $(-\infty, b)$ and $(-\infty, \infty)$.

Note that corresponding integrals in complex variable theory use limits of $[-b, b)$ as $b \to +\infty$, and do not come under the present theory, for which $a \to -\infty$ and $b \to +\infty$ *independently*.

The Harnack extension is as follows. Let F be a compact set on $\mathbb{R}$. Then F lies in a bounded interval, say $I = [u, v)$, and we can arrange that $u \in F$, so that $G = I \setminus F$ is an open set and so the union of a sequence (I_k) of disjoint open intervals. If a function f is integrable by some reasonable means over F and the separate I_k with

$$M_k \equiv \max_J \left| \int_J f \, dx \right|$$

for all intervals $J \subseteq I_k$, and if the series of M_k is convergent, then the Harnack-extended integral over I is the sum of the integrals over F and the separate I_k.

We now tease out the implications of this definition. First, the convergence of the sum of moduli means that, given $\varepsilon > 0$, there is an integer N with

(2.10.14)
$$\sum_{k=N}^{\infty} M_k \leq \varepsilon$$

i.e. if $J_k \subseteq I_k$ for a finite number of distinct $k \geq N$, then

(2.10.15)
$$\left| \sum \int_{J_k} f \, dx \right| \leq \varepsilon.$$

Conversely, if (2.10.15) is true and if f is real-valued, we take the positive integrals in one sum, the negative integrals in another, to give

$$\sum \left| \int_{J_k} f \, dx \right| \leq 2\varepsilon.$$

As usual, for complex-valued f we finish with 4ε on the right, for any finite number of $k \geq K$, and so for all $k \geq K$, giving (2.10.14) with 2ε or 4ε on the right. Thus the following theorem shows that the generalized Riemann integral includes the Harnack extension process, and this extension gives nothing new.

Theorem 2.10.7. *For a compact set $F \subseteq [a, b]$ with $a, b \in F$, let $G = (a, b)\setminus F$, an open set and so the union of disjoint intervals $[u_j, v_j)$ with $[u_j, v_j] \subseteq G$ $(j = 1, 2, \ldots)$. Let $(\mathbb{R}, \mathcal{T}, \mathbf{A})$ be a decomposable additive division space with $[a, b)$ an elementary set. Given $h\colon \mathcal{U}^1 \to \mathbb{C}$, let*

$$(2.10.16) \qquad \int_{[a, b)} \chi(F; \cdot)\, dh, \quad \int_{[u_j, v_j)} dh$$

exist. If, given $\varepsilon > 0$, there are an integer J and a $\mathcal{U} \in \mathbf{A}\,|\,[a, b)$ such that for every finite collection $\mathcal{D}$ of disjoint partial sets $[u, v)$ of $[a, b)$, with some $(I, t) \in \mathcal{U}$, some $t \in F$, and $[u, v) = [u_j, v_j) \cap I$ for some $j > J$, and no two intervals $[u, v)$ lying in the same $[u_j, v_j)$, we have

$$(2.10.17) \qquad \left|(\mathcal{D})\sum \int_{[u, v)} dh\right| < \varepsilon,$$

then there exists

$$(2.10.18) \qquad \int_{[a, b)} dh = \int_{[a, b)} \chi(F; \cdot)\, dh + \sum_{j=1}^{\infty} \int_{[u_j, v_j)} dh.$$

Proof. Subtracting the first integral in (2.10.16) from both sides of (2.10.18), we need only prove that

$$(2.10.19) \qquad \int_{[a, b)} \chi(G; \cdot)\, dh = H_2([a, b)) \quad \text{where}$$

$$H_2(E) \equiv \sum_{j=1}^{\infty} \int_{[u_j, v_j) \cap E} dh$$

for each elementary set $E \subseteq [a, b)$. As E is a finite union of intervals each of which has two frontier points, $[u_j, v_j) \cap E$ is either empty or is $[u_j, v_j)$, for all but a finite number of j. Hence by (2.10.17) the sequence of partial sums of the infinite series for $H_2(E)$ is fundamental and so convergent, and H_2 exists.

To prove (2.10.19) we use the proof of Theorem 2.10.6 for the intervals $[u_j, v_j)$ $(j = 1, 2, \ldots)$, while for the set F, $\chi(G; t) = 0$ there, and t does not lie in $[u_j, v_j)$ $(j = 1, \ldots, J)$. Taking these from $[a, b)$, we are left with an elementary set E_1, a finite union of disjoint intervals, and the same is true of $[a, b)\setminus E_1$. These provide a partition of $[a, b)$ and we can assume a $\mathcal{U}_0 \in \mathbf{A}\,|\,[a, b)$ such that every division from $\mathcal{U}_0$ refines that partition. Thus by

2.10 SPECIAL RESULTS

decomposability we can find a $\mathscr{U} \in \mathbf{A}|[a, b)$ and complete the proof as in Theorem 2.10.6 with $v = b$.

Combining Theorems 0.1.1, 2.10.6, and 2.10.7, then at each stage of construction of the special Denjoy integral from the Lebesgue integral, the construction gives the gauge integral. Thus at the end of the process, in a countable, though possibly transfinite, number of steps, the special Denjoy integral is included in the gauge integral, and the converse holds when $h(I, t) = f(t)m(I)$ for $m(I)$ the length of I. The gauge integral is wider as it can integrate more general $h(I, t)$ than $f(t)m(I)$. In Section 7.1 we show that in the $f(t)m(I)$ case the gauge integral is equivalent to the Perron integral, so that we have another proof of the equivalence of the Perron and special Denjoy integrals on $\mathbb{R}$.

Note that all the results in one dimension for this section appear in Henstock (1963c) for the gauge integral.

Example 2.10.1 In Exercise 0.1.2 each division $\mathscr{E}$ of $[-1, 1)$ from $\mathscr{U}$ must include two intervals each with associated point 0, assuming that in (I, t), t is at an end of I. Then

$$(\mathscr{E})\sum |(\Delta f)(\Delta g)| = 1, \quad \bar{V}((\Delta f)(\Delta g); \mathbf{A}; [-1, 1)) = 1,$$

and integration by parts fails.

Exercise 2.10.2 For the gauge integral, let $\varepsilon > 0$ and take

$$f(t) = 1/(t \log t) \quad (0 < t < 1), \; f(0) = 0, \; g(t) = t.$$

Show that as $\varepsilon \to 0$, neither integral

$$\int_{[\varepsilon, 1/2]} f \, dg = \Delta(\log \log (1/t); [\varepsilon, 1/2)),$$

$$\int_{[\varepsilon, 1/2]} g \, df = \Delta(1/\log t - \log \log(1/t); [\varepsilon, 1/2))$$

tends to a limit, so that by Theorem 2.5.4(2.5.13) the integrals do not exist over $[0, 1/2)$. However,

$$|f(t) - f(0)||g(t) - g(0)| \to 0 \quad \text{as } t \to 0+,$$

so that on using Theorem 2.4.1(2.4.1), (2.10.2) is true. Another proof is given in Henstock (1973a).

Example 2.10.3 Note that when both f and g are differentiable in (a, b) and continuous at the ends, as in the calculus, (2.10.2) is easily shown true. Thus (2.10.2) never appears in the calculus.

CHAPTER 3

LIMITS UNDER THE INTEGRAL SIGN, FUNCTIONS DEPENDING ON A PARAMETER

3.1 INTRODUCTION AND NECESSARY AND SUFFICIENT CONDITIONS

It is well known that the Lebesgue integral handles with ease many problems of the commutability of operations of integration and the taking of a limit, partially resolving C. Jordan's question quoted by de la Vallée Poussin (1892a).

> Give a rigorous theory of differentiation under the integral sign of definite integrals, with precise conditions which limit Leibnitz's rule, principally for unbounded regions of integration or unbounded functions, and particularly many celebrated definite integrals.

After a century of partial results we go beyond Lebesgue to necessary and sufficient conditions, proofs in division space theory being transparent.

The simplest limit concerns a measure m over an elementary set E and sequences $(f_n(t))$ tending pointwise to a finite limit $f(t)$ m-almost everywhere, given that each $f_n m$ is integrable over E. We examine two properties, the integrability of fm over E, and

$$(3.1.1) \qquad \lim_{n \to \infty} \int_E f_n \, dm = \int_E \lim_{n \to \infty} f_n \, dm = \int_E f \, dm.$$

We now replace n by a variable y over an uncountable set Y with a limit process, symbolically written '$y\to$'. For example, the integral's continuity in y at c, is studied taking '$y\to$' as $y \to c$. Such limit processes are considered in this chapter, while Chapter 4 deals with differentiation relative to y, (3.1.1) becoming one- or n-dimensional differentiation, e.g.

$$(3.1.2) \qquad \frac{d}{dy} \int_E f(y, t) \, dm = \lim_{h \to 0} \int_E \frac{f(y+h, t) - f(y, t)}{h} \, dm$$

$$= \int_E \lim_{h \to 0} \frac{f(y+h, t) - f(y, t)}{h} \, dm = \int_E \frac{\partial f(y, t)}{\partial y} \, dm.$$

3.1 NECESSARY AND SUFFICIENT CONDITIONS

In this area is the inversion of order of repeated integrals over $E \otimes H$ (Chapter 5),

$$(3.1.3) \quad \int_H \left\{ \int_E g(y,t)\,dm \right\} dM = \lim_{\delta \to 0+} (\mathscr{E}_y) \sum \int_E g(y,t)\,dm \times M(v,w)$$

$$= \int_E \lim_{\delta \to 0+} (\mathscr{E}_y) \sum g(y,t) M(v,w)\,dm = \int_E \int_H g(y,t)\,dM\,dm,$$

in a reasonable notation. Either integral is the limit process, the other being the integral in the generalization of (3.1.1).

Supposing a decomposable division space for (3.1.1), when the linear space K (such as a Banach space) has a norm $\|\cdot\|$, we can use properties (i) $\bar V(m; \mathbf{A}; E) < \infty$, (ii) $\|f_n - f\| \to 0$ m-almost everywhere in E, (iii) Fm and $f_n m$ ($n = 1, 2, \ldots$) integrable on E, (iv) $m \geq 0$ and $\|f_n\| \leq F$ ($n = 1, 2, \ldots$), the Arzelà–Lebesgue conditions in K. But (i), which forbids many applications to Feynman integration, and (iv) restrict the test. However, Section 3.3 gives a condition depending on Riemann sums, that is usually equivalent to (iv), and this condition can be generalized more easily. Further, not all topological groups K have even a group norm; and the restriction to $f_n(t)m(I,t)$ and $f(y;t)m(I,t)(y \in Y)$ is a weakness. Generalizing to $h_n(I,t)$ and $h(y;I,t)$ respectively, a problem is highlighted by Exercises 2.3.5 and 2.3.6 and avoided by a slight change of Henstock (1969), p. 527.

We first look at the question of what corresponds to (ii). For $(K, \mathscr{G})$ a topological linear space, to each $(I, t) \in \mathscr{U}^1$ let there be a sequence $(Z^j(I,t))$ of sets of K containing the zero z, such that for each $\mathscr{G}$-neighbourhood G of z, a positive integer j and a $\mathscr{U} \in \mathbf{A}|E$ exist and for all divisions $\mathscr{E}$ over E from $\mathscr{U}$,

$$(3.1.4) \quad (\mathscr{E}) \sum Z^j(I,t) \equiv \{(\mathscr{E}) \sum z^j(I,t) : z^j(I,t) \in Z^j(I,t)\} \subseteq G.$$

This is invariant under the action of real continuous linear functionals, and so is a useful idea in topological linear spaces, see (3.1.6).

Theorem 3.1.1.

(3.1.5) *For $(T, \mathscr{T}, \mathbf{A})$ a division space, if P is a proper partial set of E it can replace E in (3.1.4), divisions of P replacing the $\mathscr{E}$.*

(3.1.6) *Further, for $\mathscr{F}$ a real continuous linear functional on K and*

$$R^j(I,t) \equiv \mathscr{F}(Z^j(I,t)) \equiv \{\mathscr{F}(z^j(I,t)) : z^j(I,t) \in Z^j(I,t)\},$$

$R^j(I,t)$ can replace $Z^j(I,t)$ in (3.1.4) with $G = (-\varepsilon, \varepsilon)$ for various $\varepsilon > 0$.

Proof. $z \in Z^j(I,t)(I \subseteq E \setminus P)$ gives (3.1.5). Linearity, continuity, and a suitable $\mathscr{G}$-neighbourhood G of z give (3.1.6) with $\mathscr{F}(z) = 0$ and

$$(\mathscr{E}) \sum R^j(I,t) = \{(\mathscr{E}) \sum \mathscr{F}(z^j(I,t)) = \mathscr{F}((\mathscr{E}) \sum z^j(I,t)) : z^j(I,t) \in Z^j(I,t)\} \subseteq \mathscr{F}(G)$$

$$\subseteq (-\varepsilon, \varepsilon). \quad \square$$

122   THE GENERAL THEORY OF INTEGRATION

To correspond to (ii), for each $y \in Y$ let $h(y; I, t)$ be integrable over E, let $X \subseteq T$, $\mathcal{U}_+ \in \mathbf{A}|E$, $\mathcal{U}_+ \subseteq \mathcal{U}$, and for each integer $j \geq 1$ and each $(I, t) \in \mathcal{U}_+$ let $k(j; I, t) \in Y$. For $h: \mathcal{U}_+ \to K$ let

(3.1.7) $h(y; I, t) - h(I, t) \in Z^j(I, t) ((I, t) \in \mathcal{U}_+, t \notin X, y \geq k(j; I, t)$ in $Y)$,

(3.1.8) h and $h(y; \cdot)$ are of variation zero in X, relative to $E, \mathbf{A}$.

The integration is not affected if we take X empty and replace h by $h \cdot \chi(\backslash X; \cdot)$, $h(y; \cdot)$ by $h(y; \cdot) \chi(\backslash X; \cdot)$. We now show that in the $f_n m$ and $f(y; \cdot)m$ cases (ii) is true when $(3.1.n)(n = 4, 7, 8)$ are true, given other reasonable properties, and conversely, so that the new arrangement includes the old.

Theorem 3.1.2. *Let $(T, \mathcal{T}, \mathbf{A})$ be a decomposable division space with K normed, let*

$$h(y; I, t) = f(y; t)m(I, t), \quad h(I, t) = f(t)m(I, t),$$

let X_1 be the set of t where $f(y; t) = f(t)$ for some $J(t) \in Y$ and all $y \geq J(t)$ in Y, let $m \cdot \chi(\backslash X_1; \cdot)$ be VBG* in E with $f(y; t) \to f(t)$ except in a set X_2 of m-variation zero, and for each $t \in X_2$ let either $f(y; t)$ be bounded in y or $m(I, t) = 0$ (all $(I, t) \in \mathcal{U} \in \mathbf{A}|E$, $t \in X_2$). Then (3.1.7) holds, the $Z^j(I, t)$ being spheres $S(z, r)$ with centre the zero z and radii $r = r(I, t)$, and $S(z, 0) = \text{sing}(z)$.

Conversely, for $(T, \mathcal{T}, \mathbf{A})$ a decomposable division space, K a normed linear space satisfying Theorem 2.1.10(2.1.33) (e.g. $K = \mathbb{R}$ or $\mathbb{C}$), $Z^j(I, t) = S(z, r(I, t))$, $G = S(z, \varepsilon)$, Y the set of positive integers, and $h(n; I, t) = f_n(t)m(I, t)$, $h(I, t) = f(t)m(I, t)$, then (3.1.4), (3.1.7) give $\bar{V}(m; \mathbf{A}; E; X_2) = 0$, X_2 being the set of t where $f_n(t) \not\to f(t)$ as $n \to \infty$, $f_n(t)$ being bounded in n for each $t \in X_2$ with $m(I, t) \neq 0$ for some $(I, t) \in \mathcal{U}_+$, and $m \cdot \chi(\backslash X_1; \cdot)$ is VBG* in E. For a general Y and a sequence $(y_n) \subseteq Y$ such that for each $y \in Y$ there is an $N > 0$ with $y_n \geq y$ $(n \geq N)$, then $f(y_n; t) \to f(t)$ m-almost everywhere, with the other results.

Proof. Let mutually disjoint $X_1, X_2, X_p \subseteq T$ have union T and $\mathcal{U}_p \in \mathbf{A}|E$ have

(3.1.9)
$$0 < \bar{V}(m; \mathbf{A}; E; X_p) \leq V(m; \mathcal{U}_p; E; X_p) < 2\bar{V}(m; \mathbf{A}; E; X_p) < \infty \quad (p = 3, 4, \ldots).$$

In X_1, $f(y; t) = f(t)$ $(y \geq J(t))$ and so can be ignored. For each $t \in X_2$, either $m(I, t) = 0$ for some $\mathcal{U}_0 \in \mathbf{A}|E$ and all $(I, t) \in \mathcal{U}_0$, or, for a finite positive function φ on X_2,

$$\|f(y; t) - f(t)\| \leq \varphi(t),$$

and as in Theorem 2.2.1(2.2.4),

(3.1.10) $\bar{V}(\varphi m; \mathbf{A}; E; X_2) = 0$, $V(\varphi m; \mathcal{U}_0; E; X_2) < \tfrac{1}{2}\varepsilon$,

$$V((f(y; \cdot) - f)m; \mathcal{U}_0; E; X_2) < \tfrac{1}{2}\varepsilon.$$

3.1 NECESSARY AND SUFFICIENT CONDITIONS   123

Let $\mathcal{U}_+ \in \mathbf{A}|E$ lie in the diagonal of $(\mathcal{U}, X_1; \mathcal{U}_0, X_2; \mathcal{U}_p, X_p)$, with $k(t), r(I, t)$ so that

(3.1.11)
$$\|f(y; t) - f(t)\| \cdot \|m(I, t)\| \leq \varepsilon \cdot 2^{-p-2} \bar{V}(m; \mathbf{A}; E; X_p)^{-1} \|m(I, t)\| = r(I, t)$$

$$(t \in X_p, p = 3, 4, \ldots, y \geq k(t)).$$

From (3.1.n) ($n = 9, 10, 11$) then (3.1.4), (3.1.7) are true since

$$(\mathcal{E}) \sum \|f(y; t) - f(t)\| \cdot \|m(I, t)\| \leq$$

$$\tfrac{1}{2}\varepsilon + \sum_{p=3}^{\infty} \varepsilon \cdot 2^{-p-2} \bar{V}(m; \mathbf{A}; E; X_p)^{-1} V(m; \mathcal{U}_+; E; X_p) < \varepsilon.$$

Conversely, from the given hypotheses, (3.1.4), (3.1.7), and Theorem 2.1.10 (2.1.33),

(3.1.12) $(\mathcal{E}) \sum \|(f_{n(t)}(t) - f(t))m(I, t)\| < q\varepsilon$

for $\mathcal{E}$ over E from some $\mathcal{U} \in \mathbf{A}|E$. If for $t \in Z$ and $\eta > 0$, $\|f_n(t) - f(t)\| \geq \eta$ for an infinity of n, then

$$(\mathcal{E}) \sum \eta \|m(I, t)\| \chi(Z; t) \leq q\varepsilon, \quad V(m; \mathcal{U}; E; Z) \leq q\varepsilon/\eta, \quad V(m; \mathbf{A}; E; Z) = 0$$

since $\varepsilon > 0$ is arbitrarily small. Taking $\eta = 1, \tfrac{1}{2}, \tfrac{1}{3}, \ldots$ and using Theorem 2.2.1(2.2.1) (analogue), X_2 has m-variation zero. In (3.1.12) then either $m(I, t) = 0$ for all $(I, t) \in \mathcal{U}$, or $f_n(t)$ is bounded in $n \geq N$. As the number of n less than N is finite $(f_n(t))$ is bounded in n. For $t \in X_1$ we take $k(j; I, t) \geq J(t)$ so that $f_n(t) = f(t)$ in (3.1.7) when $n \geq k(j; I, t)$. To show that m is VBG^* in $\setminus X_1$ let Y_r ($r \geq 1$) be the set of t with

$$\|f_{n(t)}(t) - f(t)\| \geq 1/r \quad (\text{some } n(t) \geq k(j; I, t)).$$

If $t \notin X_1$ then for some integer r, $t \in Y_r$, so $\setminus X_1 = \bigcup_{r=1}^{\infty} Y_r$. From (3.1.12),

$$(\mathcal{E}) \sum \|m(I, t)\| \chi(Y_r; t)/r \leq q\varepsilon, \quad \bar{V}(m; \mathbf{A}; E; Y_r) \leq V(m; \mathcal{U}; E; Y_r) \leq qr\varepsilon,$$

so m is VB^* in $Y_r = Y_r(U)$ and so VBG^* in $\setminus X_1$. □

We now come to the two theorems giving the necessary and sufficient conditions for the integrability of the limit function h and the result (3.1.1).

Theorem 3.1.3. *Let Z^j, $h(y; \cdot)$, h satisfy (3.1.n)($n = 4, 7, 8$) with $h(y; \cdot)$ integrable over E for each $y \in Y$. Let K be complete $(\mathbf{A}; E)$ and locally compact. Then h is integrable over E if and only if there are a compact set C of arbitrarily small diameter, some $M: E^* \to Y$, some $\mathcal{U} \in \mathbf{A}|E$, and all divisions $\mathcal{E}$ of E from $\mathcal{U}$, such that*

(3.1.13) $(\mathcal{E}) \sum h(y(I, t); I, t) \in C$ $\quad$ (all $y(I, t) \geq M(I, t)$ in Y).

Proof. If h is integrable over E, there is a neighbourhood S of $H \equiv \int_E dh$ of arbitrarily small diameter, such that

(3.1.14) $\qquad (\mathscr{E})\sum h(I, t) \in S.$

By (3.1.n) ($n = 4, 7, 8$), the usual construction for sets of h-variation zero, and an arbitrary $\mathscr{G}$-neighbourhood G of z, we can choose $M(I, t)$ to satisfy

$$(\mathscr{E})\sum \{h(y(I, t); I, t) - h(I, t)\} \in G + G \quad \text{(all } y(I, t) \geqslant M(I, t) \text{ in } Y\text{)},$$

and (3.1.13) holds with $S + G + G$, of arbitrary small diameter, replacing C. As in Theorems 0.4.11, 0.4.13, we can so choose S, G, that, given an arbitrary $\mathscr{G}$-neighbourhood G_1 of H, there is a $\mathscr{G}$-neighbourhood G_2 of H whose $\mathscr{G}$-closure is compact (as K is locally compact) and lies in G_1, and

$$S + G + G \subseteq G_2 \subseteq \bar{G}_2 \subseteq G_1.$$

Then $\bar{G}_2$ can replace C in (3.1.13). Conversely, if (3.1.n) ($n = 4, 7, 8, 13$) hold, we take $M(I, t)$ to be the greater of those needed in (3.1.7), (3.1.13), and so have (3.1.14) with a suitable S of arbitrarily small diameter. Integrability follows as K is complete (**A**; E). □

Theorem 3.1.4. *Given C and the conditions of Theorem 3.1.3, with (3.1.13), a necessary and sufficient condition for the analogue of (3.1.1) is that there are a $J \in Y$, a compact set C_1 of arbitrarily small diameter, either containing a $\mathscr{G}$-neighbourhood of $H \equiv \int_E dh$, or such that $C \cap C_1$ is not empty, and a $\mathscr{U}_y \in \mathbf{A}|E$ with*

(3.1.15) $\qquad (\mathscr{E})\sum h(y; I, t) \in C_1$

for all divisions $\mathscr{E}$ of E from $\mathscr{U}_y$ with $y \geqslant J$ in Y.

Proof. As H and $H_y = \int_E dh(y; \cdot)$ exist for $y \in Y$, we have the analogue of (3.1.1) if and only if $H_y \in H + G$ for each neighbourhood G of z and all $y \geqslant J$ in Y, where J depends on G. The result follows from this, the definitions of the integrals H_y and H, and local compactness of K. □

The difference between (3.1.13) and (3.1.15) is that y varies with (I, t) in the first and is constant in the second. Neither condition contains the other, unless other conditions are imposed. When theorems 3.1.3, 3.1.4 are specialized to the case of repeated integrals the two conditions are interchangeable, see Chapter 5.

It is worth while to give the necessary and sufficient conditions in the simple case when $h_n = f_n m$, $h = fm$, using pointwise convergence, in one theorem with $K = \mathbb{C}$, just as it is given (with slight rearrangements) in the only other place where it has appeared, namely, Henstock (1988a), pp. 105–7, Theorems 11.1, 11.2.

3.1 NECESSARY AND SUFFICIENT CONDITIONS

Theorem 3.1.5. *Let $(T, \mathcal{T}, \mathbf{A})$ be a division space, let K be $\mathbb{C}$, let $m: \mathcal{U}^1 \to \mathbb{C}$ be VBG*, and let $f_n: E^* \to \mathbb{C}$ with $f_n m$ integrable in E $(n = 1, 2, \ldots)$, such that $f_n \to f$ pointwise in E^* as $n \to \infty$. Then the necessary and sufficient conditions for the integrability of fm over E and the result*

$$(3.1.16) \qquad \lim_{n \to \infty} \int_E f_n \, dm = \int_E \lim_{n \to \infty} f_n \, dm$$

are that

$$(3.1.17) \qquad (\mathscr{E}) \sum f_{n(t)}(t) m(I, t) \in S$$

for some compact set S of arbitrarily small diameter, some $\mathcal{U} \in \mathbf{A} | E$, some positive function M on E^*, all positive integer-valued functions $n \geq M$ on E^*, and all divisions $\mathscr{E}$ of E from $\mathcal{U}$ (for the integrability) and

$(3.1.18)$ *a positive integer J and a $\mathcal{U}_n \in \mathbf{A} | E$ for each $n = 1, 2, \ldots$, such that*

$$(\mathscr{E}) \sum f_n(t) m(I, t) \in S_1$$

for all divisions $\mathscr{E}$ of E from $\mathcal{U}_n$ and all $n \geq J$, where S_1 has arbitrarily small diameter and where $S \cap S_1$ is not empty (for (3.1.16)).

Comparing the two accounts, clearly the above is a slight generalization of the one in Henstock (1988a).

Example 3.1.1 (D. Przeworska-Rolewicz and S. Rolewicz (1966). Also see Henstock (1969) pp. 532–3). Let

$$K = L^p[0, 1] \quad (0 < p < 1), \, m([u, v)) = v - u,$$

$$f_n(t; y) = n \cdot \chi([k-1)/n, k/n); y) \quad (k - 1 \leq nt < k, \, k = 1, 2, \ldots, n),$$

$$\|f_n(t; \cdot)\|_p = \int_{[0,1]} |f_n(t; y)|^p \, dy = n^{p-1} \to 0 \quad \text{as } n \to \infty,$$

$$\int_{[0,1]} f_n(t; y) \, dt = \sum_{k=1}^n n \cdot \chi([(k-1)/n, k/n); y)/n = \chi([0, 1); y),$$

non-zero and independent of n. Thus (3.1.1) is false here. Note that for $\mathscr{E}$ a division over E with arbitrarily small norm $N = \max m(I)$ for $(I, t) \in \mathscr{E}$,

$$(\mathscr{E}) \sum \|f_n(t; \cdot) m(I)\|_p = n^{p-1} (\mathscr{E}) \sum \|m(I)\|_p$$
$$= n^{p-1} (\mathscr{E}) \sum m(I) m^{p-1}(I) \geq n^{p-1} (\mathscr{E}) \sum m(I) N^{p-1}$$
$$= m(E)(nN)^{p-1} \to \infty \quad (N \to 0).$$

Exercise 3.1.2 (Henstock (1988a) pp. 107–8). On the real line let $f_j = 1$ in $[0, 1)$, $f_j = (-1)^n n(n+1)$ in $2 - 1/n < t < 2 - 1/(n+1)$ $(n = 1, 2, \ldots, j)$, $f_j = 0$ in $2 - 1/(n+1) < t < 2$. Then f_j is gauge-integrable in $[0, 2)$. As

$j \to \infty$ the limit f satisfies $f = 1$ in $[0, 1)$, $(-1)^n n(n+1)$ in $2 - 1/n < t < 2 - 1/(n+1)$ $(n = 1, 2, \ldots)$, $f(2) = 0$.

In the respective intervals take a function $M \geq 1$, $M \geq n$, with $M(2) \geq 1$. If $j \geq M$ then $f_j = f$. For suitable $\mathcal{U} \in \mathbf{A}|[0, 2)$ the sums oscillate between 0 and 1 while

$$\int_{[0, 2 - 1/n)} f \, dt = 1 + \sum_{n=1}^{j} (-1)^n$$

and does not tend to a limit as $j \to \infty$. By Theorem 2.5.4(2.5.13) the integral has the Cauchy limit property if it exists. So the limit function is not integrable over $[0, 2)$.

Show that (3.1.17) is not satisfied by, say, taking even values for $n(t)$.

3.2 MONOTONE SEQUENCES AND FUNCTIONS

For each j we can have $h_j(I, t)$ integrable over an elementary set E and tending to $h(I, t)$ as $j \to \infty$ without $h(I, t)$ being integrable; or if integrable, the integral of h need not be the limit of the integral of h_j. See Exercises 2.3.5, 2.3.6. However, we make progress when h_j has the form $h_j(I, t) = f_j(t)k(I, t)$. One major point is that it has been assumed by the majority of mathematical analysts that the Lebesgue limit theorems, which are the subject of this and the following section, follow because Lebesgue measure is countably additive. For example, see Luxemburg (1971), p. 971. However, proof of these theorems by the methods of this book, and in particular the property of decomposability, were given in detail in Henstock (1974) without ever using countable additivity of the measure, this now being $k(I, t)$ which need not be even finitely additive. The proofs in Henstock (1974) and in these two sections can be carefully checked to confirm these remarks. Note that some definitions are slightly altered. The earlier non-additive division space has now become the division space while the earlier division space is now the additive division space, two better titles for the respective spaces.

We begin with the weak monotone convergence theorem, due to Levi (1906) in the Lebesgue form.

Theorem 3.2.1. *Let $(T, \mathcal{T}, \mathbf{A})$ be a decomposable division space, $k: \mathcal{U}^1 \to \mathbb{R}^+$, $f_j: T \to \mathbb{R}$ $(j = 1, 2, \ldots)$, and let each of $f_j(t)k(I, t)$ $(j = 1, 2, \ldots)$ be integrable over a fixed elementary set E. If for each t, $(f_j(t))$ is a bounded monotone increasing sequence in j, and if $(H_j(E))$ is bounded above, where*

(3.2.1) $$H_j(P) \equiv \int_P f_j \, dk \quad (j = 1, 2, \ldots),$$

then for $f(t) = \lim_{j \to \infty} f_j(t)$, $f(t)k(I, t)$ is integrable over E to $H(E) = \lim_{j \to \infty} H_j(E)$.

3.2 MONOTONE SEQUENCES AND FUNCTIONS

Proof. Theorems 2.5.2(2.5.4) and 2.3.4(2.3.9) show that $(H_j(P))$ exists and is monotone increasing. When $P = E$ it is bounded above and so tends to its supremum M. Thus, given $\varepsilon > 0$, an integer n exists with

(3.2.2) $\qquad M - \varepsilon < H_j(E) \leqslant M \quad \text{(all } j \geqslant n\text{)}.$

Using Theorem 2.3.3(2.3.7) to replace f_j by $f_j - f_1 \geqslant 0$, we see that taking $f_j \geqslant 0$ causes no loss of generality. By Theorem 2.5.5(2.5.15) there is a $\mathcal{U}_j \in \mathbf{A}|E$ with

(3.2.3) $\qquad (\mathcal{E}) \sum |f_j(t)k(I, t) - H_j(I)| \leqslant \varepsilon \cdot 2^{-j} \quad (j = 1, 2, \ldots)$

for each division $\mathcal{E}$ of E from $\mathcal{U}_j$. As $\mathbf{A}$ is decomposable and directed in the sense of divisions, given a function $m(t)$ of t with integer values not less than n, there is a $\mathcal{U}(m) \in \mathbf{A}|E$ for which

(3.2.4) $\qquad \mathcal{U}(m)[\text{sing}(t)] \subseteq \mathcal{U}_{m(t)}[\text{sing}(t)] \cap \mathcal{U}.$

This may appear to need full decomposability, but as the points t with equal $m(t)$ can be grouped together, only a countable number of sets is required. As the integrals are finitely additive over co-divisional partial sets (Theorem 2.5.2(2.5.6)) and as (3.2.1) is monotone increasing in j, if $\mathcal{E}$ is a division of E from $\mathcal{U}(m)$, with u, v the least and greatest values of $m(t)$ for the finite number of $(I, t) \in \mathcal{E}$, so that $n \leqslant u \leqslant v$, we have

$$M - \varepsilon < H_u(E) = (\mathcal{E}) \sum H_u(I) \leqslant (\mathcal{E}) \sum H_{m(t)}(I) \leqslant (\mathcal{E}) \sum H_v(I) = H_v(E) \leqslant M.$$

Putting together the (I, t) with equal $m(t)$, and using (3.2.3), (3.2.4),

(3.2.5) $\qquad M - 2\varepsilon < (\mathcal{E}) \sum f_{m(t)}(t) k(I, t) < M + \varepsilon$

for each division $\mathcal{E}$ of E from $\mathcal{U}(m)$. We use two functions m.

First let $r(t)$ be the least integer greater than 1, for which

(3.2.6) $\qquad f_{r(t)}(t) > 0.$

If for some t there is no such $r(t)$, we put $r(t) = 1$ and note that for this t,

(3.2.7) $\qquad f_j(t) = 0 \ (j = 1, 2, \ldots), \quad f(t) = \lim_{j \to \infty} f_j(t) = 0.$

Secondly there is a least integer $J = J(t) \geqslant n$ for which, even when $r(t) = 1$,

(3.2.8) $\qquad f(t) \geqslant f_j(t) \geqslant f_J(t) \geqslant f(t) - \varepsilon f_{r(t)}(t) \quad (j \geqslant J(t) \geqslant r(t)).$

By (3.2.4) with $m = r$ and $m = J$, and as $\mathbf{A}$ is directed in the sense of divisions, there is a $\mathcal{U}^* \in \mathbf{A}|E$ with $\mathcal{U}^* \subseteq \mathcal{U}(r) \cap \mathcal{U}(J)$. If $\mathcal{E}$ is a division of E from $\mathcal{U}^*$, then by (3.2.5), (3.2.8), and the monotonicity of f_j in j,

$$M - 2\varepsilon < (\mathcal{E}) \sum f(t) k(I, t) \leqslant (\mathcal{E}) \sum f_{J(t)}(t) k(I, t) + \varepsilon(\mathcal{E}) \sum f_{r(t)}(t) k(I, t)$$
$$< (M + \varepsilon)(1 + \varepsilon).$$

It is true for each $\varepsilon > 0$ and each division $\mathscr{E}$ of E from $\mathscr{U}^* \in \mathbf{A}|E$ which depends on ε. Hence fk is integrable and Theorem 3.2.1 is true with

$$\int_E f\,dk = M = \lim_{j \to \infty} \int_E f_j\,dk.$$

If k is of bounded variation, $r(t)$ can be omitted and in (3.2.8) we can use

$$f_J(t) \geqslant f(t) - \varepsilon. \quad \square$$

Next comes the strong monotone convergence theorem, due in the Lebesgue version to Vitali (1907).

Theorem 3.2.2. *In Theorem 3.2.1, if we omit only the boundedness of $(f_j(t))$ in j for each $t \in E^*$, there is a set $X \subseteq T$ of k-variation zero such that $f_j(t)$ tends to a finite limit for $t \in E^* \setminus X$. Defining the finite-valued function*

$$f(t) \equiv \lim_{j \to \infty} f_j(t)\chi(\setminus X; t),$$

$f(t)k(I, t)$ *is integrable over E to $H(E) = \lim_{j \to \infty} H_j(E)$.*

Proof. As the monotone increasing sequence $(H_j(E))$ is bounded above, it tends to a finite limit as $j \to \infty$. Thus, taking a subsequence if necessary, we assume that

(3.2.9) $\qquad 0 \leqslant H_{j+1}(E) - H_j(E) \leqslant 4^{-j} \quad (j = 1, 2, \ldots).$

As $H_{j+1} - H_j \geqslant 0$ is finitely additive over divisions of E, if X_j is the set of t where

$$f_{j+1}(t) - f_j(t) \geqslant 2^{-j},$$

and writing $V(X) \equiv \bar{V}(k; \mathbf{A}; E; X)$, Theorem 2.5.5(2.5.16), (3.2.9), and Theorem 2.2.1 (analogue) give

$$2^{-j}V(X_j) \leqslant V((f_{j+1} - f_j)k; \mathbf{A}; E) = V(H_{j+1} - H_j; \mathbf{A}; E)$$
$$= H_{j+1}(E) - H_j(E) \leqslant 4^{-j}, \quad V(X_j) \leqslant 2^{-j},$$

$$X^N \equiv \bigcup_{j=N}^{\infty} X_j, \quad X \equiv \bigcap_{N=1}^{\infty} X^N, \quad V(X) \leqslant V(X^N) \leqslant \sum_{j=N}^{\infty} V(X_j) \leqslant 2^{1-N}, \quad V(X) = 0.$$

If $t \notin X$ then for some N, $t \notin X^N$, and so $t \notin X_j$ $(j \geqslant N)$, and $f_j(t)$ tends to a finite limit since

$$0 \leqslant f_{j+1}(t) - f_j(t) \leqslant 2^{-j} \quad (j \geqslant N).$$

As $f_j(t)$ is monotone increasing in j for each fixed $(I, t) \in \mathscr{U}$, the original sequence tends to the same limit as the subsequence when $t \notin X$, and so k-almost everywhere. Replacing $f_j(t)$ by $f_j(t)\chi(\setminus X; t)$, Theorems 2.4.1, 3.2.1 finish the proof.

3.2 MONOTONE SEQUENCES AND FUNCTIONS

There is an easy extension to functions monotone increasing with respect to a real number $y \to \infty$, obtained by noting that for $[y]$ denoting the integer part of y,

$$f(t, [y]) \leq f(t, y) \leq f(t, [y] + 1), \quad H(E, [y]) \leq H(E, y) \leq H(E, [y] + 1).$$

It is now trivial to replace $y \to \infty$ by $y \to c$, for some real number c, and similar alterations. A more difficult case is that of $f_d(t) (d \in D)$ where D is a directed set.

Theorem 3.2.3. *Let $(T, \mathcal{T}, \mathbf{A})$ be a decomposable division space, D a set directed upwards, $k: \mathcal{U}^1 \to \mathbb{R}^+$, $f_d: T \to \mathbb{R}\, (d \in D)$. If f_d is monotone increasing in the direction of D, if $f_d(t) k(I, t)$ is integrable to $H_d(E)$ over E for each $d \in D$, and if*

$$H(E) \equiv \sup_{d \in D} H_d(E)$$

is finite, then there is an $f: T \to \mathbb{R}$ such that $f = \lim_{n \to \infty} f_{d(n)}$ k-almost everywhere, for each monotone increasing sequence $(d(n))$ in D with $\lim_{n \to \infty} H_{d(n)}(E) = H(E)$, and fk is integrable to $H(E)$ over E. But we can have $f \neq \lim_{d \in D} f_d$ everywhere. See Example 0.5.1.

Proof. Given the monotone increasing sequence $(d(n)) \subseteq D$ with $H_{d(n)} \to H$, by Theorem 3.2.2 there is an $f: T \to \mathbb{R}$ with $f_{d(n)} \to f$ k-almost everywhere, fk integrable, and

$$\int_E f \, dk = \lim_{n \to \infty} \int_E f_{d(n)} \, dk = \lim_{n \to \infty} H_{d(n)}(E) = H(E).$$

To show that f is independent of the chosen sequence $(d(n))$, modulo values in sets of k-variation zero, we replace $(d(n))$ by $(e(n)) \subseteq D$ with $f_{e(n)} \to f_0$ k-almost everywhere. Then by direction in D and Theorem 2.5.5(2.5.16) there is an $(r(n)) \subseteq D$ with $f_{r(n)} \to f_*$ k-almost everywhere, $r(n) \geq d(n)$, $r(n) \geq e(n)$, $r(n) \geq r(j) \, (1 \leq j \leq n)$,

$$f_* \geq f, \quad f_* \geq f_0,$$

$$\bar{V}((f_* - f)k; \mathbf{A}; E) = \bar{V}\left(\int (f_* - f) \, dk; \mathbf{A}; E\right) = \int_E (f_* - f) \, dk = 0,$$

$$f_* - f \leq 1/n$$

k-almost everywhere ($n = 1, 2, \ldots$), and $f_* \leq f$ (and so $f_* = f$) k-almost everywhere, by Theorem 2.2.1 (analogue). Similarly $f_* = f_0$ k-almost everywhere, so that $f_0 = f$ k-almost everywhere, and the limit is independent of the particular sequence $(d(n))$ used, modulo values in sets of k-variation zero. We write f as $\lim_d^* f_d(t)$. □

The form of the Radon–Nikodym Theorem 2.9.6 is only suitable for absolute integration and real- or complex-valued functions. We are now able to give a result suitable for non-absolute integrals.

Theorem 3.2.4. *Let $(T, \mathcal{T}, \mathbf{A})$ be an infinitely divisible decomposable additive division space, let $h: \mathcal{U}^1 \to \mathbb{C}$, $k: \mathcal{U}^1 \to \mathbb{C}$, with k integrable over an elementary set E, and such that $\mathcal{F}(E: P)$ has k-variation zero for each proper partial set P of E. Let (X_j) be a sequence of mutually disjoint sets of T with union T, such that*

$$x_j(I, t) \equiv h(I, t)\chi(X_j; t), \quad y_j(I, t) \equiv k(I, t)\chi(X_j; t)$$

are integrable and of bounded variation in E with y_j absolutely continuous in E with respect to x_j ($j = 1, 2, \ldots$). Then there is a function $f: T \to \mathbb{C}$ with fh integrable, such that for every partial set P of E,

(3.2.10) $$\int_P dk = \int_P f(t)\, dh.$$

Further, if all but a finite number of the X_j are empty, then h and k are of bounded variation and $|f(t)|\bar{V}(h; \mathbf{A}; I)$ is integrable over E (with a finite integral).

Note that k is ACG* with respect to h, while h, k are VBG* in E. The integrability of x_j, y_j reduces to the h- and k-measurability of X_j in the Lebesgue integral case. Further, if, given $\delta > 0$, there is a partition $\mathscr{P}$ of E with the variation of x_j in each $I \in \mathscr{P}$ less than δ, then y_j is of bounded variation by the property of absolute continuity, and one hypothesis can then be removed, and x_j is also continuous.

Proof of Theorem 3.2.4. When all but a finite number of X_j are empty, the union T then satisfies the same conditions as the X_j, and h and k obey the conditions of Theorem 2.9.6 while k is absolutely continuous in E relative to h. By Theorem 2.7.3, h and $V(I) \equiv \bar{V}(h; \mathbf{A}; I)$ are variationally equivalent, and k and $V_1(I) \equiv \bar{V}(k; \mathbf{A}; I)$ are variationally equivalent. By Example 2.3.8, V_1 is absolutely continuous in E relative to V. Thus when all but a finite number of X_j are empty, the proof of the result for $h = V$, $k = V_1$, plus Theorem 2.9.6, give what is required.

We put $W = V_1 - bV$ in Theorem 2.5.9, where b is a non-negative constant, so that

$$\bar{V}(V_1 - bV; \mathbf{A}; E) \leq V_1(E) + bV(E) < \infty.$$

Given $\varepsilon > 0$, there is a partial set P of E with

$$V_1(P_1) < bV(P_1) + \varepsilon, \quad V_1(P_2) > bV(P_2) - \varepsilon$$

for all partial sets P_1 of P, P_2 of $E \setminus P$. First, for $b = 2^{-n}$, $\varepsilon = 2^{-2-n}$, let P be P_3. Then suppose that $P_3, \ldots, P_m$ have been defined. If their union is E, we

3.2 MONOTONE SEQUENCES AND FUNCTIONS

stop there. Otherwise we replace E by $E\setminus(P_3 \cup P_4 \cup \cdots \cup P_m)$, we take $b = (m-1)2^{-n}$, $\varepsilon = 2^{-m-n}$, and let P be P_{m+1}, continuing the inductive definition. For all partial sets P_0 of P_m,

(3.2.11)
$$V(P_0)(m-3)2^{-n} - 2^{-m-n+2} < V_1(P_0) < V(P_0)(m-2)2^{-n} + 2^{-m-n+1} \quad (m \geq 3),$$

(3.2.12)
$$V(E\setminus(P_3 \cup \cdots \cup P_m)) < 2^n\{V_1(E\setminus(P_3 \cup \cdots \cup P_m)) + 2^{-m-n+1}\}/(m-2)$$
$$\leq 2^n\{V_1(E) + 1\}/(m-2).$$

By Exercise 2.9.1 and (3.2.11), taking $f_{nN}(t) = (m-2)2^{-n}$ in P_m ($3 \leq m \leq N$) and 0 in $E\setminus(P_3 \cup \cdots \cup P_N)$, we have for a partial set $P \subseteq P_m$,

$$\int_P f_{nN}(t)\,dV = (m-2)2^{-n}V(P), \quad \left|\int_P f_{nN}(t)\,dV - V_1(P)\right| \leq 2^{-n}V(P) + 2^{-m-n+2}.$$

By finite additivity, for all partial sets P of E we have

(3.2.13)
$$\left|\int_P f_{nN}(t)\,dV - V_1(P)\right| \leq (V(P) + 1)2^{-n} + V_1(P \cap E\setminus(P_3 \cup \cdots \cup P_N)).$$

As f_{nN} is bounded and monotone increasing in N with a bounded integral, Theorem 3.2.1 shows that the limit $f_n(t)$ as $N \to \infty$, is finitely integrable, while from (3.2.12), (3.2.13) and the absolute continuity of V_1,

(3.2.14) $\quad \left|\int_P f_n\,dV - V_1(P)\right| \leq (V(P) + 1)2^{-n}, \quad \lim_{n \to \infty} \int_P f_n\,dV = V_1(P).$

For $m > n$ and all partial sets P of E, the first result in (3.2.14) gives

(3.2.15)
$$\left|\int_P (f_m - f_n)\,dV\right| \leq (V(P) + 1)2^{1-n}.$$

By a usual argument, from each partition $\mathscr{P}$ of E we take separately all those $I \in \mathscr{P}$ over which the integral of $f_m - f_n$ is positive, giving a P, and put the rest of the $I \in \mathscr{P}$ into $E\setminus P$, (3.2.15) then producing

$$(\mathscr{P})\sum \left|\int_I (f_m - f_n)\,dV\right| \leq (V(E) + 2)2^{1-n},$$

$$\bar{V}(|f_m - f_n|V; \mathbf{A}; E) = \bar{V}\left(\left|\int_I (f_m - f_n)\,dV\right|; \mathbf{A}; E\right) \leq (V(E) + 2)2^{1-n},$$

$$\bar{V}\left(\sum_{n=1}^{\infty} |f_{n+1} - f_n|V; \mathbf{A}; E\right) \leq \sum_{n=1}^{\infty} \bar{V}(|f_{n+1} - f_n|V; \mathbf{A}; E) \leq (V(E) + 2)2$$

using Theorem 2.4.1(2.4.1). Thus by Exercise 3.2.1,

$$\sum_{n=1}^{\infty} (f_{n+1}(t) - f_n(t))$$

is V-almost everywhere an absolutely convergent series and $f_n(t)$ tends to a limit, say, $f(t)$, there. By (3.2.14) and Exercise 3.2.1 again, fV is integrable over each P with integral equal to $V_1(P)$. Thus, collecting the results together,

$$\int_I dk = \int_I g_1 \, dV_1, \quad V_1(I) = \int_I f \, dV, \quad \int_I dh = \int_I g \, dV, \quad |g| = 1,$$

$$\int_I g^{-1} \, dh = \int_I dV = V(I), \quad \int_I dk = \int_I g_1 f g^{-1} \, dh$$

and the proof is completed in the special case. Note that f is the limit of a sequence of step functions, but g, g_1 were not obtained that way. Also $|g_1 f g^{-1}| = f \geq 0$, and fV is integrable, as required.

Replacing k by $k\chi(X;t)$ we replace $g_1 f g^{-1}$ by some function r, say, and

$$\bar{V}(k\chi(X;\cdot) - rh; \mathbf{A}; E) = 0,$$

$$\bar{V}(-rh\chi(\backslash X;\cdot); \mathbf{A}; E) = \bar{V}((k\chi(X;\cdot) - rh)\chi(\backslash X;\cdot); \mathbf{A}; E) = 0,$$

and $r = 0$ h-almost everywhere in $\backslash X$. Thus we can take $r = 0$ everywhere in $\backslash X$. In the general case we replace h, k, X by x_j, y_j, X_j, and we have, for $j = 1, 2, \ldots$, a function $r_j(t)$ that is 0 in $\backslash X_j$. For r the sum of the r_j then $r = r_j$ in X_j and by Theorem 2.4.1(2.4.1),

$$\bar{V}(k - rh; \mathbf{A}; E) \leq \sum_{j=1}^{\infty} \bar{V}(k - rh; \mathbf{A}; E; X_j) = \sum_{j=1}^{\infty} \bar{V}(y_j - r_j x_j; \mathbf{A}; E; X_j) = 0.$$

As k is integrable we can replace it by the finitely additive integral, which then is also the integral of rh, which finishes the proof.

The proofs of the monotone convergence theorems using the gauge integral appear in Henstock (1963c) pp. 82–4.

Exercise 3.2.1 Let $(T, \mathcal{T}, \mathbf{A})$ be a decomposable division space, let $h: \mathcal{U}^1 \to \mathbb{R}^+$, $f_j: T \to \mathbb{R}$, and let $f_j h$, $|f_j| h$ be integrable in an elementary set, for $j = 1, 2, \ldots$. Prove that if

$$\sum_{j=1}^{\infty} \int_E |f_j| \, dh$$

is finite, $f(t) \equiv \sum_{j=1}^{\infty} f_j(t)$ exists as an absolutely convergent series h-almost everywhere in E and fh is integrable over E with

$$\int_E f \, dh = \sum_{j=1}^{\infty} \int_I f_j \, dh.$$

(*Hint*: $\sum_{j=1}^{J} f_j h$ is monotone increasing in J while $\sum_{j=1}^{J} \int_E f_j dh$ is bounded. Apply Theorem 3.2.2 and then consider $|f_j| \pm f_j \geq 0$. For complex-valued f_j split into the real and imaginary parts.)

Exercise 3.2.2 For $(T, \mathcal{T}, \mathbf{A})$ a decomposable additive division space let $h: \mathcal{U}^1 \to \mathbb{R}^+$ be integrable over an elementary set E, let $f: T \to \mathbb{R}$ with fh and $|f|h$ integrable over E, let b be a constant, and let f_0^b be the indicator of the set $X(f \geq b)$ where $f \geq b$. Prove that $f_0^b h$ is integrable over E to the value $V(h; \mathbf{A}; E; X(f \geq b))$. Similarly for the sets where $f = b$, $f > b$, $a \leq f \leq b$, etc.

(*Hint*: Consider $f_0^{ab} = \max(\min(f, b), a)$, equal to f when $a \leq f \leq b$, to a when $f < a$, and to b when $f > b$. Then $f_0^b = \lim_{a \to b-} (f_0^{ab} - a)/(b - a)$, the fraction being monotone with value lying in $[0, 1]$. Similarly the indicators $\chi(X(f > a)) = \lim_{b \to a+} f_0^b$,

$$\chi(X(f = a)) = \min\{\chi(X(f \geq a)), 1 - \chi(X(f > a))\},$$
$$(X(a \leq f < b)) = \min\{\chi(X(f \geq a)), \chi(X(f < b))\}.)$$

Exercise 3.2.3 Given $x \in [0, 1)$, let $X(x)$ be the set of all y in $[0, 1)$ with $y - x$ rational, and let $M \subseteq [0, 1)$ be such that for each $x \in [0, 1)$, $m \cap X(x)$ contains exactly one point. Assuming that such a set can be found, show that $\chi(M; t) \Delta x$ is not integrable in $[0, 1)$.

(*Hint*: Let (r_n) be the sequence of mutually disjoint rationals in $[0, 1)$, and M_n the set of all $x + r_n$, $x + r_n - 1$ that lie in $[0, 1)$ with $x \in M$. If $\chi(M; t) \Delta x$ is integrable, so is $\chi(M_n; t) \Delta x$ and

$$\int_{[0,1)} \chi(M_n; t) dt = \int_{[0,1)} \chi(M; t) dt,$$

$$n \int_{[0,1)} \chi(M; t) dt = \sum_{j=1}^{n} \int_{[0,1)} \chi(M_j; t) dt = \int_{[0,1)} \chi\left(\bigcup_{j=1}^{n} M_j; t\right) dt \leq 1,$$

$$\int_{[0,1)} \chi(M; t) = 0,$$

$$1 = \int_{[0,1)} \chi\left(\bigcup_{j=1}^{\infty} M_j; t\right) dt = \sum_{j=1}^{\infty} \int_{[0,1)} \chi(M_j; t) dt = 0,$$

a contradiction.)

Exercise 3.2.4 For $(T, \mathcal{T}, \mathbf{A})$ a decomposable additive division space, $h: \mathcal{U}^1 \to \mathbb{C}$, $f: T \to \mathbb{C}$, and $fh, |fh|, |h|$ all integrable over the elementary set E, prove that the integral F of fh is absolutely continuous relative to h and of bounded variation in E. (For $\mathscr{E}$ a division of E use Theorems 2.3.6 and

2.5.2(2.5.6).) Then

$$(\mathscr{E})\sum |F(I)| \leq (\mathscr{E})\sum \int_I |f|d|h| = \int_E |f|d|h|$$

and F is of bounded variation. For $N > 0$ let $f_N \equiv \min(|f|, N)$. Then $f_N|h|$ is integrable (Theorem 2.7.6) and is monotone increasing with N to $|f|$. Hence

$$\lim_{N \to \infty} \int_E f_N dh = \int_E |f|d|h|$$

by Theorem 3.2.1. Hence, given $\varepsilon > 0$, there is an integer N such that

$$\int_E (|f| - f_N)d|h| < \tfrac{1}{2}\varepsilon.$$

If $X \subseteq T$ satisfies $\bar{V}(h; \mathbf{A}; E; X) < \varepsilon/(2N)$ then

$$|f| = (|f| - f_N) + f_N, \quad \bar{V}(F; \mathbf{A}; E; X) = \bar{V}(fh; \mathbf{A}; E; X)$$
$$\leq \bar{V}((|f| - f_N)|h|; \mathbf{A}; E; X) + \bar{V}(f_N|h|; \mathbf{A}; E; X)$$
$$\leq \int_E (|f| - f_N)d|h| + N \cdot \bar{V}(h; \mathbf{A}; E; X) < \varepsilon.$$

(Use Theorems 2.5.5 and 2.1.1(2.1.6) analogue.)

3.3 THE BOUNDED RIEMANN SUMS TEST AND THE MAJORIZED (DOMINATED) CONVERGENCE TEST OF ARZELÀ AND LEBESGUE

For over 80 years the regular tests used in Lebesgue theory have been the two monotone convergence tests and the majorized or dominated convergence test, and these are the basis of some axiomatic accounts of Lebesgue integration. The first two are in Section 3.2, the third in Theorem 3.3.2, and we have gone further, with necessary and sufficient conditions in Section 3.1. Controlled convergence is given in Henstock (1988a).

Theorem 3.3.1. *Let $(T, \mathcal{T}, \mathbf{A})$ be a decomposable additive division space, C a compact set in the linear topological space K such that real continuous linear functionals exist to separate all points of C, and let $k: \mathcal{U}^1 \to \mathbb{R}^+$, $f_n: T \to K$, $f: T \to K$, $X \subseteq T$ of k-variation 0 relative to E, $G \in \mathcal{G}$ with $z \in G$, $\mathcal{U} \in \mathbf{A}|E$ depending on G, and $m: I \otimes T \to I$ (the set of positive integers), such that*

(3.1.4) $\quad (\mathscr{E})\sum Z^j(I, t) \equiv \{(\mathscr{E})\sum z^j(I, t): z^j(I, t) \in Z^j(I, t)\} \subseteq G$

3.3 CONVERGENCE TESTS

for all divisions $\mathscr{E}$ over E from $\mathscr{U}$,

(3.3.1) $\quad \{f_n(t) - f(t)\}k(I, t) \in Z^j(I, t) \quad ((I, t) \in \mathscr{U}, t \notin X, n \geq m(j, t)),$

(3.3.2) $\quad\quad\quad\quad\quad\quad (\mathscr{E})\sum f_{n(t)}(t)k(I, t) \in C$

for all divisions $\mathscr{E}$ of E from $\mathscr{U}$ and all choices of $n: T \to \mathscr{I}$ for $(I, t) \in \mathscr{E}$. For each fixed integer n let $f_n(t)k(I, t)$ be integrable over E to $H_n(E)$. Then fk is integrable over E to $H(E) = \lim_{n \to \infty} H_n(E)$, which limit exists.

Proof. By (3.1.4), (3.3.1), (3.3.2),

$$(\mathscr{E})\sum f(t)k(I, t) \in C + G.$$

By Theorems 0.4.14, 0.4.9, $C + G$ can be replaced by $\mathscr{G}C = C$. Then by Theorem 0.5.2(0.5.5), $\bar{S}(fk: \mathbf{A}; E)$ is not empty and lies in C. If it contains two points $a \neq b$, a real continuous linear functional $\mathscr{F}$ exists with $\mathscr{F}(a) \neq \mathscr{F}(b)$. By Theorem 3.1.1(3.1.6), $R^j(I, t) \equiv \mathscr{F}\{Z^j(I, t)\}$ can replace $Z^j(I, t)$ in (3.1.4) with $G = (-\varepsilon, \varepsilon)$ for arbitrarily small $\varepsilon > 0$, while by linearity, (3.3.2) and continuity (Theorem 0.4.7(0.4.20)),

$$(\mathscr{E})\sum \mathscr{F}\{f_{n(t)}(t)\}k(I, t) = \mathscr{F}\{(\mathscr{E})\sum f_{n(t)}(t)k(I, t)\} \in \mathscr{F}(C),$$

a compact set, so that $\mathscr{F}(f_n)k$ satisfies (3.3.2). Also by linearity,

$$(\mathscr{E})\sum \mathscr{F}\{f_n(t)\}k(I, t) - \mathscr{F}\{H_n(E)\} = \mathscr{F}\{(\mathscr{E})\sum f_n(t)k(I, t) - H_n(E)\}, \mathscr{F}(z) = 0.$$

Hence $\mathscr{F}(f_n)k$ is integrable over E to $\mathscr{F}(H_n(E))$. Thus we can take $K = \mathbb{R}$ and omit $\mathscr{F}$. By (3.3.2) and Theorem 2.7.6, for each N, Q in $Q > N \geq 0$ we have the integrability of

$$\max(f_N(t), \ldots, f_Q(t))k(I, t), \quad \min(f_N(t), \ldots, f_Q(t))k(I, t)$$

By Theorem 3.2.1 and (3.3.2) the following exist (finite) and are integrable in E, when multiplied by k.

$$\inf_{n \geq N} f_n = \lim_{Q \to \infty} \min_{N \leq n \leq Q} f_n, \quad \sup_{n \geq N} f_n = \lim_{Q \to \infty} \max_{N \leq n \leq Q} f_n,$$

$$\liminf_{n \to \infty} f_n = \lim_{N \to \infty} \inf_{n \geq N} f_n, \quad \limsup_{n \to \infty} f_n = \lim_{N \to \infty} \sup_{n \geq N} f_n.$$

For the integrals themselves we have the results

$$\int_E \liminf_{n \to \infty} f_n\, dk = \lim_{N \to \infty} \int_E \inf_{n \geq N} f_n\, dk = \lim_{N \to \infty} \lim_{Q \to \infty} \int_E \min_{N \leq n \leq Q} f_n\, dk$$

$$\leq \lim_{N \to \infty} \lim_{Q \to \infty} \min_{N \leq n \leq Q} \int_E f_n\, dk = \liminf_{n \to \infty} H_n(E)$$

$$\leq \limsup_{n \to \infty} H_n(E) \leq \int_E \limsup_{n \to \infty} f_n\, dk,$$

the proof of the part involving the lim sup being similar. By Theorem 3.1.2, in this case (3.3.1) implies that $f(t) \equiv \lim_{n \to \infty} f_n(t)$ exists k-almost everywhere so that the two outer integrals have the same value, and it is equal to the limit of $H_n(E)$, and we have the result when $K = \mathbb{R}$.

In the general case the proof contradicts $\mathscr{F}(a) \neq \mathscr{F}(b)$, so that $\bar{S}(fk; \mathbf{A}; E) = \text{sing}(a)$ for some a. By Theorem 2.3.2, fk is integrable to a. By (3.3.2) the integral $H_n(E)$ of $f_n k$ over E, lies in C, so that by Exercise 0.4.7, $(H_n(E))$ has at least one limit-point, say b. By continuity of every $\mathscr{F}$ $(\mathscr{F} H_n(E))$ has limit-point $\mathscr{F}(b)$. We have proved that $\mathscr{F} H_n(E)$ tends to a limit, which is therefore $\mathscr{F}(b)$, and it also is the integral of $\mathscr{F}(f)k$ over E and so is $\mathscr{F}(a)$. By choice of $\mathscr{F}$ we see that the only possible value of b is $b = a$, and the theorem is proved.

Let X_1 be the set of t where $f_n(t) = f(t)$ for some integer $N = N(t)$ and all $n \geq N$. Then when K is a normed linear space we can by Theorem 3.1.2 replace (3.1.4), (3.3.1) by $f_n \to f$ k-almost everywhere, with k VBG* in $\setminus X_1$. But for more general K, (3.1.4) and (3.3.1) are likely to be stronger. In the majorized (dominated) convergence test of Arzelà and Lebesgue, (3.3.2) holds and we can write the test in the following form.

Theorem 3.3.2. *Let* $(T, \mathscr{T}, \mathbf{A})$ *be a decomposable additive division space, K a normed linear space, $k: \mathscr{U}^1 \to \mathbb{R}^+$, f_n and $f: T \to K$, $g: T \to \mathbb{R}^+$, $X \subseteq T$ of k-variation zero relative to E, and $k\chi(\setminus X_1; \cdot)$ VBG* in E where X_1 is the set of t where $f_n(t) = f(t)$ for some integer $N = N(t)$ and all $n \geq N$. Let $f_n k$ and gk be integrable in E with $\|f_n\| \leq g$ and $f_n \to f$ as $n \to \infty$ except possibly in X. Then fk is integrable over E to $H(E)$, the limit as $n \to \infty$ of the integral of $f_n k$ over E.*

When $K = \mathbb{R}$, (3.3.2) is equivalent to the Arzelà–Lebesgue condition. We need only prove the following result.

Theorem 3.3.3. *Let* $(T, \mathscr{T}, \mathbf{A})$ *be a decomposable additive division space with $k: \mathscr{U}^1 \to \mathbb{R}^+$, $f_n: T \to \mathbb{R}$, and $f_n k$ satisfying (3.3.2) and integrable for each fixed n. Then $|f_n - f_1|k$ is integrable on E, and for $g(t) \equiv \sup_n |f_n(t) - f_1(t)|$ whenever this is finite, and otherwise $g(t) = 0$, gk is integrable on E.*

Proof. As $K = \mathbb{R}$ we can replace the compact set C by $[-M, M]$ for some finite $M > 0$. The arbitrary $n(t)$ for (3.3.2) is first chosen to be 1 for all $t \in E^*$, and then secondly, $n(t) = n$ when $f_n(t) \geq f_1(t)$, say in a set X, and $n(t) = 1$ in $\setminus X$. Thus

$$(\mathscr{E})\sum |f_n(t) - f_1(t)|\chi(X; t)k(I, t) \leq 2M, \quad (\mathscr{E})\sum |f_n(t) - f_1(t)|\chi(\setminus X; t)k(I, t) \leq 2M$$

similarly, so that

(3.3.3) $\qquad (\mathscr{E})\sum |f_n(t) - f_1(t)|k(I, t) \leq 4M,$

and $(f_n - f_1)k$ is of bounded variation in E. By Theorem 2.7.3, $|f_n - f_1|k$ is integrable over E. Clearly we can replace $f_n - f_1$ by $f_{n(t)} - f_1$ or $f_{n(t)} - f_n$, and we can go further. If $k(I, t) = 0$, $gk = 0$ for any g. If

$$k(I, t) > 0, \quad g(t) \equiv \sup_n |f_n - f_1| \leqslant 4M/k(I, t) < +\infty$$

(3.3.4) $$(\mathscr{E}) \sum g(t) k(I, t) \leqslant 4M$$

The integrability of gk over E follows from (3.3.4), Theorem 2.7.6 for the maximum when $1 \leqslant n \leqslant N$, and Theorem 3.2.1 for the limit as $N \to \infty$. □

Thus it follows that the Arzelà–Lebesgue test on $f_n - f_1$ and the bounded Riemann sums test are equivalent when $K = \mathbb{R}$. For more general K the bounded Riemann sums test has a wider application.

A result for a one-sided boundedness when $K = \mathbb{R}$ corresponds to Fatou's lemma.

Theorem 3.3.4. *Let $(T, \mathscr{T}, \mathbf{A})$ be a decomposable additive division space with $k: \mathscr{U}^1 \to \mathbb{R}^+$, $f_n: T \to \mathbb{R}$, $f: T \to \mathbb{R}$, $B \in \mathscr{R}$, $\mathscr{U} \in \mathbf{A}|E$, and $X \subseteq T$ of k-variation 0 relative to E, such that for all divisions $\mathscr{E}$ of an elementary set E from $\mathscr{U}$ and all choices of $n: T \to \mathscr{I}$ for $(I, t) \in \mathscr{E}$,*

(3.3.5) $$(\mathscr{E}) \sum f_{n(t)}(t) k(I, t) \geqslant B.$$

If for each fixed integer n, $f_n(t) k(I, t)$ is integrable over E to $H_n(E)$ and if $\liminf_{n \to \infty} H_n(E)$ is finite, then $(\liminf_{n \to \infty} f_n) k$ is integrable over E and

(3.3.6) $$\int_E \liminf_{n \to \infty} f_n \, dk \leqslant \liminf_{n \to \infty} \int_E f_n \, dk.$$

If (3.3.5) is replaced by

(3.3.7) $$(\mathscr{E}) \sum f_{n(t)}(t) k(I, t) \leqslant V$$

for some $V \in \mathscr{R}$ with finite $\limsup_{n \to \infty} H_n(E)$, then $(\limsup_{n \to \infty} f_n) k$ is integrable over E and

(3.3.8) $$\int_E \limsup_{n \to \infty} f_n \, dk \geqslant \limsup_{n \to \infty} \int_E f_n \, dk.$$

If (3.3.5) and (3.3.7) hold with f_n tending to a finite limit except possibly in X, we are back with Theorem 3.3.1 and $K = \mathbb{R}$.

Proof. We follow the proof of Theorem 3.3.1 when $K = \mathbb{R}$, omitting $\mathscr{F}$. For the first part we look at the minimum and infimum and see that (3.3.5) and the finiteness of $\liminf_{n \to \infty} H_n(E)$ are sufficient hypotheses for the proof to go through, omitting the part involving the lim sup. The second part is then a mirror image of the first. □

If gk is integrable over E and $f_n \geq g$ for all positive integers n, then (3.3.5) follows. In Lebesgue theory this is exactly *Fatou's lemma*. When $K = \mathbb{R}$, Theorem 3.3.1 follows from Theorem 3.3.4. This is the pattern followed in Henstock (1963c, 1968b, 1969, 1974, 1988a). Here we have a more general K and a different pattern. However, we return to the previous pattern on turning to functions $f_d(t)$ depending on a parameter d that lies in a directed set D that is more than a sequence. The most usual extension is to the real line or a subset of it such as a one-sided or two-sided neighbourhood of a real number. Writing the results in terms of $y \in \mathbb{R}$ and $y \to \infty$, we can obtain other results from it, and we can see the difficulties. Thus $\lim_{y \to c+} f(x, y)$ results from $\lim_{y \to \infty} f(x, c + 1/y)$, etc. We have to avoid measurability difficulties in the generalization of Fatou's lemma.

Theorem 3.3.5. *Let $(T, \mathcal{T}, \mathbf{A})$ be a decomposable additive division space, let $k: \mathcal{U}^1 \to \mathbb{R}^+$, $f(\cdot, y): T \to \mathbb{R}$ (each $y \in \mathbb{R}^+$) and let $f(t, y)k(I, t)$ be integrable over E to $H(E, y)$ for each $y \in \mathbb{R}^+$. First, if for each real Y, Z in $0 \leq Y \leq Z$,*

(3.3.9)
$$\inf_{Y \leq y \leq Z} f(t, y)k(I, t)$$

is integrable in E and, given $\mathcal{U} \in \mathbf{A}|E$, if there is a real number L such that

(3.3.10)
$$(\mathcal{E}) \sum f(t, y(t))k(I, t) \geq L$$

for each division $\mathcal{E}$ of E from $\mathcal{U}$ and for each $y(t): T \to \mathbb{R}^+$, then for a finite right side, the left side exists in the following and

(3.3.11)
$$\int_E \liminf_{y \to \infty} f(\cdot, y) \, dk \leq \liminf_{y \to \infty} \int_E f(\cdot, y) \, dk.$$

Secondly, if for each real Y, Z in $0 \leq Y \leq Z$

(3.3.12)
$$\sup_{Y \leq y \leq Z} f(t, y)k(I, t)$$

is integrable in E and if, given $\mathcal{U} \in \mathbf{A}|E$, there is a real number M such that

(3.3.13)
$$(\mathcal{E}) \sum f(t, y(t))k(I, t) \leq M$$

for each division $\mathcal{E}$ of E from $\mathcal{U}$ and for each $y(t): T \to \mathbb{R}^+$, then for a finite left side, the right side exists in the following and

(3.3.14)
$$\limsup_{y \to \infty} \int_E f(\cdot, y) \, dk \leq \int_E \limsup_{y \to \infty} f(\cdot, y) \, dk.$$

If $f(t) \equiv \lim_{y \to \infty} f(t, y)$ exists for t k-almost everywhere, we do not need the integrability of (3.3.9) nor of (3.3.12). Further, if (3.3.10), (3.3.13) hold, so that (3.3.2) holds, fk is integrable over E and

(3.3.15)
$$\int_E f \, dk = \lim_{y \to \infty} \int_E f(\cdot, y) \, dk.$$

3.3 CONVERGENCE TESTS

Proof. For fixed Y, (3.3.9) is monotone decreasing as $Z \to \infty$. Taking Z an integer and applying Theorem 3.2.2 to (3.3.9) with (3.3.10), so that the integral of (3.3.9) is not less than L, then $\inf_{y \geq Y} f(t, y) k(I, t)$ exists for t k-almost everywhere. By monotonicity the limit, when it exists, as $Z \to \infty$ through the integers, is the same as the limit for unrestricted $Z \to \infty$, that limit is integrable, and

(3.3.16)

$$\int_E \inf_{y \geq Y} f(\cdot, y) \, dk = \lim_{Z \to \infty} \int_E \inf_{Y \leq y \leq Z} f(\cdot, y) \, dk \leq \inf_{y \geq Y} \int_E f(\cdot, y) \, dk.$$

Further, $\inf_{y \geq Y} f(\cdot, y) k(I, t)$ is monotone increasing in Y, and the right side of (3.3.11) is the limit, as $Y \to \infty$, of the right side of (3.3.16), given as finite. Hence by Theorem 3.2.2 again,

$$\liminf_{y \to \infty} f(t, y) k(I, t) = \lim_{Y \to \infty} \inf_{y \geq Y} f(t, y) k(I, t)$$

is finite k-almost everywhere and is integrable with (3.3.16), giving (3.3.11). The mirror image result gives (3.3.14).

If $f(t) \equiv \lim_{y \to \infty} f(t, y)$ exists k-almost everywhere, then for some $(y_n) \to +\infty$,

$$\liminf_{y \to \infty} \int_E f(\cdot, y) \, dk = \lim_{n \to \infty} \int_E f(\cdot, y_n) \, dk \leq \int_E \lim_{n \to \infty} f(\cdot, y_n) \, dk$$

$$= \int_E \lim_{y \to \infty} f(\cdot, y) \, dk$$

and similarly for (3.3.14), without using the integrability of (3.3.9), (3.3.12). We have used Theorem 3.3.4(3.3.6). Then (3.3.11), (3.3.14) together, with the convergence of $f(t, y)$ to $f(t)$ k-almost everywhere, give (3.3.15). □

(3.3.10) and (3.3.13) can be respectively replaced by $f(t, y) \geq g(t)$ and $f(t, y) \leq g(t)$ k-almost everywhere, with gk integrable over E.

We can avoid the assumptions of measurability in (3.3.9), (3.3.12), and at the same time replace $y \in \mathbb{R}$ by $d \in D$, a general set directed upwards, on using the limit

$$f = \lim_{d}{}^* f_d(t)$$

of Theorem 3.2.3 where, for a suitable sequence $(d(n)) \subseteq D$,

$$f(t) = \lim_{n \to \infty} f_{d(n)}(t)$$

k-almost everywhere; and variations on the same theme.

As in (3.3.10) we assume that there are a real number L and a $\mathcal{U} \in \mathbf{A}|E$ such that

(3.3.17) $$(\mathcal{E}) \sum f_{g(t)}(t)k(I, t) \geq L$$

for each division $\mathcal{E}$ of E from $\mathcal{U}$ and for each function $g: T \to D$.

If $u, v \in D$ with $u < v$ in the order of D let $[u, v)$ or $u \leq d < v$ denote the $d \in D$ (if any) that satisfy both $u \leq d$ and $d < v$. Then we can consider

$$F(d(n); u, v; t) \equiv \inf_{u \leq d(n) \leq v} f_{d(n)}(t)$$

for a sequence $(d(n)) \subseteq [u, v)$. Using (3.3.17) we can integrate F_k over E. If M is the infimum of

$$\int_E F(d(n); u, v; t) \, dk$$

over such sequences $(d(n))$, then by (3.3.17), $M \geq L$, and there is a sequence of such sequences for which the integral tends to M. Combining the sequences into one sequence by the usual diagonal process we have a $(d(n))$, and then an $F(d(n); u, v; t)$ whose integral over E is not greater than the separate integrals, and so is M. This particular F is denoted by

$$\inf_{u \leq d \leq v} {}^* f_d(t).$$

To show that it is unique modulo values in sets of k-variation zero we use a similar argument to that for Theorem 3.2.3. Let $(d(n))$ and $(d_1(n))$ lie in $[u, v)$, the integrals of the corresponding F being M. If $(d_2(n))$ is the combined sequence, then

$$F(d_2(n); u, v; t) \leq F(d(n); u, v; t), \quad F(d_2(n); u, v; t) \leq F(d_1(n); u, v; t)$$

and, multiplying by k and integrating over E, we have the minimum M each time. Thus k times the non-negative differences of the F for the d_2 and d have integrals 0 and so, k-almost everywhere,

$$F(d_2(n); u, v; t) = F(d(n); u, v; t), \quad F(d_2(n); u, v; t) = F(d_1(n); u, v; t),$$
$$F(d_1(n); u, v; t) = F(d(n); u, v; t),$$

and F is unique modulo values in sets of k-variation zero. Similarly for sup*. With these constructions we can consider the usual majorized convergence and other results, defining

$$\inf_{u \leq d} {}^* f_d(t) = \lim_v {}^* \inf_{u \leq d \leq v} {}^* f_d(t), \quad \liminf_d {}^* f_d(t) = \lim_u {}^* \left\{ \inf_d {}^* f_d(t) \right\},$$

and similar definitions involving sup*.

It is enough to give the (generalized) Fatou result. The rest of the theorem corresponding to Theorem 3.3.5 can then easily be written.

Theorem 3.3.6 *Let $(T, \mathcal{T}, \mathbf{A})$ be a decomposable additive division space, let $k: \mathcal{U}^1 \to \mathbb{R}^+$, $f_d: T \to \mathbb{R}$ (all $d \in D$) and $f_d k$ integrable in E. If (3.3.17) is true.*

$$(3.3.18) \qquad \int_E \liminf_d {}^* f_d(t)\, dk \leq \liminf_d \int_E f_d(t)\, dk.$$

Proof. Use Theorem 3.3.4 since the starred limits are based on sequences.

CHAPTER 4

DIFFERENTIATION

4.1 THE DIFFERENTIATION OF STRONG VARIATIONAL INTEGRALS

In the calculus on the real line, integration and differentiation are inverse operations. This chapter studies what corresponds in division spaces $(T, \mathcal{T}, \mathbf{A})$ to differentiation. For an elementary set E, a fixed point $t \in T$, and two functions $F(I, t)$ and $h(I, t)$, the analogue is the limit of $F(I, t)/h(I, t)$ for $(I, t) \in \mathcal{U}$ as $\mathcal{U}$ shrinks in $\mathbf{A}|E$. The limit, say $f(t)$, is called the *derivative* $D = D(F, h; \mathbf{A}; E; t)$ *of F with respect to h, $\mathbf{A}$, E, at t.* Thus for real- or complex-valued functions, given $\varepsilon > 0$, there is a $\mathcal{U} \in \mathbf{A}|E$ such that

(4.1.1) $\qquad |F(I, t)/h(I, t) - f(t)| \leq \varepsilon,$

(4.1.2) $\qquad |F(I, t) - f(t)h(I, t)| \leq \varepsilon |h(I, t)|$

for all $(I, t) \in \mathcal{U}$ with $h(I, t) \neq 0$. If $h(I, t) = 0$ in (4.1.2) then $F(I, t)$ would also have to be 0 if the inequality holds, and then the ratio in (4.1.1) would be the indeterminate form 0/0. If this behaviour occurs for some $t \in T$, some $\mathcal{U} \in \mathbf{A}|E$, and all $(I, t) \in \mathcal{U}$, i.e.

(4.1.3) $\qquad h(I, t) = 0 \quad (\text{all } (I, t) \in \mathcal{U}),$

then the derivative cannot be defined. Conversely, if the derivative is $f(t)$ and is $g(t)$, where

$$|f(t) - g(t)| > 2\varepsilon > 0,$$

then from (4.1.2) and suitable $\mathcal{U} \in \mathbf{A}|E$,

$$2\varepsilon |h(I, t)| < |f(t) - g(t)||h(I, t)| \leq 2\varepsilon |h(I, t)|, \quad h(I, t) \neq 0,$$

gives a contradiction, and (4.1.3) occurs, (4.1.2) is true for arbitrary $f(t)$, and (4.1.1) is meaningless. If the division space is fully decomposable, the set of points t satisfying (4.1.3) has h-variation 0 relative to E and so may be disregarded in the sequel. It is clear that differentiation is a pointwise process, and in order to fit together the portions of many $\mathcal{U} \in \mathbf{A}|E$ that are connected with each point in question, full decomposability is appropriate. Thus we now assume in the whole of this section that

(4.1.4) $(T, \mathcal{T}, \mathbf{A})$ *is a fully decomposable additive division space.*

4.1 DIFFERENTIATION OF STRONG VARIATIONAL INTEGRALS

When K is a Banach space, such as $\mathbb{R}$ or $\mathbb{C}$, with norm $\|.\|$, and $\mathcal{U}_0 \subseteq \mathcal{U}^1$ is a set of interval–point pairs (I, t), with indicator $\mathrm{ind}(\mathcal{U}_0; I, t)$, we write $IV(h; \mathbf{A}; E; \mathcal{U}_0)$ for $\bar{V}(\mathrm{ind}(\mathcal{U}_0; \cdot)\|h\|; \mathbf{A}; E)$, the *inner h-variation of $\mathcal{U}_0$ over E relative to $\mathbf{A}$*. Note that in $\bar{V}(h; \mathbf{A}; E; X)$, X is a set of points $t \in T$, whereas in $IV(h; \mathbf{A}; E; \mathcal{U}_0)$, $\mathcal{U}_0$ is a set of some $(I, t) \in \mathcal{U}^1$, so justifying the different notation.

In a Banach space we usually work with (4.1.2) and not (4.1.1) so that to consider fh we need to multiply together values that may be in different spaces. Let K_1, K_2, K be Banach spaces such that any value from K_1 can be multiplied on the right by any value from K_2 to give a value in K. For example, one of K_1, K_2 could be $\mathbb{R}$ or $\mathbb{C}$, and the other could be K. Or $K_1 = K_2 = K$, a Banach algebra. For simplicity we write all three norms as $\|.\|$, such that $\|fh\| = \|f\| \cdot \|h\|$.

Theorem 4.1.1. *Let $f: T \to K_1$, $h: \mathcal{U}^1 \to K_2$, with fh strong variationally integrable to $F(P)$ over the partial sets P of the elementary set E. Then for $\varepsilon > 0$ and $\mathcal{U}(\varepsilon)$ the set of $(I, t) \in \mathcal{U}^1$ for which*

(4.1.5) $$\|F(I) - f(t)h(I, t)\| > \varepsilon \|h(I, t)\|,$$

(4.1.6)
$$IV(h; \mathbf{A}; E; \mathcal{U}(\varepsilon)) = V(\mathrm{ind}(\mathcal{U}(\varepsilon); \cdot)\|h\|; \mathbf{A}; E) = \int_E d(\mathrm{ind}(\mathcal{U}(\varepsilon); \cdot)\|h\|) = 0.$$

(4.1.7) *If also $D(F, h; \mathbf{A}; E; t)$ exists equal to $g(t)$ at all points t of a set $X \subseteq E^*$ in which h is VBG^*, then $g(t) = f(t)$ h-almost everywhere in X.*

Proof. Given $\eta > 0$, there is a $\mathcal{U} \in \mathbf{A}|E$ for which every division $\mathscr{E}$ of E from $\mathcal{U}$ satisfies

(4.1.8)
$$(\mathscr{E})\sum \|F(I) - f(t)h(I, t)\| < \eta, \quad (\mathscr{E})\sum \mathrm{ind}(\mathcal{U}_0; I, t)\|h(I, t)\| < \eta/\varepsilon,$$

giving (4.1.6) as $\eta \to 0$. For (4.1.7), given $\delta > 0$, there is a $\mathcal{U} \in \mathbf{A}|E$ with

(4.1.9) $\quad \|F(I) - g(t)h(I, t)\| \leq \delta \|h(I, t)\| \quad$ (all $(I, t) \in \mathcal{U}$ with $t \in X$).

By full decomposability we can take $\mathcal{U}$ independent of t. Then from (4.1.8), (4.1.9),

$$(\mathscr{E})\sum \|f(t) - g(t)\| \|h(I, t)\| \chi(X; t) = (\mathscr{E})\sum \|f(t)h(I, t) - F(I)$$
$$+ F(I) - g(t)h(I, t)\| \chi(X; t)$$
$$< \eta + (\mathscr{E})\sum \delta \|h(I, t)\| \chi(X; t).$$

144   THE GENERAL THEORY OF INTEGRATION

If $\|f(t) - g(t)\| \geq \zeta > 0$ in $X_1(\zeta) \subseteq X$ where h is VB*, then

$$\zeta(\mathscr{E})\sum \|h(I, t)\|\chi(X_1(\zeta); t) < \eta + \delta V(h; \mathscr{U}; E; X_1(\zeta)) \to 0 \quad \text{as } \eta \to 0, \delta \to 0,$$
$$V(h; \mathbf{A}; E; X_1(\zeta)) = 0.$$

We prove (4.1.7) using decomposability with $\zeta = 1/n$ $(n = 1, 2, \ldots)$ and the splitting up of X into sets on which h is VB*. □

However, full decomposability cannot prove directly that the union of $\mathscr{U}(1/n)$ $(n = 1, 2, \ldots)$ also has inner variation zero, e.g. in the case of the gauge integral it could happen that for all I outside the neighbourhood of t with centre t and radius $1/n$, $(I, t) \in \mathscr{U}(1/n)$. The union of the $\mathscr{U}(1/n)$, would then be every (I, t), causing difficulties. To avoid such an example, given $\mathscr{U}_0 \subseteq \mathscr{U}^1$, we look at the set X_0 of all t for which, for every $\mathscr{U} \in \mathbf{A}|E$, there is a partial interval I of E with $(I, t) \in \mathscr{U} \cap \mathscr{U}_0$. Full decomposability can then act on such sets X_0.

In one dimension a lemma of Sierpinski (1923) is used to prove that if $\mathscr{U}_0$ has inner h-variation zero and if X_0 is the corresponding set, then X_0 has h-variation zero. In $\mathbb{R}^n$ for an integer $n > 1$, Vitali's covering theorem (Vitali (1908)) is used analogously on bricks satisfying a condition on the ratios of edges of intervals or bricks. As these results are excessively geometrical, with no significant obvious generalization, I will not reproduce the theory here but refer to Henstock (1988a), pp. 140–5, Theorems 15.10–12.

Theorem 4.1.2. *In Theorem 4.1.1 let K_3 be separable and $K_3 \subseteq K_1$, and let $f \in K_3$ except possibly for t in a set of h-variation zero. Let h, $\|h\|$, fh, $\|fh\|$ all be integrable in E. Then*

(4.1.10) $$\int_J \|f(t) - f(x)\| \, \mathrm{d} \|h\|/\bar{V}(h; \mathbf{A}; J) \to 0 \quad ((J, x) \in \mathscr{U})$$

as $\mathscr{U}$ shrinks in $\mathbf{A}|E$, except for a countable union of sets $\mathscr{U}_0$ of (J, x) with $IV(h; \mathbf{A}; E; \mathscr{U}_0) = 0$.

Proof. For fixed $k \in K_1$, Exercise 2.7.2 shows that $\|(f-k)h\|$ is integrable. By Theorem 2.7.3, for $V(J) \equiv \bar{V}(h; \mathbf{A}; J)$, the integrals of $\|(f-k)h\|$ and $\|f-k\|V$ are equal, for all partial sets J of E. Hence by Theorem 4.1.1, as $\mathscr{U}$ shrinks in $\mathbf{A}|E$,

(4.1.11) $$\lim \int_J \|f(t) - k\| \, \mathrm{d} \|h\|/V(J) = \|f(x) - k\| \quad ((J, x) \in \mathscr{U}),$$

except for x in the union of a countable number of sets each with inner h-variation zero. As a countable number of a countable number is still countable, (4.1.11) is true for a countable number of $k \in K_1$. Taking these k to be

4.1 DIFFERENTIATION OF STRONG VARIATIONAL INTEGRALS

everywhere dense in K_3, and noting that for arbitrary $k_1 \in K_3$,

$$|\,\|f(t) - k\| - \|f(t) - k_1\|\,| \leq \|k - k_1\|,$$

$$\left|\int_J \|f(t) - k\|\,d\|h\| - \int_J \|f(t) - k_1\|\,d\|h\|\right| \leq \|k - k_1\| V(J).$$

Replacing k by k_1 in (4.1.11), we see that for the exceptional set depending on the countable number of k but not on k_1, the upper and lower limits of the fraction lie between

$$\|f(x) - k\| \pm \|k - k_1\|$$

apart from x in the exceptional set. These upper and lower limits are independent of k, and we can take $\|k - k_1\| \to 0$, ensuring that the upper and lower limits are both equal to $\|f(x) - k_1\|$. Hence the limit exists with this value. Now, h-almost everywhere in x, we have $f(x) \in K_3$ and so $f(x)$ is such a k_1, and in (4.1.11) we can put $f(x)$ for k and have the result. □

Note that if $V(h; \mathbf{A}; E; X_0) = 0$ and if $\mathcal{U}_1 = \mathcal{U}^1[X_0]$, then $IV(h; \mathbf{A}; E; \mathcal{U}_1)$ = 0, since $h \cdot \chi(X_0; t) = h \cdot \mathrm{ind}(\mathcal{U}_1; I, t)$. If $t_0 \mathcal{U}_0$ corresponds X_0, then $\mathcal{U}_0 \subseteq \mathcal{U}_1$, so that

(4.1.12) $\qquad IV(h; \mathbf{A}; E; \mathcal{U}_0) = 0 \quad$ when $V(h; \mathbf{A}; E; X_0) = 0$.

In Section 3.1 one of the limit processes is that of differentiation by a parameter, of an integral over a fixed elementary set E, (3.1.2) giving the one-dimensional result. We can obtain necessary and sufficient conditions for (3.1.2) to hold by translating into the present mathematical language the general Theorems 3.1.3 and 3.1.4.

Theorem 4.1.3. *For a decomposable additive division space $(T, \mathcal{T}, \mathbf{A})$, E an elementary set, $m: \mathcal{U}^1 \to \mathbb{R}^+$, $f: T \times \mathbb{R} \to \mathbb{R}$, and with fm integrable over E to $H(E; y)$, so that*

(4.1.13) $\qquad \{f(t, y + h) - f(t, y)\} m(I, t)/h$

is integrable to

$$\{H(E; y + h) - H(E; y)\}/h,$$

and such that (4.1.13) tends to the limit $\partial f/\partial y$ as $h \to 0$, for t m-almost everywhere in E^*, then $\partial f/\partial y$ is integrable over E if and only if there are a compact set C of arbitrarily small diameter, some $M: \mathcal{U}^1 \to \mathbb{R}$, some $\mathcal{U} \in \mathbf{A}|E$, and all divisions $\mathcal{E}$ of E from $\mathcal{U}$, such that

(4.1.14) $\qquad (\mathcal{E}) \sum \{f(t, y + h(I, t)) - f(t, y)\} m(I, t)/h(I, t) \in C$

for all $|h(I, t)| \leq M(I, t)$. Then when these hold, the integral of $\partial f/\partial y$ is the derivative of $H(E, y)$, if and only if there are a compact set C_1 of arbitrarily small diameter such that $C \cap C_1$ is not empty, a fixed M, and a $\mathscr{U}(h) \in \mathbf{A}|E$ with

(4.1.15) $\qquad (\mathscr{E})\sum \{f(t, y + h) - f(t, y)\}m(I, t)/h(I, t) \in C_1$

for all $|h| \leq M$.

The proofs follow the proof of Theorems 3.1.3, 3.1.4. Henstock (1988a) gives the theorem for gauge integrals on pp. 135–6, Theorem 15.5, and Theorems 15.1–4, pp. 133–5 give other tests.

Exercise 4.1.1 Let the convex function $r: \mathbb{R} \to \mathbb{R}$, $\mathbb{R}$ being the union of disjoint sets X_j ($j = 1, 2, \ldots$), such that for some gauge δ on $\mathbb{R}$,

(4.1.16) $\qquad |r(y) - r(x)| \leq 2^j|y - x| \quad (x \in X_j, |y - x| < \delta(x), j = 1, 2, \ldots)$

Let $h: \mathscr{U}^1 \to \mathbb{R}^+$ be integrable to H and let $f: T \to \mathbb{R}^+$ have fh integrable to H_1, where $(T, \mathscr{T}, \mathbf{A})$ is a decomposable additive division space, E an elementary set. Then $r(f)h$ is integrable over E if and only if the set of sums

(4.1.17) $\qquad (\mathscr{E})\sum r(f)h$

is bounded above.

For let $\mathscr{U}_j \in \mathbf{A}|E$ be such that for all divisions $\mathscr{E}$ of E from $\mathscr{U}_j$,

$$(\mathscr{E})\sum |fH - H_1| < \varepsilon \cdot 4^{-j}.$$

The differentiation of integrals gives, for some $\mathscr{U}^* \in \mathbf{A}|E$,

$$|f(t) - H_1(I)/H(I)| < \delta(f(t)) \quad ((I, t) \in \mathscr{U}^*).$$

Direction and decomposability give a $\mathscr{U}^+ \in \mathbf{A}|E$ such that if W_j is the set of t with $f(t) \in X_j$, then $\mathscr{U}^+[W_j] \subseteq \mathscr{U}^* \cap \mathscr{U}_j$ ($j = 1, 2, \ldots$). If $\mathscr{E}$ is a division of E from $\mathscr{U}^+$ with $\mathscr{E}_j$ the subset of E with $t \in W_j$, show that

$$(\mathscr{E})\sum |r(f)H - r(H_1/H)H| = (\mathscr{E})\sum |r(f) - r(H_1/H)|H$$

$$\leq \sum_{j=1}^{\infty} 2^j (\mathscr{E}_j)\sum |f - H_1/H|H = \sum_{j=1}^{\infty} 2^j (\mathscr{E}_j)\sum |fH - H_1| \leq \varepsilon.$$

Next, show that $r(H_1/H)H$ is integrable when (4.1.17) is bounded above, by noting that if $\mathscr{P}$ is a partition of I formed of intervals J, then

$$(\mathscr{P})\sum H(J)/H(I) = 1,$$

$$r(H_1(I)/H(I)) = r((\mathscr{P})\sum H_1(J)/H(I))$$
$$= r((\mathscr{P})\sum (H_1(J)/H(J))(H(J)/H(I)))$$
$$\leq (\mathscr{P})\sum r(H_1(J)/H(J))H(J)/H(I),$$

$r(H_1(I)/H(I))H(I) \leq (P)\sum r(H_1(J)/H(J))H(J)$ and $r(H_1/H)H$ is finitely subadditive.

Now complete the proof. There may be an easier proof.

4.2 FURTHER RESULTS ON DIFFERENTIATION

When $p(I, t)$ is of bounded variation, or of generalized bounded variation, it has good differentiation properties. In essence the result was first proved by Lebesgue.

Theorem 4.2.1. *Let $p, q: \mathcal{U}^1 \to K$, where K is a Banach space with norm $\|.\|$. With q arbitrary and p of bounded variation or of generalized bounded variation in a set X, relative to E, and with $(T, \mathcal{T}, \mathbf{A})$ a decomposable division space, then $p(I, t)q^{-1}(I, t)$ is either of the form 00^{-1} or is bounded, for some $\mathcal{U} \in \mathbf{A}|E$ and all $(I, t) \in \mathcal{U}$ with $t \in X$, except for $t \in X_1$, X_1 being of inner q-variation zero (i.e. X_1 is the set of points associated with the $\mathcal{U}_0 \subseteq U^1$ of inner q-variation zero.)*

If $V(p; \mathbf{A}; E; X) = 0$ then $D(p, q; \mathbf{A}; E; t) = 0$ in X, except for a set of inner q-variation 0.

Proof. When p is of bounded variation, if, for a constant $M > 0$,

$$\|p(I, t)\| > M \|q(I, t)\| \quad ((I, t) \in \mathcal{U}_0 \subseteq \mathcal{U}^1),$$

and if $\mathcal{E}_0$ is a division of E from $\mathcal{U}$, we have

$$M(\mathcal{E}_0 \cap \mathcal{U}_0) \sum \|q\| \leq (\mathcal{E}_0 \cap \mathcal{U}_0) \sum \|p\| \leq (\mathcal{E}_0) \sum \|p\|$$
$$\leq V(p; \mathcal{U}; E; X) < \infty,$$

$$V(q \, \mathrm{ind} \, (\mathcal{U}_0; \cdot); \mathcal{U}; E) \leq V(p; \mathcal{U}; E; X)/M.$$

If the ratio is unbounded, we can take M arbitrarily large; showing that the set X_1 is of inner q-variation zero.

When p is of generalized bounded variation we replace X by a sequence (X_j) of sets ($j = 2, 3, \ldots$) on each of which p is of bounded variation, and the proof yields a sequence of sets of inner q-variation zero.

(4.2.1) *If (Y_j) is a sequence of sets $Y_j \subseteq T$, each of inner q-variation zero, then the union Y of the Y_j also has inner q-variation zero.*

Replacing Y_j by $Y_j/(Y_1 \cup Y_2 \cup \cdots \cup Y_{j-1})$ if necessary, we take the Y_j mutually disjoint. We then follow the proof of Theorem 2.2.1 (2.2.1), using various $\mathrm{ind}\,(\mathcal{U}_0; \cdot)$ in an added complication, thus improving various results in Section 4.1.

When $\bar{V}(p; \mathbf{A}; E; X) = 0$, then for each $M > 0$, $V(q \, \mathrm{ind}(\mathcal{U}_0; \cdot); \mathcal{U}; E)$ is as small as we please, by choice of $\mathcal{U} \in \mathbf{A}|E$, and $IV(q; \mathbf{A}; E; \mathcal{U}_0) = 0$. Taking $M = 1/n$ ($n = 1, 2, \ldots$) in turn, giving a sequence of $\mathcal{U}_0$, (4.2.1) shows that the union is still of inner q-variation zero. □

Theorem 4.2.2. *For $(T, \mathcal{T}, \mathbf{A})$ a decomposable division space and p, q, r, s four functions $\mathcal{U}^1 \to K$ such that p, r are variationally equivalent, and q, s are variationally equivalent, both relative to a set $X \subseteq T$ and an elementary set E.*

Then, for $t \in X$, either both derivatives exist and

$$D(p, q; \mathbf{A}; E; t) = D(r, s; \mathbf{A}; E; t),$$

or else neither derivative exists, except for t in a set of inner q-variation zero.

Proof. If $D = D(p, q; \mathbf{A}; E; t)$ exists for a particular $t \in X$, then as

$$\|r - D.S\| \leq \|r - p\| + \|p - D.q\| + \|D\| \|q - S\|$$

(Naturally D is in K_1 while q and s are in K_2),

$$V(r - p; \mathbf{A}; E; X) = 0 = V(q - s; \mathbf{A}; E; X).$$

We now use the second part of Theorem 4.2.1. □

The theory of Chapter 4 has not yet been clarified. The author surmises that eventually a theory not based on Vitali's and Sierpinski's covering theorems will emerge, to prove that a set of inner variation zero is of variation zero in much wider and more general circumstances than are used at present.

CHAPTER 5

CARTESIAN PRODUCTS OF A FINITE NUMBER OF DIVISION SYSTEMS (SPACES)

5.1 FUBINI-TYPE RESULTS

Taking two division systems (spaces) $(T^u, \mathcal{T}^u, \mathbf{A}^u)$ $(u = x, y)$ that could sometimes be equal, or could be extremely different, we put $T^z = T^x \otimes T^y$, the Cartesian product of T^x and T^y in that order. Let $\mathcal{T}^z$, $\mathcal{U}^{1z}$ be the respective families of Cartesian products $I^x \otimes I^y$ for all $I^u \in \mathcal{T}^u$, and $(I^x \otimes I^y, (x, y))$, written $(I^x, x) \otimes (I^y, y)$, for all $(I, u) \in \mathcal{U}^{1u}(u = x, y)$. To complete the usual three elements $(T^z, \mathcal{T}^z, \mathbf{A}^z)$ let $\mathbf{A}^z$ be some non-empty family of non-empty subsets $\mathcal{U}^z \subseteq \mathcal{U}^{1z}$, such that $\mathbf{A}^x, \mathbf{A}^y, \mathbf{A}^z$ have the Fubini property in common. Thus first, for E^u the fixed or an arbitrary elementary set in T^u $(u = x, y)$, $E^z = E^x \otimes E^y$, and $\mathcal{U}^z$ an arbitrary member of $\mathbf{A}^z | E^z$, a non-empty set, then for each $x \in (E^x)^*$ there is a $\mathcal{U}^y(x) \in \mathbf{A}^y | E^y$, and to each collection of divisions $\mathcal{E}^y(x)$ of E^y from $\mathcal{U}^y(x)$, one division for each such x, there corresponds a $\mathcal{U}^x \in \mathbf{A}^x | E^x$ such that if $(I^x, x) \in \mathcal{U}^x$, $(I^y, y) \in \mathcal{E}^y(x)$, then $(I^x, x) \otimes (I^y, y) \in \mathcal{U}^z$; and secondly the corresponding property when x and y are interchanged, except that we keep the Cartesian products in $T^x \otimes T^y$, i.e. x first and then y. Note the use of star sets as a locating device.

The space of values of functions connected with $(T^u, \mathcal{T}^u, \mathbf{A}^u)$ is taken to be a Banach space K^u $(u = x, y)$, such that if $r \in K^x$, $s \in K^y$ then $rs \in K$, another Banach space. For example, r could be real-valued with s in K, or s could be real-valued with r in K, or the scalar component could be complex-valued, or all three spaces could be the same Banach algebra. For simplicity we use the same sign for norm in the three spaces, assuming that $\|rs\| = \|r\| \cdot \|s\|$.

We now turn to a way of constructing $\mathbf{A}^z$. Let E^u be an elementary set in T^u and, for each $z \in T^z$ let $\mathcal{U}^u(z) \in \mathbf{A}^u | E^u (u = x, y)$. Let $\mathcal{U}^z$ be the family of all $(I^x, x) \otimes (I^y, y)$ with I^u a partial interval of E^u and $(I^u, u) \in \mathcal{U}^u(z)$ $(u = x, y)$. Let $\mathbf{A}^z$ be the family of finite unions of such $\mathcal{U}^z$ for all finite unions of disjoint products $E^x \otimes E^y$. Then $(T^z, \mathcal{T}^z, \mathbf{A}^z)$ is the product division system (space) of the division systems (spaces) $(T^u, \mathcal{T}^u, \mathbf{A}^u)$ $(u = x, y)$.

If $\mathbf{A}^u(u = x, y, z)$ have the Fubini property in common and if $(T^z, \mathcal{T}^z, \mathbf{A}^z)$ is a division system (space), we call it a *Fubini division system (space)*.

Theorem 5.1.1. *The product division system (space) of the fully decomposable division systems (spaces) $(T^u, \mathcal{T}^u, \mathbf{A}^u)$ $(u = x, y)$ is a fully decomposable Fubini division system (space).*

Proof. Theorem 2.9.1(2.9.4) shows that the $(T^u, \mathcal{T}^u, \mathbf{A}^u)$ $(u = x, y)$ are stable as they are fully decomposable, and when $x \in (E^x)^*$ it is lying in a set that does not vary with the $\mathcal{U}$ that have shrunk enough. Then let $\mathcal{U}^y(x) \in \mathbf{A}^y | E^y$ be contained in the diagonal relative to y of the $\mathcal{U}^y(x, y)$, with x fixed. Let $\mathscr{E}^y(x)$ be a division of E^y from $\mathcal{U}^y(x)$ with associated points $y_1(x), \ldots, y_{n(x)}(x)$, say. As $\mathbf{A}^x$ is directed in the sense of divisions, for each such x there is a $\mathcal{U}_1^x(x) \in \mathbf{A}^x | E^x$ with $\mathcal{U}_1^x(x) \subseteq \mathcal{U}^x(x, y_j(x))$ for $j = 1, 2, 3, \ldots, n(x)$. This is one of the places where it is vital that a division has only a finite number of interval–point pairs. By full decomposability we can replace the $\mathcal{U}_1^x(x)$ by a $\mathcal{U}_2^x \in \mathbf{A}^x | E^x$. If $(I^x, x) \in \mathcal{U}_2^x$ and $(I^y, y) \in \mathscr{E}^y(x)$, then for some j, $y = y_j(x)$ and $(I^x, x) \in \mathcal{U}^x(x, y_j(x))$, $(I^x, x) \otimes (I^y, y) \in \mathcal{U}^z$.

Interchanging x and y in the proof (without altering the Cartesian product $T^x \otimes T^y$) we have the other half of the Fubini property, and $\mathbf{A}^u$ $(u = x, y, z)$ have that property in common.

From the construction, clearly $(T^z, \mathcal{T}^z, \mathbf{A}^z)$ is fully decomposable and so stable. □

Let $\mathscr{E}^x$ be a division of E^x from $\mathcal{U}_2^x$, and for each associated point x in E^x take the division $\mathscr{E}^y(x)$ of E^y. Then the products of $(I^x, x) \in \mathscr{E}^x$ and $(I^y, y) \in \mathscr{E}^y(x)$ form a division of $E^x \otimes E^y$. Similarly for finite unions of disjoint products of elementary sets, so that $\mathbf{A}^z$ divides all elementary sets in T^z, or just the special elementary set if both components of the product system or space are just division systems. Other properties of the $\mathbf{A}^u$ $(u = x, y)$ carry through to $\mathbf{A}^z$, as easy proofs show, and as the two component division systems or spaces can be wildly different there may be curious one-sided properties. By construction

(5.1.1) $$(E^x \otimes E^y)^* = (E^x)^* \otimes (E^y)^*.$$

We now have the main Fubini-type theorem, in an unsymmetrical form.

Theorem 5.1.2. *Let $\mathbf{A}^u$ $(u = x, y, z)$ have the Fubini property in common; the $(T^u, \mathcal{T}^u, \mathbf{A}^u)$ being stable decomposable division spaces. Let $h^x(I^x, x)$ and $h^y(x; I^y, y)$ be defined for $(I^u, u) \in \mathcal{U}^{1u}$ $(u = x, y,$ respectively) and all $x \in T^x$, with the given value spaces.*

(5.1.2) *If* $\quad h(I^x \otimes I^y, (x, y)) = h^x(I^x, x) h^y(x; I^y, y)$

is strong variationally integrable to H in $E^z = E^x \otimes E^y$, the integral

(5.1.3) $$J(x) = \int_{E^y} dh^y(x; I^y, y)$$

exists as a strong variational integral, except for the $x \in X^x$ with $\bar{V}(h^x; \mathbf{A}^x; E^x; X^x) = 0$. Putting $J(x) = 0$ in X^x, the strong variational integral of $J(x) h^x(I^x, x)$ in E^x is H.

5.1 FUBINI-TYPE RESULTS

Note that (5.1.2) is unsymmetrical. If it also holds when x, y are interchanged, then for some ks,

$$h^x(I^x, x)h^y(x; I^y, y) = k^x(y; I^x, x)k^y(I^y, y),$$
$$h^y(x; I^y, y)/k^y(I^y, y) = k^x(y; I^x, x)/h^x(I^x, x),$$

if the divisions can take place in some appropriate sense. In the last equation the left side does not depend on I^x. Hence the same is true for the right side, and it does not depend on I^y either, so that it is a function, say $f(x, y)$, of x, y only. Thus if we can divide by $h^x(I^x, x) \cdot k^y(I^y, y)$, there results the symmetric form

(5.1.4) $$h(I^z, z) = f(x, y)h^x(I^x, x)k^y(I^y, y)$$

that is nearer to the usual product considered for Fubini's theorem.

Proof of theorem Given $\varepsilon > 0$, let $\mathcal{U}^z \in \mathbf{A}^z | E^z$ be such that for all divisions $\mathscr{E}^z$ of E^z from $\mathcal{U}^z$,

(5.1.5) $$(\mathscr{E}^z) \sum \| h(I^z, z) - H(I^z) \| < \varepsilon.$$

As in the proof of Theorem 5.1.1 the Fubini property gives a division of E^z, using divisions $\mathscr{E}^x$ and $\mathscr{E}^y(x)$, that satisfies (5.1.5). For each x and each $(I^y, y) \in \mathscr{E}^y(x)$, there is a division over I^y that is from $\mathcal{U}^y(x)$, and putting the divisions together, we have a division $\mathscr{E}_1^y(x)$ that refines $\mathscr{E}^y(x)$, for each $x \in (E^x)^*$. By the Fubini property there is a suitable $\mathcal{U}_1^x$, corresponding to $\mathcal{U}^x$, and by direction in $\mathbf{A}^x | E^x$ there is a $\mathcal{U}_2^x \in \mathbf{A}^x | E^x$ with $\mathcal{U}_2^x \subseteq \mathcal{U}^x \cap \mathcal{U}_1^x$. Taking a division $\mathscr{E}^x$ of E^x from $\mathcal{U}_2^x$, it can be used with $\mathscr{E}^y(x)$ and with $\mathscr{E}_1^y(x)$ for divisions in (5.1.5), to obtain

(5.1.6) $$(\mathscr{E}^x) \sum \| h^x(I^x, x) \| (\mathscr{E}^y(x)) \sum \| h^y(x; I^y, y)$$
$$- (\mathscr{E}_1^y(x) \cdot I^y) \sum h^y(x; J^y, y') \| < 2\varepsilon.$$

Here, $(J^y, y') \in \mathscr{E}_1^y(x) \cdot I^y$, the part of the division lying in I^y. If in the set X_n^x of x we can always find $\mathscr{E}^y(x)$ and $\mathscr{E}_1^y(x)$ from $\mathcal{U}^y(x)$ with $\mathscr{E}_1^y(x) \leq \mathscr{E}^y(x)$ so that in (5.1.6) the sum over $\mathscr{E}^y(x)$ is greater than 2^{-n}, then by (5.1.6),

$$(\mathscr{E}^x) \sum \| h^x(I^x, x) \| \chi(X_n^x, x) < 2^{n+1}\varepsilon, \quad V(h^x; \mathbf{A}^x; E^x; X_n^x) = 0.$$

By decomposability and Theorem 2.4.1(2.4.1) the union X^x of the X_n^x ($n = 1, 2, \ldots$) also has h^x-variation zero. When $x \notin X^x$, given the positive integer n and $\varepsilon = 4^{-n}$, there is a $\mathcal{U}_n^z \in \mathbf{A}^z | E^z$ with (5.1.5) such that every $\mathscr{E}_1^y(x) \leq \mathscr{E}^y(x)$ from $\mathcal{U}_n^z$ and the Fubini property give

(5.1.7) $$(\mathscr{E}^y(x)) \sum \| h^y(x; I^y, y) - (\mathscr{E}_1^y(x) \cdot I^y) \sum h^y(x; J^y, y') \| < 2^{-n}.$$

True for $n = 1, 2, \ldots$ so that $J(x)$ exists over E^y. Writing the integral over I^y as $J(x; I^y)$, and taking the limit in (5.1.7) as the sum over $\mathscr{E}_1^y(x) \cdot I^y$ tends to that integral,

(5.1.8) $\quad (\mathscr{E}^y(x)) \sum \|h^y(x; I^y, y) - J(x; I^y)\| < 2^{-n}.$

But we cannot substitute this directly into (5.1.6) as $\mathscr{E}^x$ depends on $\mathscr{E}_1^y(x)$ as well as $\mathscr{E}^y(x)$. We proceed as follows.

Given $x \in (E^x)^* \setminus X^x$ let $\mathscr{U}_n^y(x)$ be the family in $\mathbf{A}^y | E^y$ obtained from the Fubini property using $\mathscr{U}_n^z$. By direction in $\mathbf{A}^y | E^y$ we can assume that $\mathscr{U}_n^y(x)$ is monotone decreasing in n. For otherwise there is in $\mathbf{A}^y | E^y$ a $\mathscr{U}_{1n}^y(x) \subseteq \bigcap_{j=1}^n \mathscr{U}_j^y(x) \cap \mathscr{U}_{1,n-1}^y(x)$ with $\mathscr{U}_{10}^y(x) = \mathscr{U}^{1y}$, and we can replace $\mathscr{U}_n^y(x)$ by $\mathscr{U}_{1n}^y(x)$.

Let $\mathscr{E}_n^y(x)$ be a division of E^y from $\mathscr{U}_n^y(x)$ and put

$$f_n(x) = (\mathscr{E}_n^y(x)) \sum h^y(x; I^y, y).$$

For $\mathscr{E}_n^y(x)$ and $\mathscr{E}_{n+1}^y(x)$ replacing $\mathscr{E}^y(x)$, the Fubini property plus direction in $\mathbf{A}^x | E^x$ gives a $\mathscr{U}^x \in \mathbf{A}^x | E^x$ and an $\mathscr{E}^x$ dividing E^x from $\mathscr{U}^x$ such that $\mathscr{E}^x$ corresponds to $\mathscr{E}_n^y(x)$ and $\mathscr{E}_{n+1}^y(x)$, and by (5.1.6),

$$(\mathscr{E}^x) \sum \|h^x(I^x, x)\| (\mathscr{E}_n^y(x)) \sum \|h^y(x; I^y, y)$$
$$- (\mathscr{E}_{n+1}^y(x) \cdot I^y) \sum h^y(x; J^y, y')\| < 2 \cdot 4^{-n},$$

(5.1.9) $\quad (\mathscr{E}^x) \sum \|h^x(I^x, x)\| \cdot \|f_n(x) - f_{n+1}(x)\| < 2 \cdot 4^{-n}.$

By (5.1.7),

(5.1.10) $\quad \|f_n(x) - f_r(x)\| < 2^{-n} (r > n),$
$\qquad \|f_n(x) - J(x)\| \leq 2^{-n} \quad (x \in (E^x)^* \setminus X^x).$

From (5.1.9), with X_{np}^x the set in $\setminus X^x$ where

$$\|f_{n+1}(x) - f_n(x)\| > 2^{-n-p},$$
$$2^{-n-p} \bar{V}(h^x; \mathbf{A}^x; X_{np}^x) \leq \bar{V}(h^x \| f_{n+1}(x) - f_n(x)\|; \mathbf{A}^x; E^x; X_{np}^x) \leq 2 \cdot 4^{-n},$$
$$\bar{V}(h^x; \mathbf{A}^x; E^x; X_{np}^x) \leq 2^{1-n+p},$$
$$\bar{V}(h^x; \mathbf{A}^x; E^x; W_p^x) \leq 2^{1+p} \quad \left(W_p^x \equiv \bigcup_{n=1}^{\infty} X_{np}^x \right).$$

As X^x has h^x-variation zero, and putting

$$s(x) = 2^{-1-2p}(x \in W_p^x \setminus (W_1^x \cup \cdots \cup W_{(p-1)}^x))$$

(5.1.11) $\quad \bar{V}(sh^x; \mathbf{A}^x; E^x; S) \leq 1, \quad S \equiv \bigcup_{p=1}^{\infty} W_p^x \cup X^x = \bigcup_{n,p=1}^{\infty} X_{np}^x \cup X^x.$

If $x \notin S$ then $x \notin X_{np}^x$ (all n, p) and so

$$\|f_{n+1}(x) - f_n(x)\| \leq 2^{-n-p} (\text{all } n, p), \quad \|f_{n+1}(x) - f_n(x)\| = 0,$$
$$f_n(x) = f_1(x) = J(x) \quad (x \notin S, \text{ all } n)$$

By (5.1.10), given $\varepsilon > 0$, in $S \setminus X^*$ we choose the integer $t = t(x)$ so large that

(5.1.12) $$\|f_{t(x)}(x) - J(x)\| < s(x)\varepsilon,$$

while in $\setminus S$ we take $t(x) = 1$. Thus (5.1.12) is satisfied in $(E^*)^* \setminus X^*$, in which we take $\mathscr{E}_{t(x)} y_x$, and in X^* we take $\mathscr{E}_1^y(x)$. By the Fubini property there is a suitable $\mathscr{U}^*$, and by direction we can also assume that for all divisions $\mathscr{E}^*$ of E^* from $\mathscr{U}^*$,

$$(\mathscr{E}^*)\sum s(x)\|h^*(I^*, x)\| < 1 + \varepsilon,$$

$$(\mathscr{E}^*)\sum \|h^*(I^*, x)\| \cdot \|f_1(x) - J(x)\| \chi(X^*, x) < \varepsilon;$$

$$(\mathscr{E}^*)\sum \|h^*(I^*, x)\| \cdot \|f_{t(x)}(x) - J(x)\| \leq (\mathscr{E}^*)\sum s(x)\|h^*(I^*, x)\|\varepsilon + \varepsilon$$

$$< (2 + \varepsilon)\varepsilon,$$

(5.1.13) $$(\mathscr{E}^*)\sum \|J(x)h^*(I^*, x) - H(I^* \otimes E^y)\| < (3 + \varepsilon)\varepsilon.$$

Hence Jh^x is strong variationally integrable on E^x to $H(E^x \otimes E^y)$. □

The form (5.1.2) seems to be the most general for which the conclusions of Theorem 5.1.2 can be proved. However, if part of the conclusions is assumed, with a special condition, we can use general h.

Theorem 5.1.3. *Let $(T^z, \mathscr{T}^z, \mathbf{A}^z)$ be a stable decomposable Fubini division space, let $h((I^x, x) \otimes (I^y, y))$, with values in K, be defined in $\mathscr{U}^{1z}$ and strong variationally integrable to H in $E^z = E^x \otimes E^y$. For each $(I^x, x) \in \mathscr{U}^{1x}$ let the strong variational integral relative to (I^y, y),*

$$G(I^x, x) \equiv \int_{E^y} dh((I^x, x) \otimes (I^y, y)),$$

exist. Let the real-valued $m(I^x, x) \geq 0$ be VBG in E^x. If, given $\varepsilon > 0$, there are $\mathscr{U}^y(x) \in \mathbf{A}^y | E^y$ such that all divisions $\mathscr{E}^y(x)$ over E^y from $\mathscr{U}^y(x)$, and $\mathscr{U}^x \in \mathbf{A}^x | E^x$, connected by the Fubini property, satisfy*

(5.1.14) $$\|G(I^x, x) - (\mathscr{E}^y(x))\sum h((I^x, x) \otimes (I^y, y))\| \leq \varepsilon m(I^x, x)((I^x, x) \in \mathscr{U}^x),$$

then

$$H = \int_{E^x} dG(I^x, x).$$

If m is of bounded norm variation we can omit decomposability.

Note that full decomposability enables us to stitch together families of $\mathscr{U}^x$ that differ from point to point. But if they also differ as the I^x varies, the resulting families stitched together would not normally be of the same kind as the $\mathscr{U}^x \in \mathbf{A}^x$. Thus it seems unlikely that one could in general obtain an inequality like (5.1.14) with $\varepsilon m(I^x, x)$ replaced by a suitable small sum that occurs since $G(I^x, x)$ exists. The rapidity of convergence of the sum to the integral may change discontinuously as I^x changes with fixed x.

Proof of theorem. As m is VBG*, there is an m_1 of bounded variation with $0 \leq m \leq nm_1$ for the integer-valued function $n = n(x)$. By decomposability and choice of $\varepsilon > 0$ we can replace m by m_1, so that we can assume that m is of bounded variation. Then given the $\mathscr{U}^y(x)$ satisfying the Fubini property, and also that all divisions $\mathscr{E}^y(x)$ of E^y from $\mathscr{U}^y(x)$ satisfy (5.1.14), with the allied $\mathscr{E}^x$ also satisfying

$$(\mathscr{E}^x)\sum m \leq \bar{V}(m; \mathbf{A}^x; E^x) + 1,$$

$$\|(\mathscr{E}^x)\sum G - H\| \leq \|(\mathscr{E}^x)\sum G - (\mathscr{E}^x)\sum(\mathscr{E}^y(x))\sum h\|$$
$$+ \|(\mathscr{E}^x)\sum(\mathscr{E}^y(x))\sum h - H\|$$
$$\leq (\mathscr{E}^x)\sum \varepsilon m + \varepsilon \leq \{2 + \bar{V}(m; \mathbf{A}^x; E^x)\}\varepsilon,$$

and hence the theorem. □

Theorem 5.1.2 appeared first in Henstock (1961b), and the proof has been refined greatly since then. See also Kurzweil (1973a). Theorem 5.1.3 is due to T. W. Lee (1971/1972). In Lebesgue integration Theorem 5.1.2 in the symmetrical form is essentially due to Fubini (1907), and the unsymmetrical form is due to Cameron and Martin (1941) and Robbins (1948). See also Bosanquet (1930a).

5.2 TONELLI-TYPE RESULTS ON THE REVERSAL OF ORDER OF DOUBLE INTEGRALS

We now come to the inverted limits of (3.1.3) and the specialization of Theorems 3.1.3, 3.1.4 to the case of repeated integrals, noting also Theorem 3.1.2. We use two fully decomposable division spaces $(T_j, \mathscr{T}_j, \mathbf{A}_j)$ $(j = 1, 2)$ with $T = T_1 \otimes T_2$, $\mathscr{T} = \mathscr{T}_1 \otimes \mathscr{T}_2$. We need not consider the product division space nor integration on T, though the existence of the integral over an elementary set of T is a useful criterion for the reversal of order of integration. The example of Sierpiński (1920) is not relevant here, the double integrals of the characteristic function in the plane can be reversed even though the set is non-measurable in the plane.

For elementary sets E_j relative to T_j $(j = 1, 2)$ we consider the product set $E = E_1 \otimes E_2$. The general elementary set in T is a finite union of such disjoint product sets, but its introduction here would be irrelevant to our purpose. As we are reversing the order of integration we take products of functions like (5.1.4). We take the integration using $(T_1, \mathscr{T}_1, \mathbf{A}_1)$ as the integration in Theorems 3.1.3, 3.1.4, and the integration using $(T_2, \mathscr{T}_2, \mathbf{A}_2)$ as the limit process. Then if we reverse the roles of the two integrals we will find that the conditions are interchanged. Note that real-valued functions are also complex-valued with the imaginary part 0, so that instead of saying $\mathbb{R}$ or $\mathbb{C}$ for the value space, we just say, $\mathbb{C}$.

Theorem 5.2.1. *For the arrangement outlined above let $h_j: \mathcal{U}_j \to \mathbb{C}$ be VBG* $(j = 1, 2)$ and for the complex-valued $f(t_1, t_2)$ and $(I_j, t_j) \in \mathcal{U}_j (j = 1, 2)$ let $f(t_1, t_2) h_1(I_1, t_1)$ be integrable over E_1 for each fixed $t_2 \in E_2^*$, and let $f(t_1, t_2) h_2(I_2, t_2)$ be integrable over E_2 for each fixed $t_1 \in E_1^*$, where $\mathcal{U}_j \in \mathbf{A}_j | E_i$ $(j = 1, 2)$. Then*

(5.2.1) $$\int_{E_1} \left\{ \int_{E_2} f(t_1, t_2) dh_2 \right\} dh_1 = F$$

exists if and only if there are a circle C with centre F and arbitrarily small radius, some function $M: \mathcal{U}^1 \to \mathbf{A}_2 | E_2$, some $\mathcal{U}_1 \in \mathbf{A}_1 | E_1$, and all divisions $\mathscr{E}_1$ of E_1 from $\mathcal{U}_1$, such that

(5.2.2) $$(\mathscr{E}_1) \sum (\mathscr{E}_2(I_1, t_1)) \sum f(t_1, t_2) h_1(I_1, t_1) h_2(I_2, t_2) \in C$$

(all $\mathscr{E}_2(I_1, t_1)$ from $M(I_1, t_1)$). Given (5.2.2), a necessary and sufficient condition that also there exists

(5.2.3) $$\int_{E_2} \left\{ \int_{E_1} f(t_1, t_2) dh_1 \right\} dh_2 = F$$

is that there are a $\mathcal{U}_2 \in \mathbf{A}_2 | E_2$, another circle C_1 with centre F and arbitrarily small radius, and a $\mathcal{U}_1(I_2, t_2) \in \mathbf{A}_1 | E_1$, with

(5.2.4) $$(\mathscr{E}_2) \sum (\mathscr{E}_1)(I_2, t_2)) \sum f(t_1, t_2) h_1(I_1, t_1) h_2(I_2, t_2) \in C_1$$

(all $\mathscr{E}_2$ from $\mathcal{U}_2$ and all $\mathscr{E}_1(I_2, t_2)$ from $\mathcal{U}_1(I_2, t_2)$).

Clearly, if we take the integration using $(T_2, \mathcal{T}_2, \mathbf{A}_2)$ as the integration and the integration using $(T_1, \mathcal{T}_1, \mathbf{A}_1)$ as the limit process, all that happens is that (5.2.2) and (5.2.4) are interchanged.

For a weaker test on the existence and equality in (5.2.1), (5.2.3), a good test is the existence of the integral over the Cartesian product $E_1 \otimes E_2$, see the Fubini-type Theorem 5.1.2. For then the repeated integral one way is equal to the integral over the Cartesian product, which in turn is equal to the repeated integral the other way.

Considering other tests, monotonicity does not seem relevant, but the analogue of the bounded Riemann sums test seems helpful.

Theorem 5.2.2. *Let $h_1: \mathcal{U}_1^1 \to \mathbb{R}$, $h_2: \mathcal{U}_2^1 \to \mathbb{R}$, $f: T \to \mathbb{R}$, and for each fixed t_2, fh_1 is integrable over E_1, while for each fixed t_1, fh_2 is integrable over E_2. Let $B < C$ be two real numbers, and let $\mathcal{U}_j \in \mathbf{A}_j | E_j (j = 1, 2)$. Then (5.2.1) and (5.2.3)* *exist and are equal when for every division $\mathscr{E}_1$ of E_1 from $\mathscr{U}_1$ and every division $\mathscr{E}_2(I_1, t_1)$ of E_2 from $\mathscr{U}_2$ satisfy*

(5.2.5) $\quad B \leq (\mathscr{E}_1) \sum (\mathscr{E}_2(I_1, t_1)) \sum f(t_1, t_2) h_1(I_1, t_1) h_2(I_2, t_2) \leq C$

and the similar inequalities on changing $\mathscr{E}_1$ to $\mathscr{E}_2$ and $\mathscr{E}_2(I_1, t_1)$ to $\mathscr{E}_1(I_2, t_2)$.

This theorem is very similar to Theorem 5.2.1, except that the $\mathscr{U}_1, \mathscr{U}_2$ are fixed and do not vary according to divisions in the other space, and also $[B, C]$ is fixed and not arbitrarily small.

When $h_j \geq 0$ ($j = 1, 2$) these conditions hold if for $g \colon T_2 \to \mathbb{R}^+$ and $k \colon T_1 \to \mathbb{R}^+$ with gh_2 integrable over E_2, kh_1 integrable over E_1 and $|f(t_1, t_2)| \leq g(t_2)$ (all $t_1 \in E_1^*$), $|f(t_1, t_2)| \leq k(t_1)$ (all $t_2 \in E_2^*$).

The necessary and sufficient conditions first appeared in Henstock (1988a). Conditions in Lebesgue form appear in Tonelli (1923, 1926).

CHAPTER 6

INTEGRATION IN INFINITE-DIMENSIONAL SPACES

6.1 INTRODUCTION

In countable Cartesian product spaces (sequence spaces) a result of Jessen (1934) holds, so that the integral is the limit of the corresponding integral over the first n terms of the sequences. The origins of integration in uncountable Cartesian product spaces (function spaces) are in Einstein (1906) and Smoluchowski (1906), examples being given by Chandrasekhar (1943). Two integrals vital for theoretical physicists and theoretical chemists are in Wiener (1930) and Feynman (1948). It may be that the integral of this chapter gives mathematical expression to the latter; other expressions used are far too complicated. Again we can sometimes prove that the integral is the limit of integrals over spaces of finite dimensions. See also the account given in Muldowney (1987). This chapter goes into more detail of the pure mathematics.

We use paths $x(b)$ parameterized by a real variable b lying in an interval $[a, c)$ where c is often $+\infty$, and for generalized interval we use the set of those paths that, when $b = b_j$, lie in the range

$$u^*(b_j) = u(j) \leqslant x(b_j) < v^*(b_j) = v(j),$$
$$u(j) < v(j)(1 \leqslant j \leqslant n), \qquad a < b_1 < \cdots < b_n < c.$$

A function of such generalized intervals I is the integral

(6.1.1) $$P(I) \equiv \int_{u(1)}^{v(1)} \cdots \int_{u(n)}^{v(n)} p(x_1, \ldots, x_n; C)\,dx_1 \cdots dx_n$$

$$(C = (b_1, \ldots, b_n)),$$

(6.1.2) $$p(x_1, \ldots, x_n; C) \equiv q(x_1; b_1 - a|x_0)$$
$$\times q(x_2; b_2 - b_1|x_1) \cdots q(x_n; b_b - b_{n-1}|x_{n-1}),$$

with q obeying a consistency condition of Smoluchowski, Chapman, and Kolmogorov, namely that for each s in $0 < s < t$,

(6.1.3) $$q(y; t|x) = \int_{-\infty}^{+\infty} q(z; s|x)q(y; t - s|z)\,dz.$$

For example, the probability measure for the path of a free spherical Brownian particle starting at time $b = a$ from $x = 0$, can be constructed taking $x_0 = 0$ and

(6.1.4) $\quad q(y; t|x) = g(y - x; 4Dt), \quad g(x; t) = (\pi t)^{-1/2} \exp(-x^2/t).$

The diffusion constant D varies with the viscosity and temperature of the medium and the radius of the Brownian particle. The probability measure is only an approximation and should not be pushed too far. For $D = \frac{1}{2}$, $P(I) = W(I)$, the Wiener measure of I. Feynman (1948) used complex-valued q, one being given by (6.1.4) with $4D = i$. The integral is then very difficult to develop, so highly complex along each dimension b, with a continuum of b. Limits under the integral sign are difficult to find, though there may be hope that the test involving bounded Riemann sums might be useful. However, there is no difficulty with Wiener measure and integration, for which q is positive. The intervals at $b = b_1, \ldots, b_n$ are just like the edges of an n-dimensional rectangle or brick, and the whole collection of paths is the definition of the Cartesian product of $(-\infty, +\infty)$, one real axis for each b.

6.2   DIVISION SPACE INTEGRATION

To put such details in the setting of this book, we suppose that for each b of an index set B we have a stable fully decomposable division space $(T(b), \mathcal{T}(b), \mathbf{A}(b))$. For the Cartesian product T of $T(b)$ $(b \in B)$ we need a suitable $(T, \mathcal{T}, \mathbf{A})$. For each $C \subseteq B$ let

$$\prod(X(b); C), \quad X(b) \subseteq T(b) \, (b \in C), \quad X(b) = T(b) \, (b \in B \backslash C),$$

be the Cartesian product of the given $X(b)$. Let $X(C)$ be the Cartesian product of $X(b)$ for $c \in C$ alone, so that $T(C)$ comes from $T(b)$. We take the intervals of T to be $I = \prod (I(b); C)$ for all finite sets $C \subseteq B$ and all $I(b) \in \mathcal{T}(b)$ $(b \in C)$, while $\mathcal{U}^1(C)$, $\mathcal{U}^1$ are the Cartesian products of $(I(b), f(b))$ for all $b \in C$, all $b \in B$, respectively. Further, with $f: B \to K$ we write $f^S: S \to K$ when $S \subseteq B$ and $f^S(b) = f(b)(b \in S)$.

(6.2.1) $\quad\quad\quad\quad\quad E(C)^* = E^*(C) \quad (\text{finite } C \subseteq B),$

from (5.1.1), used repeatedly, when $E(b)$ is an elementary set or is $T(b)$ for each $b \in B$. For simplicity we assume that

(6.2.2) $T(b)$ is an elementary set relative to $\mathcal{T}(b)$, for each $b \in B$.

Section 5.1 gives the product division space $(T(C), \mathcal{T}(C), \mathbf{A}(C))$ for each finite $C \subseteq B$ while in T we construct families $\mathcal{U}$ dividing elementary sets E, associating a finite set $C_1(\mathcal{U}, f) \subseteq B$ with each f with $f(b) \in T(b)^*$ $(b \in B)$, and

6.2 DIVISION SPACE INTEGRATION

a $\mathcal{U}_1(C; \mathcal{U}; f) \in A(C)$ for each finite $C \subseteq B$. Then $\mathcal{U}$ is the family of all (I, f) with

(6.2.3) $\quad I = \prod(I(b); C) \subseteq E$, C finite, $C_1(\mathcal{U}, f) \subseteq C \subseteq B$;

(6.2.4) $\quad \bigotimes_c (I(b), f(b)) \in \mathcal{U}_1(C; \mathcal{U}; f)$;

(6.2.5) $\quad f(b) \in T(b)^* \quad (b \in B \setminus C)$.

For **A** the family of all such $\mathcal{U}$, then $(T, \mathcal{T}, \mathbf{A})$ is the product division space of $(T(b), \mathcal{T}(b), \mathbf{A}(b))$ $(b \in B)$. Note that there might not be order in B, and the $b \in B$ can be rearranged without affecting the construction. When we take limits in B, it appears that convergent series of real or complex terms can be rearranged without affecting the convergence, so that they must be absolutely convergent (by a result of Dirichlet) and convergent series of Banach space values have to be unconditionally convergent (unless a reasonable order is given in B).

Theorem 6.2.1. *The above $(T, \mathcal{T}, \mathbf{A})$ is fully decomposable and so stable, and*

$$I = \prod(I(b); C) \text{ implies } I^* = \bigotimes_B I(b)^* \equiv I^*(B).$$

Proof. Full decomposability follows since $(T(C), \mathcal{T}(C), \mathbf{A}(C))$ is fully decomposable. By Theorem 2.9.1(2.9.4), stability follows. Now let $f \in I^*(B)$. Given $\mathcal{U}$, by (6.2.5) we need only consider $b \in C$. By (6.2.n) ($n = 1, 3, 4$) and the stability of $\mathbf{A}(b)$, there are $J(b) \in \mathcal{T}(b)$ with $J(b) \subseteq I(b)$,

$$\prod(J(b); C) \subseteq I, \bigotimes_c (J(b), f(b)) \in \mathcal{U}_1(C, \mathcal{U}, f),$$
$$(\prod(J(b); C), f) \in \mathcal{U}, \quad f \in I_{\mathcal{U}}^*.$$

As true for all $\mathcal{U}$ of the above kind, $f \in I^*$. Conversely, let $\mathcal{U}(b)$ be such that $I_{\mathcal{U}(b)}^*(b) = I^*(b)$ by stability, let $\mathcal{U}_1(C; \mathcal{U}; f) \subseteq \bigotimes_c \mathcal{U}(b)$, and let $\mathcal{U}$ be formed from these $\mathcal{U}_1(C; \mathcal{U}; f)$. If $f \in I^*$ there is a

$$J = \prod(J(b); C_2) \subseteq (I(b); C)$$

with $(\prod(J(b); C_2), f) \in \mathcal{U}$; thus $C_2 \supseteq C \cup C_1(\mathcal{U}, f)$,

$$J(b) \subseteq I(b).$$

By (6.2.n) ($n = 1, 4, 5$),

$$f(b) \in J(b)^* \subseteq I(b)^* \quad (b \in B), \ f \in I^*(B), \text{ so } I^*(B) = I_{\mathcal{U}}^*$$

independent of $\mathcal{U}$, and $I^* = I^*(B)$. □

Next, assuming an abstract version of continued bisection, we prove that $\mathcal{U}$ divides every interval and so every elementary set.

Theorem 6.2.2. Let $J = \prod(J(b); C) \in \mathcal{T}$ and let a sequence $(\mathcal{P}_n(b))$ of partitions of $T(b)$ exist for each $b \in B$ independent of $\mathcal{U}$, such that

(6.2.6) if $I(b) \in \mathcal{P}_1(b)$ then either $I(b) \subseteq J(b)$ or $I(b) \cap J(b)$ is empty;

(6.2.7) $\mathcal{P}_{n+1}(b)$ is a refinement of $\mathcal{P}_n(b)$ ($\mathcal{P}_{n+1}(b) \leqslant \mathcal{P}_n(b)$) for $n = 1, 2, \ldots$;

(6.2.8) for each $f \in T$ and each $C_2 \supseteq C_1(\mathcal{U}, f)$, there is an integer N depending on $\mathcal{U}, f$, and $\mathcal{U}_1(C_2; \mathcal{U}; f)$, and so on $\mathcal{U}, f, C_2$, such that if $n \geqslant N$, $I(b) \in \mathcal{P}_n(b)$, $I(b) \subseteq J(b), f(b) \in I(b)^*$, then $I(b)$ is an interval $L(b)$ or a finite union of disjoint intervals $L(b)$ ($b \in C_2$) with

$$\bigotimes_{C_2}(L(b), f(b)) \in \mathcal{U}_1(C_2; \mathcal{U}; f);$$

(6.2.9) let $(T(b), \mathcal{T}(b), \mathbf{A}(b))$ be strongly $\mathcal{T}(b)$-compatible with every partial set of $T(b)$ ($b \in B$).

Then $\mathcal{U}$ divides J.

Proof. Supposing J not divided by $\mathcal{U}$, we use

(6.2.10) if $\mathcal{P}$ is a partition of an elementary set E not divided by $\mathcal{U}$, an $I \in \mathcal{P}$ is not divided by $\mathcal{U}$.

Otherwise the union of divisions from $\mathcal{U}$ of the $I \in \mathcal{P}$ would be a division of E. For finite C_2 in $C \subseteq C_2 \subseteq B$, fixed n, and all $I(b) \in \mathcal{P}_n(b)$ ($b \in C_2$) the $\prod(I(b); C_2)$ form a partition $\mathcal{P}_n(C_2)$ of T. By (6.2.6), (6.2.7), $\mathcal{P}_{n+1}(C_2) \leqslant \mathcal{P}_n(C_2)$, and each $I = I(C_2) \otimes T(B \setminus C_2)$ with $I(C_2) \in \mathcal{P}_n(C_2)$, has $I \subseteq J$ or $I \cap J$ empty. By (6.2.10) with $E = J$, there is an $I \in \mathcal{P}_1(C_2)$ not divided by $\mathcal{U}$, with $I \subseteq J$. If $I \in \mathcal{P}_n(C_2)$ not divided by $\mathcal{U}$, for some n, with $I \subseteq J$, there is by (6.2.10) an $E \in \mathcal{P}_{n+1}(C_2)$ not divided by $\mathcal{U}$, with $E \subseteq I \subseteq J$. Thus by mathematical induction the set $\mathcal{Q}_n(C_2)$ of all $I \in \mathcal{P}_n(C_2)$ not divided by $\mathcal{U}$, is not empty, and that for a finite number of n and C_2, the intersection of the $\mathcal{Q}_n(C_2)$ is a non-empty elementary set. Now by (6.2.9), $(T(b), \mathcal{T}(b), \mathbf{A}(b))$ is a division space weakly $\mathcal{T}(b)$-compatible with every partial set of $T(b)$, and so by Theorem 2.9.1(2.9.3) every pair of disjoint partial sets which are therefore co-partitional, also satisfy

$$(P_1(b) \cup P_2(b))^* = P_1(b)^* \cup P_2(b)^*$$

for those two partial sets. Thus if $Q_n(C_2)$ is the union of those $I \in \mathcal{Q}_n(C_2)$ then the corresponding finite number of $Q_n(C_2)^*$ also has a non-empty intersection. So by (6.2.9) and Theorem 1.9.4, $T(b)$ is compact in the intrinsic topology for each $b \in B$. By Tychonoff's theorem on Cartesian products, T is compact in the intrinsic topology, and hence for all n, C_2, the intersection of the $Q_n(C_2)^*$ is not empty and contains g, say. Thus there is a $C_1(\mathcal{U}, g)$, (6.2.8) applies, and we have a contradiction. Hence the theorem. $\square$

Theorem 6.2.3. In Theorem 6.2.2, $(T, \mathcal{T}, \mathbf{A})$ is a stable fully decomposable division space, additive if every $(T(b), \mathcal{T}(b), \mathbf{A}(b))$ is additive.

6.2 DIVISION SPACE INTEGRATION

Proof. After Theorems 6.2.1, 6.2.2, for stability, full decomposability, and the division of each partial set of T, we show the direction in the sense of divisions. It is given for each $\mathbf{A}(b)|E(b)$, and so holds for each $\mathbf{A}(C)|E(C)$ for finite $C \subseteq B$. If $\mathcal{U}_j \in \mathbf{A}|E$ with $C_1(\mathcal{U}_j, f) \subseteq B$ ($j = 2, 3$) take

$$C_1(\mathcal{U}_4, f) = C_1(\mathcal{U}_2, f) \cup C_1(\mathcal{U}_3, f),$$
$$\mathcal{U}_1(C, \mathcal{U}_4, f) \subseteq \mathcal{U}_1(C, \mathcal{U}_2, f) \cap \mathcal{U}_1(C, \mathcal{U}_3, f).$$

Then (6.2.3), (6.2.4) show that $\mathbf{A}|E$ is directed in the sense of divisions. The restriction property and that involving partial sets of an elementary set E, are true for each $b \in B$, and easy proofs give the result for T. Similarly for additivity if true for each $b \in B$. □

Theorem 6.2.4. *In Theorem 6.2.2 let $B = H \cup K$, $H \cap K$ empty, H and K not empty, and put*

$$V = \bigotimes_H T(b), \quad W = \bigotimes_K T(b).$$

Then $(T, \mathcal{T}, \mathbf{A})$ is a Fubini division space for V, W.

Proof. Although the $b \in H$ may interleave with the $b \in K$ we can rearrange the b so that the $b \in H$ are first, and then the $b \in K$. As noted before, this does not affect the construction. For fixed $v \in V$, some $w \in W$, and points $f = (v, w)$ there are intervals $I_v \subseteq V$, $I_W \subseteq W$ with $(I_v \otimes I_W, (v, w)) \in \mathcal{U}$. Here, for some finite $C_2 \supseteq C_1$ $(\mathcal{U}, f)$ with $C_2 \subseteq B$ and $C_2 \cap H$, $C_2 \cap K$ non-empty, (6.2.4), and Cartesian products $\prod_H$ over H, $\prod_K$ over K,

$$I_V = \prod_H (I(b); C_2 \cap H), \quad I_W = \prod_K (I(b); C_2 \cap K).$$

As the (I_W, w) form a $\mathcal{U}_W$ for W like $\mathcal{U}$ for T, Theorem 6.2.2 ensures a division $\mathscr{E}(v)$ of W with intervals $J_W \subseteq W$ and associated points, say, $w_1, \ldots, w_p$. The union of $C_1(\mathcal{U}; (v, w_j))$ ($j = 1, \ldots, p$) is a finite subset $C_3(\mathcal{U}, v)$ of B, with a $\mathcal{U}_3(C; v) \in \mathbf{A}(C)|T(C)$ and lying in the intersection of the $\mathcal{U}_1(C; \mathcal{U}; (v, w_j))$ ($j = 1, \ldots, p$). Hence we have a $\mathcal{U}_V$ of (I_V, v) like $\mathcal{U}$ for T, with

$$I_V = \prod_H (I(b); C_4 \cap H), \quad \bigotimes_{C_4} (I(b), v(b)) \in \mathcal{U}_3(C_4; v)$$

for a finite $C_4 \subseteq B$, $C_4 \supseteq C_3(\mathcal{U}; v)$, $C_4 \cap H$ and $C_4 \cap K$ not empty. Again by Theorem 6.2.2, $\mathcal{U}_V$ divides V and the first Fubini property holds. Similarly for the second.

To use Theorem 5.1.2 for every pair H, K of sets in Theorem 6.2.4 we need (5.1.4), extended to more than two variables. Thus if the finite $C \subseteq B$, $I = \prod(I(b); C)$, and $C(I)$ the smallest subset of B with $I = \prod(I(b); C(I))$,

h needs to have the form

$$g(C(I); f^{C(I)}, f^{B\setminus C(I)})k(C(I); I(C(I)), f^{C(I)})h^*(I(B\setminus C(I)) \cdot f^{B\setminus C(I)}),$$

(6.2.11) $$k(C; I(C), f^C) \equiv \prod_{b \in C} k^b(I(b), f(b)).$$

The k of (6.2.11) can have C infinite if the product of k^b has a reasonable meaning. But $I(B\setminus C(I)) = T(B\setminus C(I))$ and h^* is just a point function of $B\setminus C(I)$ and can go into g, and the most general form is

(6.2.12) $$h(I, f) = g(C(I); f)k(C(I); I(C(I)), f^{C(I)}).$$

For uniqueness, if $b \notin C$ and $I(b) = T(b)$ let $D = C \cup \text{sing}(b)$. Then

$$I \equiv \prod (I(b); C) = \prod (I(b); D),$$

$$\int_I dk(C; I, f) = \int_I dk(D; I, f)$$

$$= \int_I \left\{ \int_{T(b)} dk^b(I(b), f(b)) \right\} dk(C; I, f),$$

(6.2.13) $$\int_{T(b)} dk^b(I(b), f(b)) = 1,$$

the Smoluchowski–Chapman–Kolmogorov relation, $k(C; I, f)$-almost everywhere in $T(C)$. We have used Theorems 5.1.2, 4.1.1(4.1.7). As the integral in (6.2.13) is independent of $T(C)$, (6.2.13) must be true everywhere in $T(C)$, except in the very special case when the variation of $k^b(I(b), f(b))$ is 0 over $T(b)$ ($b \in C$), when the integral of (6.2.12) is 0 over every interval, which case we will ignore.

For K a space of values, a function $F: T \to K$ is said to be *cylindrical of order* $C \subseteq B$, if $F(f) = F(g)$ when $f^{B\setminus C} = g^{B\setminus C}$; and F is cylindrical of every finite order, if cylindrical of order C for each finite $C \subseteq B$.

Theorem 6.2.5. *In Theorem 6.2.2 with (6.2.n) ($n = 11, 12, 13$),*

(6.2.14) *a function F on T that is cylindrical of every finite order, with Fk strong variationally integrable, is constant k-almost everywhere;*

(6.2.15) *if the indicator χ of a set $Y \subseteq T$ is cylindrical of every finite order, with χk integrable, then the k-variation of either Y or $\setminus Y$, but not both, is 0;*

(6.2.16) *let K contain a sequence $(\mathscr{X}_j)$ of countable families of mutually disjoint sets such that for each j, K is the union of sets $X_{jk} \in \mathscr{X}_j$ and such that if $m \neq p$ in K there are j, k such that X_{jk} contains m but not p. A function $F: T \to K$ that is cylindrical of every finite order, with χk integrable for the indicator χ of each set $F^{-1}(X_{jk})$, is constant k-almost everywhere.*

Proof. With G, M the integrals of Fk, k, respectively, and $I = I(C) \otimes T(B \setminus C)$ for a finite $C \subseteq B$, Theorem 5.1.2 (Fubini) and (6.2.13) give

$$G(I) = M(I) \int_{T(B \setminus C)} F \, dk(B \setminus C; I, f), \quad M(T) = 1,$$

$$G(T) = \int_{T(B \setminus C)} F \, dk(B \setminus C; I, f),$$

(6.2.17) $\quad G(I) = M(I)G(T),$

since F is cylindrical of order C. As C can be any finite subset of B, and by (6.2.17) $G(I) - G(T)M(I)$ has k-variation zero. By variational equivalence, $(F(f) - G(T))k$ has variation zero and $F(f) - G(T) = 0$ k-almost everywhere, giving (6.2.14). By this, in (6.2.15), $\chi = 0$ k-almost everywhere or $\chi = 1$ k-almost everywhere, while $K(T) = 1$. But if the k-variation over Y and $\setminus Y$ are both 0, then the k-variation over T is 0 and $K(T) = 0$, contradiction. Hence (6.2.15). Thus in (6.2.16) there is a $k(j)$ for each j such that $\setminus F^{-1}(X_{jk(j)})$ has k-variation 0. Thus

$$\bigcup_{j=1}^{\infty} \setminus F^{-1}(X_{jk(j)}) = \setminus \bigcap_{j=1}^{\infty} F^{-1}(X_{jk(j)}) = \setminus F^{-1} \bigcap_{j=1}^{\infty} X_{jk(j)}$$

has k-variation 0, and the last intersection contains a point $m \in K$. If $p \neq m$ there are j, k such that X_{jk} contains m but not p. For each j the X_{jk} are mutually disjoint in k. Hence $k = k(j)$, p is not in the intersection, and the intersection contains m alone. Thus $F = m$ k-almost everywhere. □

We now look at the integral over finite dimensions, given that the integral over the whole space T exists, with the aim to have a limiting result like that of Jessen (1934).

Theorem 6.2.6. *Let $(6.2.n)$ $(n = 11, 12, 13)$ and the conditions of Theorem 6.2.2 hold, with J replaced by T. Let h be strong variationally integrable to G. Then for finite $C \subseteq B$, the strong variational integral of h over any interval J of $T(C)$, written $G(B \setminus C; J, f)$, exists for $f^{B \setminus C} k(B \setminus C; I(B \setminus C), f^{B \setminus C})$-almost everywhere. If we put $G(B \setminus C; J, f) = 0$ where it is otherwise undefined, then the integral of $G(B \setminus C; T(C), f)k(B \setminus C; I(B \setminus C), f^{B \setminus C})$ ever $T(B \setminus C)$ is equal to $G(T)$.*

Proof. Apply Theorem 5.1.2 (Fubini). □

Here, $H(C, f) \equiv G(B \setminus C; T(C), f)$ is an integral over the finite dimensional space $T(C)$, the integral being called a *backwards martingale* when k is a probability measure. For if C is finite and

$$S \subset C \subseteq B, \quad H(C, f) = \int_{T(C \setminus S)} H(S, f) \, dk(C(I), I(C(I)), f^{C(I)}),$$

by Theorem 5.1.2 (Fubini) again. Thus $H(C,f)$ is the conditional expectation of $H(S,f)$. Probability theory studies the convergence of $H(C,f)$ to $G(T)$ as C rises in B, the first paper being Jessen (1934), pp. 273–5. As already noted, if B is uncountable so that the members of any countably infinite subset can be taken in any order, convergence along that countable subset must have an absoluteness corresponding to absolutely convergent series. Thus the restrictions imposed may not be as drastic as may otherwise appear.

Theorem 6.2.7. *Assuming (6.2.2), that T is an elementary set, and taking $K = \mathbb{R}$, so that G is real-valued, let gk and $|gk|$ be integrable over T where (6.2.n) ($n = 11, 12, 13$) hold and the conditions of Theorem 6.2.2 hold, with J replaced by T. Then*

(6.2.18) $$\lim_{C \subseteq B}{}^* G(B\setminus C; T, f) = G(T)$$

for f k-almost everywhere in T^*.

Proof. Let $\mathcal{U}_j \in \mathbf{A}$ be such that every division $\mathcal{E}$ over T from $\mathcal{U}_j$ satisfies

(6.2.19) $$|(\mathcal{E})\sum gk - G(T)| < 1/j.$$

By direction in $\mathbf{A}$ we can take the $\mathcal{U}_j$ monotone decreasing in j. By mathematical induction we can choose a sequence $(\mathcal{E}_j)$ of divisions of T such that $\mathcal{E}_j$ comes from $\mathcal{U}_j$, and that $\mathcal{E}_{j+1}$ refines $\mathcal{E}_j$. Let $C_j = C_0(\mathcal{E}_j)$ be the finite set of all $b \in B$ for which at least one $(I, f) \in \mathcal{E}_j$ has $b \in C(I)$. Then by the refinement property, (C_j) is monotone increasing with $B^* \equiv \bigcup_j C_j$, at most countable. For fixed j the divisions $\mathcal{E}_j^+$ of T obtained from $\mathcal{U}_j$ by restricting the $C(I)$ to lie in B^* (This restricts the kinds of I but not the f) include $\mathcal{E}_j$ and are a subset of all divisions of T from $\mathcal{U}_j$, and so satisfy (6.2.19). The corresponding restricted integral has the same value $G(T)$. Writing the restricted integral with (A^*) before the integral sign, and B^* as (b_j), $C^n = \{b_1, \ldots, b_n\}$, $G_n(J, f) = G(B\setminus C^n; J, f)$, we use Jessen's proof in this more general situation. Let $X_N(M)$ be the set of $f \in T$ for which

$$\sup_{1 \leq n \leq N} G_n(T, f) > M,$$

let Y_n be the set of f where $G_n(T, f) > M$, and let

$$Z_N = Y_N, \quad Z_n = Y_n \setminus (Y_N \cup \cdots \cup Y_{n+1}) \quad (1 \leq n < N).$$

Then each Y_n is cylindrical of order C^n, and the same is true of Z_n as $N, N-1, \ldots, n+1$ are all greater than n. Clearly the Z_n are disjoint with $Z_n \subseteq Y_n$, and by the usual proofs they are k-measurable ($n = 1, \ldots, N$).

6.2 DIVISION SPACE INTEGRATION

Hence for $V(n) = T(B\setminus C^n)$,

$$\int_T \chi(Z_n;f)g\,dk = \int_{V(n)} \chi(Z_n;f)G_n(T,f)\,dk$$

$$\geq M \int_{V(n)} \chi(Z_n;f) = M \int_T \chi(Z_n;f)\,dk.$$

$$X_N(M) = Y_N \cup Y_{N-1} \cup \cdots \cup Y_1 = Z_N \cup Z_{N-1} \cup \cdots \cup Z_1,$$

and by addition for $n = 1, \ldots, N$ we have

(6.2.20) $$\int_T \chi(X_N(M);f)g\,dk \geq M \int_T \chi(X_N(M);f)\,dk$$

If $X(M)$, $X^+(M)$ are the sets of $f \in T$ for which, respectively,

$$\sup_n G_n(T,f) > M, \quad \limsup_{n \to \infty} G_n(T,f) > M,$$

then $X_N(M)$ is monotone increasing to $X(M)$ as $N \to \infty$. As gk and $|gk|$ are integrable (the only point where this is used) the integral of gk is AC* relative to k, so that we can let $N \to \infty$ in (6.2.20) to obtain

(6.2.21) $$\int_T \chi(X(M);f)g\,dk \geq M \int_T \chi(X(M);f)\,dk.$$

This is true for all real values of M. Replacing **A** by **A***, **B** by **B***, as the **A**-integral exists, so does the **A***-integral, so that from (6.2.21) we have

(6.2.22) $$(\mathbf{A}^*)\int_T \chi(X(M);f)g\,dk \geq M \cdot (\mathbf{A}^*)\int_T \chi(X(M);f)\,dk.$$

The last shows that $X(M)$ is suitably k-measurable using **A***, and clearly it contains $X^+(M)$, a set that is cylindrical of order **B***, so that by Theorem 6.2.5 for **B***, **A***, $\limsup_{n \to \infty} G_n(T,f)$ is constant, say at H, k-almost everywhere, using **A***. Taking $M < H$, $\setminus X(M)$ has k-variation zero using **A***, and (6.2.22) becomes

$$G(T) = (\mathbf{A}^*)\int_T g\,dk \geq M \cdot (\mathbf{A}^*)\int_T dk = M,$$

$$G(T) \geq H \equiv \limsup_{n \to \infty} G_n(T,f)$$

k-almost everywhere, using **A***. Replacing g by $-g$ in the proof,

(6.2.23) $$G(T) \leq \liminf_{n \to \infty} G_n(T,f), \quad \lim_{n \to \infty} G_n(T,f) = G(T),$$

k-almost everywhere, using $\mathbf{A}^*$. If B^* is replaced by a countable B^{**} in $B^* \subseteq B^{**} \subseteq B$, then (6.2.23) follows again. Thus we have

(6.2.24) $$\lim_{C \subseteq B}{}^* G(B \backslash C; T, f) = G(T)$$

k-almost everywhere. □

R. Johnson, ICI research fellow at the Queen's University, Belfast, alerted me to the problem of applying convergence tests to Feynman integration in 1962 or so, asking whether the new generalized Riemann integral would be useful. The problem was exceedingly difficult, but eventually progress was made, see Henstock (1973b, 1980b, 1988–89) and Muldowney (1987, 1988–89).

CHAPTER 7

PERRON-TYPE, WARD-TYPE, AND CONVERGENCE-FACTOR INTEGRALS

7.1 PERRON-TYPE INTEGRALS IN FULLY DECOMPOSABLE DIVISION SPACES

Let $h: \mathcal{U}^1 \to \mathbb{R}^+, f: T \to \mathbb{R}$. In Perron-type integral theory it is common to allow f to be infinite in a set of measure zero. So may I emphasize that here f does not take infinite values — such values would wreck the theory. To avoid infinite values all we have to do is to replace them by 0, and the integral is unchanged.

We aim to integrate fh by using derivates, and so have to go beyond the beginning of Chapter 4. There it is pointed out that full decomposability is appropriate, and that we can disregard points t where for some $\mathcal{U} \in \mathbf{A}|E$

(4.1.3) $\qquad h(I, t) = 0 \quad (\text{all } (I, t) \in \mathcal{U})$

An h is defined to be continuous in an elementary set E if, given $\varepsilon > 0$, there is a $\mathcal{U} \in \mathbf{A}|E$ such that $|h(I, t)| < \varepsilon$ for all $(I, t) \in \mathcal{U}$.

Let $k: \mathcal{U}^1 \to R$. Then the *lower derivate of k at $t \in E^*$ with respect to h, $\mathbf{A}$, E, is*

$$\underline{k}' = \liminf k(I, t)/h(I, t)((I, t) \in \mathcal{U}),$$

as $\mathcal{U}$ shrinks in $\mathbf{A}|E$. The *upper derivate of k at $t \in E^*$ with respect to h, $\mathbf{A}$, E, is* similarly

$$\bar{k}' = \limsup k(I, t)/h(I, t)((I, t) \in \mathcal{U}).$$

Let $M: \mathcal{P} \to \mathbb{R}$ where $\mathcal{P}$ is the family of partial sets of E. Then M is defined to be *finitely superadditive in E* if for each pair P_1, P_2 of disjoint partial sets of E,

$$M(P_1) + M(P_2) \leq M(P_1 \cup P_2).$$

Naturally the setting for this is a *division space*, which we assume throughout. Such a superadditive continuous function M is an **A**-*Perron major function* of fh *in* E if its lower derivate $\underline{M}' \geq f$, h-almost everywhere in E^*, with $\underline{M}' > -\infty$ except for a countable set X in E^*.

Later, other kinds of major functions are defined. For each of these definitions m is a minor function of fh if $-m$ is a major function of $-fh$. For such a definition to be useful we need that if M, m are respectively a major and a minor function of fh over E then $M(E) \geq m(E)$. For $N(E)$ the greatest lower bound of $M(E)$ for all major functions M of fh in E then $N(E)$ is called the *upper integral* of fh on E. For $n(E)$ the least upper bound of $m(E)$ for all minor functions m of fh in E, then $n(E)$ is called the *lower integral* of fh on E. If $n(E) = N(E)$, their common value is called the integral of fh over E of that particular kind of integral that is being considered. These remarks save much space in further definitions.

Strengthening the definitions, a real-valued superadditive continuous function M of the partial sets P of E is a *strong* **A**-*Perron major function of* fh *on* E if $\underline{M}' \geq f$ everywhere in E^*. The pattern giving other definitions now follows.

An unresolved question is as follows. Are the integral and the strong integral the same? The question is answered in an interesting special case.

Theorem 7.1.1. *When $T = \mathbb{R}$ and* **A** *is for the gauge integral, then (dropping the '***A***-' from the names of entities) the upper and lower Perron and strong upper and lower Perron integrals are the same, so that if $f, g: T \to \mathbb{R}$ with $f = g$ h-almost everywhere, and $h([u, v); t) = v - u$ the upper and lower Perron integrals of fh and gh are the same.*

Proof. Using the continuity of the major function M of fh in E, we remove the sequence (s_n) of points of X. Let $E \subseteq [a, b)$ and $M(t) = M([a, t))$, so that for $a \leq t < u$, by superadditivity,

(7.1.1) $$M(u) - M(t) \geq M([t, u)).$$

Define

(7.1.2) $$L(t, y) \equiv \sup_{0 < |x| \leq y} |M(t + x) - M(t)|,$$

$$L(t, 0) = 0, \quad L(t, -y) = -L(t, y).$$

For each fixed t, $L(t, y)$ is bounded and continuous in y and monotone decreasing to 0 as $y \to 0+$. Thus, given $\varepsilon > 0$, there are a $y_n > 0$ and $L_n(t)$ such that

(7.1.3) $$L(s_n, y_n) \leq \varepsilon \cdot 2^{-n-1} \quad (n = 1, 2, \ldots),$$

(7.1.4) $$L_n(s_n + t) = \begin{cases} L(s_n, t) - L(s_n, -y_n) & (|t| \leq y_n) \\ L(s_n, y_n) - L(s_n, -y_n) & (t \geq y_n), \\ 0 & (t \leq -y_n). \end{cases}$$

Then for $0 < x \leq y_n$, using (7.1.n) ($n = 1, 2, 3, 4$),

$$L_n(s_n + x) - L_n(s_n) \geq |M([s_n, s_n + x))|,$$

$$L_n(s_n) - L_n(s_n - x) \geq |M(s_n - x, s_n))|,$$

$$0 \leq \sum_{n=1}^{\infty} L_n \leq \varepsilon, \text{ and } M_1 \equiv M + \sum_{n=1}^{\infty} L_n$$

7.1 PERRON-TYPE INTEGRALS

is continuous in $[a,b]$ with $M_1' \geq 0$ at each s_n. As a countable set is of variation zero, M_1 satisfies the conditions with X empty.

Next we remove the exceptional set Y of variation zero where $M_1' < f$. Thus given $\varepsilon > 0$, for some gauge δ giving rise to $\mathcal{U} \in \mathbf{A}|E$, $V(h; \mathcal{U}; E; \bar{Y}) < \varepsilon$. The intervals $(y - \delta(y), y + \delta(y))$ for $y \in Y$, form a set G open in the metric topology and in the intrinsic topology, and by Sierpinski's lemma (Sierpinski (1923)), two sets of intervals (left-hand and right-hand) cover G apart from a set of variation zero, the intervals in each set being disjoint, so that the total sum for these intervals is $< 2\varepsilon$. Replacing ε by $\varepsilon \cdot 4^{-n}$, we replace G by G_n and define

$$z_n(t) \equiv 2^n \chi(G_n; t), \quad Z_n(t) \equiv \int_a^t z_n(t) \, dh, \quad Z(t) = \sum_{n=1}^{\infty} Z_n(t),$$

$$z_n(b) < \varepsilon \cdot 2^{-n}, \quad z(b) < \varepsilon.$$

Then $Z_n(t)$ and $Z(t)$ are continuous and monotone increasing in $[a,b]$. Putting

$$M_2(t) = M_1(t) + Z(t), \quad M_2' \geq f \quad (t \in [a,b] \setminus Y), \quad M_2' = +\infty \quad (t \in Y),$$

since $M_1' > -\infty$, and M_2 is a strong Perron major function of fh on E; and as

$$M(b) < M_2(b) < M(b) + 2\varepsilon,$$

together with similar results for minor functions, the theorem is proved. □

Reverting to the fully decomposable division spaces, we have the following:

Theorem 7.1.2. *If M, m are strong **A**-Perron major and minor functions, respectively, of fh in E, then $M-m$ is finitely superadditive on the partial sets P of E, with $M(E) - m(E) \geq 0$, and the strong **A**-Perron upper integral of fh over P is not less than the strong **A**-Perron lower integral of fh over P. If the two are equal for $P = E$, they are equal for all partial sets P of E. For $I(P)$ the common value, the strong **A**-Perron integral of fh over P, then*

(7.1.5) $$M(P) - I(P) \geq 0, \quad I(P) - m(P) \geq 0$$

when h is of bounded variation on E.

Proof. We have $\underline{M}' \geq f \geq \bar{m}'$ so that, given $\varepsilon > 0$, there is a $\mathcal{U} \in \mathbf{A}|E$ for which

$$M(I)/h(I,t) > f(t) - \varepsilon, \quad m(I)/h(I,t) < f(t) + \varepsilon,$$

$$M(I) - m(I) > -2\varepsilon h(I,t) \quad ((I,t) \in \mathcal{U}).$$

As M is finitely superadditive, m finitely subadditive, and $h \geq 0$ of bounded variation there are an $N > 0$ and a division $\mathscr{E}$ of P from $\mathscr{U}$, such that

$$M(P) - m(P) \geq (\mathscr{E})\sum (M(I) - m(I)) \geq -2\varepsilon(\mathscr{E})\sum h(I,t) \geq -2\varepsilon N,$$
$$M(P) - m(P) \geq 0,$$

from which follow the other results since M and m can vary independently. We let M tend to the upper integral and/or m to the lower integral to obtain (7.1.5) and that the upper integral is finitely superadditive, the lower integral finitely subadditive, so that if the two are equal for E, they are equal for all partial sets $P \subseteq E$, and the integral is finitely additive. □

Just before Theorem 2.5.5 the variational and strong variational integrals of $h(I,t)$ are defined as a function $H(P)$ that is finitely additive for copartitional partial sets P, with $h(I,t) - H(I)$ of variation zero in an elementary set E, respectively using variation sets and the p-variation. Here the space of values is the real line $\mathbb{R}$, for which the Theorem 2.1.10(2.1.33) with $q = 2$ shows that the two definitions are equivalent. This leads to the next theorem.

Theorem 7.1.3. *Let* $h\colon \mathscr{U}^1 \to \mathbb{R}^+, f\colon T \to \mathbb{R}$. *Then the strong **A**-Perron integral of fh is exactly equivalent to the strong variational integral of fh and to the integral of fh for the given division space* $(T, \mathscr{T}, \mathbf{A})$.

Proof. If the strong variational integral H of fh exists, then

$$Q(\mathscr{U}, P) \equiv \sup_{\mathscr{E}} (\mathscr{E})\sum |f(t)h(I,t) - H(I)|$$

for all divisions $\mathscr{E}$ of E from $\mathscr{U}$, is as small as we please, by choice of $\mathscr{U} \in \mathbf{A}|E$.

Clearly Q is finitely superadditive over partial sets of E, and so is $H + Q$, whereas $H - Q$ is finitely subadditive. Also if

$$(I,t) \in \mathscr{U}, H(I) + Q(\mathscr{U}, I) \geq H(I) + f(t)h(I,t) - H(I) = f(t)h(I,t).$$

Dividing by $h(I,t)$, we see that $H + Q$ is a strong **A**-Perron major function of fh, and $H - Q$ a strong **A**-Perron minor function of fh, with H also the strong variational integral. Conversely, if the strong **A**-Perron integral H of fh exists in E, we write $Q = M - H$ for a strong **A**-Perron major function M, $q = H - m$ for a strong **A**-Perron minor function M of fh. Then $Q \geq 0$, $q \geq 0$, and

$$Q(\mathscr{U}, I) \geq f(t)h(I,t) - H(I), q(\mathscr{U}, I)$$
$$\geq H(I) - f(t)h(I,t), Q(\mathscr{U}, I) + q(\mathscr{U}, I) \geq |H(I) - f(t)h(I,t)|$$
$$((I,t) \in U), \quad Q(U, E) + q(U, E) \geq \bar{V}(fh - H; A; E),$$

and H is also the strong variational integral of fh over E.

Perron integration was developed in Perron (1914), Bauer (1915), Hake (1921), Looman (1925), Saks (1937) pp. 186–203, McShane (1944) p. 313, Sarkhel (1978), and Henstock (1983).

7.2 WARD INTEGRALS IN DECOMPOSABLE DIVISION SYSTEMS AND SPACES

Let $h: \mathcal{U}^1 \to \mathbb{R}$. We wish to integrate h by using increments and not derivates, in a decomposable division system for an elementary set E. A finitely superadditive function F of the partial sets of E is an **A**-*Ward major function* of h in E, if

(7.2.1) $$F(I) \geq h(I, t) \quad ((I, t) \in \mathcal{U})$$

for some $\mathcal{U} \in \mathbf{A}|E$. Other definitions follow the pattern given in Section 7.1. The definition of **A**-Ward integral is wider than that of Ward (1936a) only in that the division system need not be on the real line, nor need **A** be that of Sections 1.4, 1.5 for the gauge integral. However, on the real line F need only be finitely additive since if $a \leq u < v \leq b$ with $g(a) = 0, g(v) = F([a, v))$, then

$$g(v) \geq F([a, u)) + F([u, v)), \quad g(v) - g(u) \geq F([A, v)),$$

$$g(b) - g(a) = F([a, b)),$$

Theorem 7.2.1. *For a division space* $(T, \mathcal{T}, \mathbf{A})$ *and K the real line, the* **A**-*Ward integral is equivalent to the variational and generalized Riemann integrals over E.*

Proof. If the finitely additive variational integral H of h exists, $\mathcal{U} \in \mathbf{A}|E$ and Theorem 2.5.6(2.5.18) give $V \equiv V(h - H; \mathcal{U}; P)$ finitely superadditive in P, so that $V + H$ are finitely superadditive and, for $(I, t) \in \mathcal{U}$,

$$H(I) - V(h - H; \mathcal{U}; I) \leq H(I) - |H(I) - h(I, t)| \leq h(I, t)$$

$$\leq H(I) + |h(I, t) - H(I)|$$

$$\leq H(I) + V(h - H; \mathcal{U}; I)$$

and $H + V$ and $H - V$ are **A**-Ward major and minor functions of h in E. By Theorem 2.5.5(2.5.15), $2V$ is arbitrarily small by choice of $\mathcal{U} \in \mathbf{A}|E$, so that the major and minor functions are arbitrarily close, and the **A**-Ward integral exists and is $H(E)$. Conversely, if the **A**-Ward integral W exists, with $\mathscr{E}$ a division of E from $\mathcal{U}$, by the finite superadditivity of the **A**-Ward major functions F of h, $-F_1$ of $-h$, we have

$$F_1(E) \leq (\mathscr{E})\sum F_1(I) \leq (\mathscr{E})\sum h(I, t) \leq (\mathscr{E})\sum F(I) \leq F(E),$$

and the limit of the sums of h over E is $W(E)$. The equivalence of the variational and generalized Riemann integrals finishes the proof. □.

Theorem 7.2.2. *When $h: \mathcal{U}^1 \to \mathbb{R}^+$ is of bounded variation and $f: T \to \mathbb{R}$, the **A**-Ward integral of fh is equivalent to the strong **A**-Perron integral.*

Proof. $M' \geq f$ implies that, given $\varepsilon > 0$, there is a $\mathcal{U} \in \mathbf{A}|E$ such that $M(I) > (f(t) - \varepsilon) h(I, t)$ for all $(I, t) \in \mathcal{U}$, while $M(I) > (f(t) - \varepsilon) h(I, t)$ (all $(I, t) \in \mathcal{U}$ depending on $\varepsilon > 0$) implies $\underline{M}' \geq f$. The result follows since for some $N > 0$,

$$\varepsilon(\mathscr{E}) \sum h(I, t) \leq \varepsilon N.$$

The range of application of the **A**-Ward integral is wider than that of the **A**-Perron integral, since h need not be non-negative-valued, and narrower than that of the generalized Riemann integral since the **A**-Ward integral is restricted to real-valued functions.

For $h: \mathcal{U}^1 \to \mathbb{R}$, $f: T \to \mathbb{R}$, $g: T \to \mathbb{R}$ with fh and gh **A**-Ward integrable, if $f = g$ at a countable set X everywhere dense in E^*, and if, as usually happens, $\bar{V}(h; \mathbf{A}; E; X) = 0$, then the equality can be disregarded. But if $T = \mathbb{R}$ with the gauge integral division space, writing 'Ward' instead of '**A**-Ward', a heavy restriction occurs in the following case.

Theorem 7.2.3. *Let $f, g, j: \mathbb{R} \to \mathbb{R}$, $h([u, v), t) = g(v) - g(u)$, $k([u, v), t) = j(v) - j(u)$ with fh, fk Ward integrable over $[a, b)$ $(a < b)$. Let $g = j$ in a set X everywhere dense in $[a, b]$. Then*

(7.2.2) $$\int_{[u,v)} f \, dh = \int_{[u,v)} f \, dk + f(v)\{g(v) - j(v)\} - f(u)\{g(u) - j(u)\}$$

$$(a \leq u < v \leq b),$$

(7.2.3) *Given $\varepsilon > 0$, and Y_ε the set of t with $|g(t) - j(t)| \geq \varepsilon$, if (7.2.2) holds then $\bar{V}(\Delta f: \mathbf{A}; E; \bar{Y}_\varepsilon) = 0$, or, alternatively,*

(7.2.4) *f is continuous and VBG^* on $\bar{Y}_\varepsilon$ with $mf(\bar{Y}_\varepsilon) = 0$ (Lebesgue measure).*

(7.2.5) *If (7.2.2) holds and if $\bar{Y}_\varepsilon$ contains an interval $[r, s]$ for some $\varepsilon > 0$, then f is constant in $[r, s]$.*

Proof. It is enough to take j identically 0, so that $g = 0$ at points in X everywhere dense in $[a, b]$. In the proof of Theorem 1.4.1(1.4.2) we choose the points y_j from X. Taking the inequality (7.2.1) and the corresponding inequality for minor functions, for $[u_j, y_j)$ and the previous $[y_{j-1}, u_j)$, we have inequalities for $[y_{j-1}, y_j)$ $(j = 2, 3, \ldots, m-1)$, with those for $[a, y_1)$ and $[y_{m-1}, b)$, taking $a = u_1$ and $u_m = b$, as can be arranged. Thus for F and F_1 the major and minor functions,

$$F_1([y_{j-1}, y_j)) \leq 0 \leq F([y_{j-1}, y_j)), \quad F_1([a, y_1))$$
$$\leq f(a)\{j(a) - g(a)\} \leq F([a, y_1)),$$
$$F_1([y_{m-1}, b)) \leq f(b)\{g(b) - j(b)\} \leq F([y_{m-1}, b)),$$

from which (7.2.2) follows. If it holds, then for each $\varepsilon > 0$, and using Theorem 2.5.5(2.5.15), with t the associated point in $[u, v)$, equal to u or v, and s the other end-point,

(7.2.6) $f(t)\{g(v) - g(u)\} - \{f(v)g(v) - f(u)g(u)\} = \{f(u) - f(v)\}g(s)$

Ward integration was developed in Ward (1936a), Henstock (1955b, pp. 277–9; 1957a; 1960a, b).

7.3 CONVERGENCE-FACTOR INTEGRALS IN FULLY DECOMPOSABLE DIVISION SPACES

Some convergence-factor integrals are defined by an inductive process beginning with the Perron or strong Perron integral, the N_0-*integral* in the inductive chain. Given N_j-integrals for $j = 0, 1, 2, \ldots, r - 1$, the N-integral or N_r-integral is defined using the S-integral or N_{r-1}-integral. For $T = \mathbb{R}$, the real line, given real numbers $a < b$ and each u, v in $a \leq u < v \leq b$ we have two functions $N_s(u, v, t)$ ($s = +, -, u \leq t \leq v$) that are monotone increasing in t with

(7.3.1) $\qquad N_s(n, v, v) - N_s(u, v, u) = 1 (s = +, -),$

$$N_+(u, v, u+) = N_+(u, v, u),$$

$$N_-(u, v, v-) = N_-(u, v, v).$$

We prove various properties for the N-integral that are true for the N_0-integral, and by induction we can assume that the S-integral has these properties. We use

(7.3.2) $\qquad I(S, N_s, F, u, v) = (S) \int_u^v F(t) d_t N_s(u, v, t) \quad (u < v),$

$$I(S, N_s, F, u, u) = 0 \quad (s = +, -).$$

The first was Burkill's Çesaro–Perron integral of order $\rho > 0$, with

$$\partial N_+/\partial t = \rho(v - t)^{\rho - 1}(v - u)^{-\rho}, \quad \partial N_-/\partial t = \rho(t - u)^{\rho - 1}(v - u)^{-\rho}$$

the S-integral being the Çesaro–Perron integral of order max $(\rho - 1, 0)$, so that the r in N_r can be ρ when ρ is an integer, and otherwise r is the next higher integer. See J. C. Burkill (1932, 1935, 1936a, b, 1951a, b) Bosanquet (1940), H. Burkill (1951, 1952, 1957, 1977), Bose (1980a).

Then Marcinkiewicz and Zygmund (1936), James and Gage (1946), James (1950, 1954, 1955), and Taylor (1955) produced similar integrals.

The next stage was the general definition of Jeffery and Miller (1945) that used Lebesgue integrals of positive functions n_+, n_-,

$$N_+(u, v, t) = (L)\int_u^t n_+(u, v, w)\,dw,$$

$$N_-(u, v, t) = (L)\int_t^v n_-(u, v, w)\,dw,$$

together with the assumed non-zero limits

$$\lambda_+(u)^{-1} = \lim_{v \to u+} \frac{1}{(v-u)N_+(u,v,v)}(L)\int_u^v n_+(u, v, t)(t-u)\,dt,$$

$$\lambda_-(v)^{-1} = \lim_{u \to v-} \frac{1}{(v-u)N_-(u,v,u)}(L)\int_u^v n_-(u, v, t)(v-t)\,dt.$$

Let $F: [a, b] \to \mathbb{R}$ be such that for $t \in [u, v] \subseteq [a, b]$, $n_\pm(u, v, t)F(t)$ is integrable in some suitable sense. Define

$$G_s(F, u, v) = \frac{1}{N_s(u, v, w)}\int_u^v n_s(u, v, t)F(t)\,dt$$

($w = v$ when $s = +$, $w = u$ when $s = -$).

Then the *upper and lower G-derivatives of F*, GD^*F and GD_*F, are the upper and lower limits as $v - u \to 0+$ of the ratio

$$\{G_s(F, u, v) - F(x)\}\lambda_s(x)/(v-u)$$

($x = u$ when $s = +$, $x = v$ when $s = -$).

The function F is a *Jeffery and Miller major function of* f if $F(a) = 0$, $G_s(F, u, v) \to F(x)$ as $v - u \to 0+$, $[u, v] \subseteq [a, b]$, $GD_*F \geq f$ almost everywhere, and $GD_*F > -\infty$ everywhere. The pattern of Section 7.1 is now followed.

We could allow GD_*F to be $-\infty$ in an at most countable set X, and then remove X using the G-continuity of F, i.e. $G_s(F, u, v) \to F(x)$ as $v - u \to 0+$ ($x = u$ when $s = +$, $x = v$ when $s = -$). The constructions in the proof of Theorem 7.1.1 can be used to show that we can assume that $GD_*F \geq f$ everywhere in (a, b), and in particular we can replace f by f^* if $f = f^*$ almost everywhere. Henstock (1983), Theorem 5, pp. 397–8, gives detail for a more general case.

To simplify the theory of Jeffery and Miller we combine the definitions with those of Ward as in Section 7.2, replacing $n_\pm$ by $d_t N(u, \pm h; t)$ in a Stieltjes integral form, in an inductive process beginning with Ward integration $(W) = (N_0)$. By the induction, whatever we prove for the N-integral we can assume for the S-integral. We assume at least an additive division space $(T, \mathcal{T}, \mathbf{A})$ with an elementary set E, and take $\mathcal{U} \in \mathbf{A}|E$. For a useful construc-

7.3 CONVERGENCE-FACTOR INTEGRALS

tion we turn functions of generalized integrals into functions of points by a geometrical device, namely,

(7.3.3) If (I, t), $(J, u) \in \mathcal{U}$ and $J \subseteq I$, there is an interval $K(t, u)$ uniquely defined by t, u, with

$$(K(t, u), t) \in \mathcal{U}, \quad K(t, u) \subseteq I.$$

We use the $\mathcal{U} \in \mathbf{A} | E$, a $\mathcal{U}_1(I, t) \in \mathbf{A} | I$ for each $(I, t) \in \mathcal{U}$, and a convergence factor $N(I, t; J)$, finitely additive in J, and defined for all $(I, t) \in \mathcal{U}$, all $(J, u) \in \mathcal{U}_1(I, t)$ with two further properties.

(7.3.4) If a real finitely superadditive function $L(P)$ of partial sets P satisfies

$$(S) \int_I L(K(t, u)) \, dN(I, t; J) \geq 0 \quad (\text{all } (I, t) \in \mathcal{U}),$$

then $L(E) \geq 0$.

This is equivalent to the condition that the N-integral (defined later) of 0 is 0.

(7.3.5) If $f(u) \geq 0$ and the S-integral below exists, then

$$(S) \int_I f(u) \, dN(I, t; J) \geq 0 \quad (\text{all } (I, t) \in \mathcal{U})$$

This is part of Theorem 7.3.1(7.3.15) later, for the S-integral. In a sense the two conditions are converses. We deduce as much as possible from them and also look at the implications on the N.

To define the N-integral of h we require the S-integrability of

(7.3.6) $h(K(t, u); t) N(I, t; J)$

over I, for each $(I, t) \in \mathcal{U}$, and for short we then say that hN is S-integrable.

A finitely superadditive function L of partial sets of E is an N-major function of h in E, relative to $\mathbf{A}$, if there is a $\mathcal{U} \in \mathbf{A} | E$ such that hN and LN are S-integrable with

(7.3.7)

$$(S) \int_I L(K(t, u)) \, dN(I, t; J) \geq (S) \int_I h(K(t, u); t) \, dN(I, t; J) \quad ((I, t) \in \mathcal{U}).$$

The other definitions follow the pattern in Section 7.1, so that if a finitely subadditive function M of partial sets of E is an N-minor function of h in E, relative to $\mathbf{A}$, then $L - M$ is finitely superadditive and by Theorem 7.3.1(7.3.13) for the S-integral, $L - M$ satisfies (7.3.4), whence

(7.3.8) $L(E) - M(E) \geq 0$.

Thus over all such L, M, the infimum $UN(h; \mathbf{A}; E)$ of $L(E)$ and supremum $LN(h; \mathbf{A}; E)$ of $M(E)$ satisfy

(7.3.9) $\quad L(E) - LN(h; \mathbf{A}; E) \geq 0, \quad UN(h; \mathbf{A}; E) - M(E) \geq 0,$

$$UN(h; \mathbf{A}; E) - LN(h; \mathbf{A}; E) \geq 0.$$

If $UN(h; \mathbf{A}; E) = LN(h; \mathbf{A}; E)$, this value is the N-integral of h in E relative to $\mathbf{A}$,

$$N(h; \mathbf{A}; E) = (\mathbf{A}, N) \int_E dk = (N) \int_E dh.$$

Normally we omit $\mathbf{A}$ in $(\mathbf{A}, N)$. By (7.3.9) the N-integral is uniquely defined if it exists, and we say that h is N-integrable over E, relative to $\mathbf{A}$. In particular, if $h(I, t) = f(t)k(I, t)$ we write as usual,

$$N(fk; \mathbf{A}; E) = (\mathbf{A}, N) \int_E f \, dk = (N) \int_E f \, dk.$$

Here, (7.3.7) becomes

(7.3.10)

$$(S) \int_I L(K(t, u)) dN(I, t; J) \geq f(t)(S) \int_I k(K(t, u); t) dN(I, t; J),$$

needing only the S-integrability of LN and kN. Relative to S-integration, $f(t)$ is a constant, and so can be taken outside by Theorem 7.3.1(7.3.14) for S-integrals.

Theorem 7.3.1.

(7.3.11) If $H(P)$ is finitely additive in E with HN S-integrable, the N-integral of H over E exists and is equal to $H(E)$.

(7.3.12) For $h_j: \mathcal{U}^1 \to \mathbb{R}$ with $h_j N$ S-integrable ($j = 1, 2$) then

$$UN(h_1; \mathbf{A}; E) + UN(h_2; \mathbf{A}; E) \geq UN(h_1 + h_2; \mathbf{A}; E)$$
$$\geq UN(h_1; \mathbf{A}; E) + LN(h_2; \mathbf{A}; E)$$
$$\geq LN(h_1 + h_2; \mathbf{A}; E)$$
$$\geq LN(h_1; \mathbf{A}; E) + LN(h_2; \mathbf{A}; E).$$

(7.3.13) In particular, for N-integrable h_2 we therefore have

$$UN(h_1 + h_2; \mathbf{A}; E) = UN(h_1; \mathbf{A}; E) + N(h_2; \mathbf{A}; E)$$

and similarly for LN, so that if both h_1 and h_2 are N-integrable then so is $h_1 + h_2$, its integral being the sum of the other two.

(7.3.14) If in (7.3.13) c is a constant, then also ch_2 is N-integrable to $cN(h_2; \mathbf{A}; E)$

7.3 CONVERGENCE-FACTOR INTEGRALS

(7.3.15) If h_2 is N-integrable, $h_1 \geq h_2$, and $h_1 N$ is S-integrable then $LN(h_1; \mathbf{A}; E) \geq N(h_2; \mathbf{A}; E)$.

Proof. As H is finitely additive in (7.3.11) we can put $L = h = H$ in (7.3.7) and the two sides are the same. Hence the N-integral is $H(E)$. For (7.3.12), given $\varepsilon > 0$, let $L = L_j$ satisfy (7.3.7) with $h = h_j$ and $L_j(E) < UN(h_j; \mathbf{A}; E) + \tfrac{1}{2}\varepsilon$ ($j = 1, 2$). Then as the S-integral satisfies (7.3.13),

$$(S)\int_I (L_1(K(t,u)) + L_2(K(t,u)))\,dN(I,t;J)$$

$$= (S)\int_I L_1\,dN + (S)\int_I L_2\,dN \geq (S)\int_I h_1\,dN + (S)\int_I h_2\,dN$$

$$= (S)\int_I (h_1(K(t,u);t) + h_2(K(t,u);t))\,dN(I,t;J).$$

Hence by definition,

$$UN(h_1 + h_2; \mathbf{A}; E) \leq L_1(E) + L_2(E) < UN(h_1; \mathbf{A}; E) + UN(h_2; \mathbf{A}; E) + \varepsilon,$$

giving the first inequality. The second inequality follows from the first since

$$UN(h_1 + h_2; \mathbf{A}; E) - LN(h_2; \mathbf{A}; E) = UN(h_1 + h_2; \mathbf{A}; E)$$
$$+ UN(-h_2; \mathbf{A}; E)$$
$$\geq UN(h_1 + h_2 - h_2; \mathbf{A}; E).$$

Then the others follow, including (7.3.13). □

For (7.3.14) we use the result for the S-integral, first taking $c \geq 0$. Then for L_2 a major function of h_2,

$$(S)\int_I cL_2\,dN = c(S)\int_I L_2\,dN \geq c(S)\int_I h_2\,dN = (S)\int_I ch_2\,dN$$

and cL_2 is a major function for ch_2. Similarly for minor functions, and the result when $c \geq 0$. When $c < 0$, the result is true for $|c|$, and for -1 where major functions become minor functions and vice versa.

For (7.3.15) we use (7.3.5) with (7.3.13) for the S-integral, on $h_1 - h_2$, to have

$$(S)\int_I h_1(K(t,u);t)\,dN(I,t;J) \geq (S)\int_I h_2(K(t,u);t)\,dN(I,t;J),$$

so that M_2 is also an N-minor function of h_1, and

$$M_2(E) \leq LN(h_1; \mathbf{A}; E), \quad N(h_2; \mathbf{A}; E) \leq LN(h_1; \mathbf{A}; E),$$

178   THE GENERAL THEORY OF INTEGRATION

These results are analogues of the results for generalized Riemann integration on division systems, which latter results begin the inductions. For the analogues of division space results for generalized Riemann integration we have the following:

Theorem 7.3.2.

(7.3.16) Let h be N-integrable over E. Then the N-integral exists over the partial sets of E and is finitely additive, when $(T, \mathcal{T}, \mathbf{A})$ is additive.

(7.3.17) When $N \geqslant 0$ and h N-integrable to H over partial sets of E, then HN is S-integrable.

(7.3.18) If E_1, E_2 are disjoint elementary sets and $N(h; \mathbf{A}; E_j)$ ($j = 1, 2$) exist, then $N(h; \mathbf{A}; E_1 \cup E_2)$ exists equal to $N(h; \mathbf{A}; E_1) + N(h; \mathbf{A}; E_2)$, when $(T, \mathcal{T}, \mathbf{A})$ is additive.

Proof. For P a proper partial set of E, $E \backslash P$ is a co-divisional partial set. Let (7.3.4) hold for all $(I, t) \in \mathcal{U} . P$. Then $L_0(P_1) \equiv L(P \cap P_1)$, or 0 for empty $P \cap P_1$, is finitely superadditive. As $(T, \mathcal{T}, \mathbf{A})$ is additive we can arrange that $\mathcal{U}$ satisfies $\mathcal{U} = (\mathcal{U} . P) \cup (\mathcal{U} . (E \backslash P))$, and as $K(t, u) \subseteq I$, (7.3.4) is true for L_0 and all $(I, t) \in \mathcal{U}$. From (7.3.4), $L(P) = L_0(E) \geqslant 0$, in particular, $L(K(t, u)) \geqslant 0$, and we have (7.3.4) holding for P. Now for L we can substitute the difference $L - M$ of N-major and N-minor functions of h and so have, for P and $E \backslash P$, and then taking limits,

$$L(P) - M(P) \geqslant 0,$$

$$UN(h; \mathbf{A}; P) - LN(h; \mathbf{A}; P) \geqslant 0,$$

$$UN(h; \mathbf{A}; E \backslash P) - LN(h; \mathbf{A}; E \backslash P) \geqslant 0,$$

$$L(E) \geqslant L(P) + L(E \backslash P) \geqslant UN(h; \mathbf{A}; P) + UN(h; \mathbf{A}; E \backslash P),$$

$$UN(h; \mathbf{A}; E) \geqslant UN(h; \mathbf{A}; P) + UN(h; \mathbf{A}; E \backslash P),$$

$$UN(h; \mathbf{A}; E) - LN(h; \mathbf{A}; E) \geqslant UN(h; \mathbf{A}; P) - LN(h; \mathbf{A}; P)$$
$$+ UN(h; \mathbf{A}; E \backslash P) - LN(h; \mathbf{A}; E \backslash P)$$

each difference being non-negative. Thus

(7.3.19) $UN(h; \mathbf{A}; E) - LN(h; \mathbf{A}; E) \geqslant UN(h; \mathbf{A}; P) - LN(h; \mathbf{A}; P)$.

Since $N(h; \mathbf{A}; E)$ exists, so does $N(h; \mathbf{A}; P)$ by (7.3.19), proving (7.3.16) since the UN are finitely superadditive, the LN finitely subadditive. For (7.3.17) and L, M the N-major and N-minor functions of h, then $L \geqslant H \geqslant M$, and for the S-integral in (7.3.15),

$$S(LN; \mathbf{A}; E) \geqslant US(HN; \mathbf{A}; E) \geqslant LS(HN; \mathbf{A}; E) \geqslant S(MN; \mathbf{A}; E),$$

$$S(LN; \mathbf{A}; E) - S(MN; \mathbf{A}; E) = S((L - M)N; \mathbf{A}; E),$$

and $N \geqslant 0$ is S-integrable and $0 \leqslant L - M \leqslant L(E) - M(E)$, as small as we please,

$$US(HN; \mathbf{A}; E) - LS(HN; \mathbf{A}; E) \leqslant (L(E) - M(E))S(N; \mathbf{A}; E),$$

and the S-integral of HN exists. This is a useful result for (7.3.11).

For (7.3.18), if L_j, M_j and $\mathcal{U}_j \in \mathbf{A}|E_j$ are chosen suitably for E_j ($j = 1, 2$),

$$L(J) = L_1(J \cap E_1) + L_2(J \cap E_2),$$
$$M(J) = M_1(J \cap E_1) + M_2(J \cap E_2), \quad \mathcal{U} \subseteq \mathcal{U}_1 \cup \mathcal{U}_2,$$

the last as $(T, \mathcal{T}, \mathbf{A})$ is additive, and L and $-M$ are N-major functions of h and $-h$, respectively, in $E_1 \cup E_2$, giving (7.3.18) on taking $L - M$ arbitrarily small. □

A function $h: \mathcal{U}^1 \to \mathbb{C}$, is *of bounded N-variation over E, relative to* $\mathbf{A}$, if there are $\mathcal{U} \in \mathbf{A}|E$ and a finitely superadditive $L(P) \geqslant 0$ such that hN and LN are S-integrable and

(7.3.20) $\left|(S)\int_I h(K(t, u); t)\,dN(I, t; J)\right| \leqslant (S)\int_I L(K(t, u))\,dN(I, t; J)$

$$((I, t) \in \mathcal{U}).$$

The *N-variation* $NV(h; \mathbf{A}; E)$ of h over E, relative to $\mathbf{A}$, is the infimum of $L(E)$ for all such L. If no such L exists we write symbolically, $NV(h; \mathbf{A}; E) = +\infty$. If $NV(h; \mathbf{A}; E) = 0$, we say that h is *of N-variation zero in E, relative to* $\mathbf{A}$. If a finitely additive $H(P)$ satisfies $NV(h - H; \mathbf{A}; E) = 0$, we say that H is *the N-variational integral of h in E, relative to* $\mathbf{A}$. Also, if $X \subseteq T$ we write $NV(h; \mathbf{A}; E; X)$ for $NV(h.\chi(X;\cdot); \mathbf{A}; E)$. For $t \in E^*$ we say that h is *N-continuous at t relative to E, $\mathbf{A}$*, if, given $\varepsilon > 0$, there are a $\mathcal{U} \in \mathbf{A}|E$ and a finitely superadditive function $L \geqslant 0$ with $L(E) < \varepsilon$ and, for that t,

(7.3.21) $\left|(S)\int_I h(K(t. u); t)\,dN(I, t; J)\right| \leqslant (S)\int_I L(K(t, u))\,dN(I, t; J)$

$$((I, t) \in \mathcal{U}).$$

(7.3.22) *Clearly, if h is of N-variation zero in E then h is N-continuous at each $t \in E^*$.*

Theorem 7.3.3. *Let hN be S-integrable. For $(T, \mathcal{T}, \mathbf{A})$ an additive division space,*

(7.3.23) *$NV(h; \mathbf{A}; P)$ is finitely additive in the partial sets P of E, if finite.*

(7.3.24) *If $NV(h; \mathbf{A}; E)$ is finite, then, given $\varepsilon > 0$, there are a $\mathcal{U} \in \mathbf{A}|E$ and a non-negative finitely superadditive function L_1 on the partial sets of E with $L_1(E) < \varepsilon$ and*

(7.3.25)
$$\left| (S) \int_I h(K(t, u), t) \, d_u N \right| \leq NV(h; \mathbf{A}; I) + (S) \int_I L_1(K(t, u)) \, d_u N$$

(7.3.26) $\quad -NV(h; \mathbf{A}; E) \leq LN(h; \mathbf{A}; E) \leq UN(h; \mathbf{A}; E) \leq NV(h; \mathbf{A}; E)$.

(7.3.27) *If h is N-integrable, $|N(h; \mathbf{A}; E)| \leq NV(h; \mathbf{A}; E)$.*

(7.3.28) *If $NV(h; \mathbf{A}; E) = 0$ then $N(h; \mathbf{A}; E)$ exists and is 0.*

(7.3.29) *If H is finitely additive with HN S-integrable then $|H(E)| \leq NV(H; \mathbf{A}; E)$.*

(7.3.30) *If $H \geq 0$ is finitely additive with HN S-integrable, then $H(E) = NV(H; \mathbf{A}; E)$.*

(7.3.31) *If $h_j N$ is S-integrable ($j = 1, 2$) with c a constant,*

$$NV(h_1 + h_2; \mathbf{A}; E) \leq NV(h_1; \mathbf{A}; E) + NV(h_2; \mathbf{A}; E),$$
$$NV(ch_1; \mathbf{A}; E) = |c| NV(h_1; \mathbf{A}; E)$$

Proof. For (7.3.23) let P_1, P_2 be disjoint partial sets of E, let $\varepsilon > 0$, and let the finitely superadditive $L_j(P)$ have hN and $L_j N$ S-integrable, satisfying

(7.3.32) $\quad L_j(P) = L_j(P \cap P_j) < NV(h; \mathbf{A}; P_j) + \tfrac{1}{2}\varepsilon$

and (7.3.20) for $(I, t) \in \mathcal{U}_j$ and some $\mathcal{U}_j \in \mathbf{A} | P_j$ ($j = 1, 2$). Then as $(T, \mathcal{T}, \mathbf{A})$ is additive, there is a $\mathcal{U} \in \mathbf{A} | P_1 \cup P_2$ with $\mathcal{U} \subseteq \mathcal{U}_1 \cup \mathcal{U}_2$, and $L_1(P) + L_2(P)$ is finitely superadditive and satisfies (7.3.20) for $(I, t) \in \mathcal{U}$. Thus

$$NV(h; \mathbf{A}; P_1 \cup P_2) \leq L_1(E) + L_2(E) = L_1(P_1) + L_2(P_2)$$
$$< NV(h; \mathbf{A}; P_1) + NV(h; \mathbf{A}; P_2) + \varepsilon.$$

Next, there is a finitely superadditive $L(P)$ with (7.3.20) for $P = P_1 \cup P_2$, $\mathcal{U} \in \mathbf{A} | P_1 \cup P_2$, and

$$NV(h; \mathbf{A}; P_1 \cup P_2) + \varepsilon > L(P \cap (P_1 \cup P_2)) \geq L(P \cap P_1) + L(P \cap P_2)$$
$$\geq NV(h; \mathbf{A}; P_1) + NV(h; \mathbf{A}; P_2).$$

As $\varepsilon > 0$ is arbitrary we finish with (7.3.23) in the form

$$NV(h; \mathbf{A}; P_1) + NV(h; \mathbf{A}; P_2) = NV(h; \mathbf{A}; P_1 \cup P_2).$$

For (7.3.24), the L satisfying (7.3.20) for E will satisfy also the conditions for a partial set $P \subseteq E$ on taking a restriction of $\mathcal{U}$ to P. Hence

$$L(P) \geq NV(h; \mathbf{A}; P), \quad L_1(P) \equiv L(P) - NV(h; \mathbf{A}; P) \geq 0.$$

As the N-variation is finitely additive by (7.3.23), $L_1(P)$ is finitely superadditive. Using Theorem 7.3.1(7.3.11) for S-integration,

$$(S) \int_I L(K(t, u)) \, dN = NV(h; \mathbf{A}; I) + (S) \int_I L_1(K(t, u)) \, dN$$

7.3 CONVERGENCE-FACTOR INTEGRALS

and we have (7.3.25). For (7.3.26), L in (7.3.20) is an N-major function of h and of $-h$, so that by definition

$$-L(E) \leqslant LN(h; \mathbf{A}; E) \leqslant UN(h; \mathbf{A}; E) \leqslant L(E).$$

Thus (7.3.27), (7.3.28) also. For (7.3.29), by Theorem 7.3.1(7.3.11) H is N-integrable to $H(E)$ and (7.3.27) gives the result. For (7.3.30) we use (7.3.29) and also put $L = h = H$ in (7.3.20), so that by definition, $NV(H; \mathbf{A}; E) \leqslant H(E)$. For (7.3.31) we use properties of the S-integral with

$$(S)\int_I (L_1 + L_2)\,dN = (S)\int_I L_1\,dN + (S)\int_I L_2\,dN$$

$$\geqslant \left|(S)\int_I h_1\,dN\right| + \left|(S)\int_I h_2\,dN\right|$$

$$\geqslant \left|(S)\int_I (h_1 + h_2)\,dN\right|,$$

$$\left|(S)\int_I ch_1\,dN\right| = |c|\left|(S)\int_I h_1\,dN\right|. \quad \square$$

Theorem 7.3.4.

(7.3.33) When $N \geqslant 0$, $S \geqslant 0$, the N-integral H of h in E is also the N-variational integral of h in E, and conversely if HN is S-integrable. Thus the N-variational integral is uniquely defined.

(7.3.34) Further, $h - H$ is N-continuous at each point of E^*,

(7.3.35) and $NV(H; \mathbf{A}; E) = NV(h; \mathbf{A}; E)$.

(7.3.36) If also $h \geqslant 0$ then $N(h; \mathbf{A}; E) = NV(h; \mathbf{A}; E)$.

Proof. (7.3.17) gives the S-integrability of HN when $N \geqslant 0$. Using Theorem 7.3.1(7.3.11), (7.3.13) for the S-integral, we have for N-major functions L, $-M$ of h, $-h$, respectively,

$$(S)\int_I (h - H)\,dN \leqslant (S)\int_I (L - H)\,dN,$$

$$(S)\int_I (H - h)\,dN \leqslant (S)\int_I (H - M)\,dN,$$

$$\left|(S)\int_I (h - H)\,dN\right| \leqslant (S)\int_I \{(L - H) + (H - M)\}\,dN$$

$$= (S)\int_I (L - M)\,dN,$$

$$L(E) - M(E) < \varepsilon,$$

$NV(h - H; \mathbf{A}; E) = 0$. Conversely, by Theorems 7.3.3(7.3.28), 7.3.1(7.3.11), (7.3.13), $h - H$ is N-integrable to 0, H is N-integrable to H, and h is N-integrable to H. (7.3.34) follows, and (7.3.35) uses Theorem 7.3.3(7.3.31) with $H = (H - h) + h$, $h = (h - H) + H$. Theorems 7.3.3(7.3.29) and 7.3.2(7.3.17) give (7.3.36). □

Theorem 7.3.5. *For* $(T, \mathcal{T}, \mathbf{A})$ *a decomposable additive division space,* $\mathcal{U} \in \mathbf{A}|E$, *and* $h: \mathcal{U} \to \mathbb{R}$, *with* hN *S-integrable over* E. *For* (X_j) *a sequence of sets of T with union* X,

(7.3.37) $$NV(h; \mathbf{A}; E; X) \leq \sum_{j=1}^{\infty} NV(h; \mathbf{A}; E; X_j),$$

(7.3.38) $$NV(h; \mathbf{A}; E; \liminf_{j \to \infty} X_j) \leq \liminf_{j \to \infty} NV(h; \mathbf{A}; E; X_j)$$

(7.3.39) *In particular, if* (X_j) *is monotone increasing,*

$$NV(h; \mathbf{A}; E; X) = \lim_{j \to \infty} NV(h; \mathbf{A}; E; X_j).$$

(7.3.40) *If* $NV(h - k; \mathbf{A}; E) = 0$ *and if* $N(f(t)h(I, t); \mathbf{A}; E)$ *exists, so does* $N(f(t)k(I, t); \mathbf{A}; E)$ *and the N-integrals are equal.*

(7.3.41) *For* $(T, \mathcal{T}, \mathbf{A})$ *a stable division space let* $h: \mathcal{U}^1 \to \mathbb{R}$ *with* $NV(h; \mathbf{A}; E; \mathcal{F}(E; P)) = 0$ *for each proper partial set P of the elementary set E. Let $\mathcal{E}$ be a division of E from a* $\mathcal{U} \in \mathbf{A}|E$. *Let* $r: T \to \mathbb{R}$ *be such that for each* $(I, t) \in \mathcal{E}$ *and some numbers* $g(I)$, $r(u) = g(I)$ $(u \in I^* \setminus \mathcal{F}(E; I))$, *r arbitrary in* $I^* \cap \mathcal{F}(E; I)$, *so that r is a step function based on E. Let $f: T \to \mathbb{R}$ be bounded in E^* with fh N-integrable over E, and let r be bounded in E^*. Then there exists*

(7.3.42) $$N(frh; \mathbf{A}; E) = (\mathcal{E}) \sum g(I)(N) \int_I f \, dh.$$

If $(T, \mathcal{T}, \mathbf{A})$ *is also decomposable, then f, r need not be bounded but only finite.*

Proof. The indicators are functions of t but not of u, see (7.3.10), so that proofs are simplified. However, the analogues, Theorems 2.4.1, 2.8.1 are proved using divisions, and we need proofs of (7.3.37), (7.3.38) that follow Perron integration proofs. First we can assume that the infinite series is convergent.

(7.3.43)   If $X \subseteq Y \subseteq T$ then $NV(h: \mathbf{A}; E; X) \subseteq NV(h; \mathbf{A}; E; Y)$.

Using this obvious result we can take the X_j mutually disjoint. Then to each $\varepsilon > 0$ and each X_j, there are a $\mathcal{U}_j \in \mathbf{A}|E$ and a non-negative finitely superadditive L_j on the partial sets of E with

$$L_j(E) < NV(h; \mathbf{A}; E; X_j) + \varepsilon 2^{-j},$$

$$(S)\int_I h(K(t, u), t) \, dN \leq (S)\int_I L_j(K(t, u)) \, dN$$

7.3 CONVERGENCE-FACTOR INTEGRALS

$((I, t) \in \mathcal{U}_j, t \in X_j)$. Decomposing the $\mathcal{U}_j$, there is a $\mathcal{U} \in \mathbf{A}|E$ with $\mathcal{U}[X_j] \subseteq \mathcal{U}_j$ ($j = 1, 2, \ldots$) and for this $\mathcal{U}$ with $L = \sum_{j=1}^{\infty} L_j$, the result follows from

$$NV(h; \mathbf{A}; E; X) \leq L(E) \leq \sum_{j=1}^{\infty} NV(h; \mathbf{A}; E; X_j) + \varepsilon.$$

For (7.3.38) first take (X_j) monotone increasing and put

$$M_j(P) \equiv NV(h; \mathbf{A}; P; X_j), \quad M(P) \equiv \lim_{j \to \infty} M_j(P).$$

We can assume $M(P)$ finite for each partial set P of E, or else the theorem is trivially true for that P. In (7.3.25) we replace ε by $\varepsilon \cdot 2^{-j}$, $NV(h; \mathbf{A}; P)$ by $M_j(P)$, $\mathcal{U} \in \mathbf{A}|E$ by $\mathcal{U}_j \in \mathbf{A}|E$, and L_1 by L_j ($j = 1, 2, \ldots$). Then for each j,

$$\left|(S)\int_I h(K)\,dN\right| \chi(X_j; t) \leq M_j(I) + (S)\int_I L_j(K)\,dN$$

$$\leq M(I) + \sum_{j=1}^{\infty} (S)\int_I L_j(K)\,dN$$

$$= M(I) + (S)\int_I \sum_{j=1}^{\infty} L_j(K)\,dN$$

by the monotone convergence theorem for the S-integral. Hence on the left we can let $j \to \infty$, and by definition have

$$NV(h; \mathbf{A}; E; X) \leq M(E) + \sum_{j=1}^{\infty} L_j(E) < M(E) + \varepsilon.$$

$$NV(h; \mathbf{A}; E; X) \leq M(E) = \lim_{j \to \infty} M_j(E).$$

To prove the general case of (7.3.38), follow the proof of Theorem 2.8.1(2.8.2), with NV for $\bar{V}$.

For (7.3.40) we use (7.3.31) to show that $b(h - k)$ is of N-variation zero for each real constant b. Taking $b = 1, 2, 3, \ldots$ with (7.3.37), $f(h - k)$ is of N-variation zero and so has N-integral 0. As fh is N-integrable, so is fk, with the same integrals. For (7.3.41) we change $\bar{V}$ to NV in Exercise 2.9.1. □

The theory of the N-integral, the N-variational integral, and the N-variation were developed in Henstock (1960b, c, 1961a, b, 1963b, 1983) Foglio (1963, 1968, 1970, 1985), and Foglio and Henstock (1973).

CHAPTER 8

FUNCTIONAL ANALYSIS AND INTEGRATION THEORY

8.1 INTRODUCTION

In order to study continuous linear functionals on a space S of functions f integrable in a given sense on the elementary sets E of a division space $(T, \mathcal{T}, \mathbf{A})$, relative to an integrator h, giving integrals

(8.1.1) $$F(fh:I) \equiv \int_I f\,dh \quad (I \subseteq E),$$

we need to define the continuity in S. We use a norm $\|f\|$, and this can be very different, depending on the properties of S. For example, if $h \geq 0$, and $|f| \in S$ when $f \in S$, so that we have an absolute integral, we usually take $\|f\| = F(|f|h:E)$, the integral being analogous to absolutely convergent series. Some S have $h \geq 0$ and some $f \in S$ with $|f| \notin S$, from non-absolute integrals analogous to conditionally convergent series, the kinds of integrals that, in the main, are studied in this book. For example, S can be the family of all functions integrated by the Cauchy–Riemann, or Cauchy–Lebesgue, or special Denjoy, or general Denjoy, or approximate Perron, or Çesaro–Perron integrals, etc. Here, many $f \in S$ have $F(|f|h; E) = +\infty$, so that kind of norm is useless. Instead we use

(8.1.2) $$\|f\| = \sup_I |F(fh; I)|$$

over all partial intervals I of E, the supremum norm $\|F\|_\infty$ of F, provided that this is finite for all $f \in S$. If infinite for some $f \in S$, then as $h \geq 0$, $F(|f|h; E) = +\infty$ also. Pfeffer (1988/1989a) gives an example of such an S. Now $F(fh; I) = F(gh; I)$, finite, for all partial intervals I of E if and only if one integral is finite and $f = g$ h-almost everywhere {Theorem 2.4.1(2.4.4), (2.4.5)}. Thus a linear functional $L(f)$ is also a linear functional $L_1(F)$ of the F if $L(f) = L(g)$ when $f - g = 0$ h-almost everywhere, or $\|f - g\| = 0$. If L is also continuous then $\|f - g\| = 0$ implies $L(f - g) = 0$, and $L_1(F)$ is a continuous linear functional of F using (8.1.2) for norm. Step functions in S give particularly simple F, say F_0. For continuous h, F is also continuous {Theorem 2.5.4(2.5.13)} so that the space S^* of F can be embedded in the space $C(E)$ of functions continuous on partial intervals of E, which can be approximated by the F_0. Thus by continuity L_1 extends to a continuous

8.2 YOUNG'S INEQUALITY AND ORLICZ SPACES

For $c > 0$ let $\varphi: [0, c] \to \mathbb{R}$ be monotone increasing. As $+\infty$ is not in $\mathbb{R}$, we imply that all values of φ are finite. Then φ is of bounded variation, every difference being non-negative, and the sum of moduli of differences of φ-values over $[0, c]$ is $\varphi(c) - \varphi(0)$. For convenience we assume that $\varphi(t) = \varphi(t-)$ $(0 < t \leq c)$ and we define $\psi: [\varphi(0), \varphi(c)] \to [0, c]$ by

$$\psi(u) = \inf\{t: 0 \leq t \leq c, \varphi(t) \geq u\}.$$

In this case we say that ψ is the *Young inverse function* of φ.

Theorem 8.2.1.

(8.2.1) *For φ, ψ as given, ψ is finite and monotone increasing and*

(8.2.2) $\quad \psi(u) = \psi(u-) \quad (\varphi(0) < u \leq \varphi(c))$,

(8.2.3) $\quad \psi(\varphi(t)) = \inf\{v: \varphi(v) = \varphi(t), \ 0 \leq v \leq t\} \ (0 \leq t \leq c)$.

(8.2.4) *If also φ is continuous at $\psi(u)$ and $\varphi(0) < u < \varphi(c)$ then $\varphi(\psi(u)) = u$.*

Proof. If $\varphi(0) \leq u < v \leq \varphi(c)$ then $\psi(u) \leq \psi(v)$ by the monotonicity of φ. If for fixed k and some u in $\varphi(0) < u \leq \varphi(c)$, $\psi(u-) < k < \psi(u)$, then for every $w < u$, $\psi(w) < k$, $w \leq \varphi(k) < u$. But if we take $w = \frac{1}{2}\{\varphi(k) + u\}$ the result is falsified. Hence (8.2.2). For (8.2.3),

$$\psi(\varphi(t)) = \inf\{z: \varphi(z) \geq \varphi(t), \ 0 \leq z \leq c\}$$

while one value of z is t. For (8.2.4), if $t > \psi(u)$ then $\varphi(t) \geq u$, so $\varphi(\psi(u)+) \geq u$. If $t < \psi(u)$, $\varphi(t) < u$, $\varphi(\psi(u)-) \leq u$. Continuity of φ at $\psi(u)$ gives (8.2.4). □

Theorem 8.2.2. *In Theorem 8.2.1, if $0 \leq a < b \leq c$, then for Riemann integrals,*

(8.2.5) $\quad \displaystyle\int_{[a,b]} \varphi \, dt + \int_{[\varphi(a), u]} \psi \, dv = bu - a\varphi(a) \quad (\varphi(b) \leq u \leq \varphi(b+))$.

Proof. φ and ψ are bounded and monotone increasing, so that the two Riemann integrals exist. Let $a = u_0 < u_1 < \cdots < u_m = b$ be a partition with $u_j - u_{j-1} < \varepsilon$ $(j = 1, 2, \ldots, m)$ for given $\varepsilon > 0$. If $u_j \in (v, w] \subseteq [a, b]$ with $\varphi(w) = \varphi(v)+$ but $\varphi(t) < \varphi(v)$ $(t < v)$, we replace u_j by v. Then several u_j may be replaced by the same v. Also the mesh may be altered. But as φ is constant in $(v, u_j]$ the mesh increase is not important. If pedantically necessary we can insert extra division points to bring the mesh down. The second partition is the transform of the first by φ,

(8.2.6) $$\varphi(a) = \varphi(u_0) < \varphi(u_1) < \cdots < \varphi(u_m) = \varphi(b) \leq u.$$
$$\psi(\varphi(u_j)) = u_j (j = 1, \ldots, m-1), \psi(u) = b$$

by Theorem 8.2.1(8.2.3) and by definition of ψ. Thus we have sums for the two integrals,

$$\sum_{j=1}^{m} \varphi(u_{j-1})(u_j - u_{j-1}) + \sum_{j=1}^{m-1} \psi(\varphi(u_j))$$
$$\times \{\varphi(u_j) - \varphi(u_{j-1})\} + \psi(u)\{u - \varphi(u_{m-1})\}$$
$$= \sum_{j=1}^{m} \varphi(u_{j-1})(u_j - u_{j-1}) + \sum_{j=1}^{m-1} u_j \{\varphi(u_j) - \varphi(u_{j-1})\} + b\{u - \varphi(u_{m-1})\}$$
$$= bu - a\varphi(a)$$

on cancellation. The mesh of (8.2.6) need not tend to 0, because of the discontinuities of φ, which introduce intervals on which ψ is constant and where we can take extra division points without affecting the sum over $[\varphi(a), u]$, so that the mesh then tends to 0. Hence (8.2.5). □

By these means we have prepared the ground for the proof of Young's inequality, which is in the next theorem.

Theorem 8.2.3. *In Theorem 8.2.2, if $0 \leq a < c$, $0 = \varphi(0) \leq b \leq \varphi(c)$, then*

(8.2.7) $$\Phi(a) + \Psi(b) \geq ab \quad \text{where}$$

$$\Phi(a) \equiv \int_{[0,a]} \varphi(t) dt \ (a > 0), \quad \Phi(0) = 0,$$

$$\Psi(b) \equiv \int_{[0,b]} \psi(u) du \ (b > 0), \quad \Psi(0) = 0.$$

(8.2.8) *Equality occurs in (8.2.7) if, and only if, $\varphi(a) \leq b \leq \varphi(a+)$.*

Proof. We use Theorem 8.2.2(8.2.5) repeatedly, particularly for the case of equality. Since $\varphi(0) = 0$, ψ is defined in $[0, \varphi(c)]$, and the results are obvious if $a = 0$ or $b = 0$. If $\varphi(a+) < t \leq c$, then by definition of ψ, $\psi(t) > a$. Hence if $a > 0, b > 0, b > \varphi(a+)$,

$$\Phi(a) + \Psi(b) = \Phi(a) + \Psi(\varphi(a+)) + \int_{[\varphi(a+),b]} \psi \, du$$
$$> a\varphi(a+) + a(b - \varphi(a+)) = ab.$$

8.2 YOUNG'S INEQUALITY AND ORLICZ SPACES

If $0 < b < \varphi(a)$, there is a t in $0 \leq t \leq a$ with

$$\varphi(t) \leq b \leq \varphi(t+), \quad \varphi(x) > b \;(t < x \leq a),$$

$$\Phi(a) + \Psi(b) = \Phi(t) + \Psi(b) + \int_{[t,a]} \varphi \, dx$$

$$> tb + b(a-t) = ab.$$

Hence if equality holds in (8.2.7) then the range of b is in (8.2.8). Conversely, if that range holds, (8.2.5) gives equality in (8.2.7) and (8.2.8) is true. □

We now use Young's inequality to prove some elementary inequalities involving Φ and Ψ remembering that $f: T \to \mathbb{R}^+$ is h-measurable if $\min(f, m)h$ is integrable for every integer m.

Theorem 8.2.4. *Let $(T, \mathcal{T}, \mathbf{A})$ be an additive division space, let $h: \mathcal{U}^1 \to \mathbb{R}^+$, $f: T \to \mathbb{R}^+$, $g: T \to \mathbb{R}^+$, f and g being h-measurable, and let a, b be positive constants. Then*

$$(8.2.9) \qquad \int_E \Phi(af) \, dh + \int_E \Psi(bg) \, dh \geq ab \int_E fg \, dh,$$

with equality if and only if

$(8.2.10) \quad \varphi(af(t)) \leq bg(t) \leq \varphi(af(t)+), \text{ for } t \text{ h-almost everywhere}.$

If for $a = a_0$, $b = b_0$, the left side of (8.2.9) is 1, then

$$(8.2.11) \qquad \int_E fg \, dh \leq a_0^{-1} b_0^{-1}.$$

Proof. As $\Phi(f)$, $\Psi(f)$ and $r(f) = f^2$ are convex functions of f we use Exercise 4.1.1 to ensure that because of the h-measurability of f and g, $\Phi(af)h$ and $\Psi(bg)h$ are integrable if and only if, for some $\mathcal{U} \in \mathbf{A}|E$ and all divisions $\mathcal{E}$ of E from $\mathcal{U}$, the sets of sums

$$(8.2.12) \qquad (\mathcal{E}) \sum \Phi(af)h, \quad (\mathcal{E}) \sum \Psi(bg)h,$$

respectively, are bounded above. To prove (8.2.9) we can assume that the left side is finite, so that for some $\mathcal{U} \in \mathbf{A}|E$ the sums in (8.2.12) are bounded for all divisions $\mathcal{E}$ of E from $\mathcal{U}$. Hence by (8.2.7).

$$(\mathcal{E}) \sum (fg) h$$

is bounded above, independent of the divisions $\mathcal{E}$ over E from $\mathcal{U}$. Hence the right side of (8.2.9) exists and the inequality follows from (8.2.7), rearranged as

$$p(t) \equiv \Phi(af(t)) + \Psi(bg(t)) - abf(t)g(t) \geq 0.$$

If equality in (8.2.9) then the integral of ph over t is 0, so that as $ph \geq 0$, the integral over every partial set of E is 0. By Theorem 2.6.1(2.6.1), $p = 0$ h-almost everywhere, so that (8.2.10) follows by Theorem 8.2.3(8.2.8). Conversely, (8.2.10) gives $p(t) = 0$ h-almost everywhere, and equality occurs in (8.2.9). Finally, (8.2.11) is obvious, with (8.2.10) for equality when $a = a_0$, $b = b_0$.

An example of the inequalities is given in Example 8.2.1, but the most familiar φ is $\varphi(t) = t^{p-1}$ where $p > 1$ is fixed. For $(q - 1)(p - 1) = 1$ or $q = p/(p - 1)$ or $1/p + 1/q = 1$, $\psi(u) = u^{q-1}$, $\Phi(t) = t^p/p$, $\Psi(u) = u^q/q$, and (8.2.9),

$$\int_E a^p f^p \, dh/p + \int_E b^q g^q \, dh/q \geq ab \int_E fg \, dh,$$

$$a_0^{-p} = \int_E f^p \, dh, \quad b_0^{-q} = \int_E g^q \, dh,$$

(8.2.13)
$$\int_E fg \, dh \leq \left(\int_E f^p \, dh \right)^{1/p} \left(\int_E g^q \, dh \right)^{1/q}$$

$(f \geq 0, g \geq 0, p > 1, q > 1, 1/p + 1/q = 1)$

with equality if and only if $a^{p-1} f^{p-1} = bg$, $a^p f^p = b^q g^q$, h-almost everywhere; i.e.

(8.2.14) $f^p/g^q = $ constant h-almost everywhere.

Thus we have Hölder's inequality for integrals, with the condition for equality, Hölder (1889). Replacing f^p by f, g^q by g, $1/p$ by s, we have

(8.2.15) $$\int_E f^s g^{1-s} \, dh \leq \left(\int_E f \, dh \right)^s \left(\int_E g \, dh \right)^{1-s} \quad (f \geq 0, g \geq 0, 0 < s < 1).$$

See Theorem 2.7.7.

We denote by L^p the space of all h-measurable functions f on E^* (i.e. those f for which $f_m \equiv \min\{m, \max(m - 1, f)\}$ and $f_m h$ is integrable over E,) with $N(f, 0)_p$ finite, where $p > 0$ is fixed and for $m = 0, \pm 1, \pm 2, \pm 3, \ldots,$

$$N(f, g)_p \equiv \int_E |f(t) - g(t)|^p \, dh.$$

For $p > 1$, $\|f\|_p \equiv N(f, 0)_p^{1/p}$ is a seminorm, which follows from Minkowski's inequality. When $0 < p \leq 1$, $N(f, 0)_p$ itself is a seminorm.

Theorem 8.2.5.

(8.2.16) *If f, g are in L^p, so is $f + g$, and (Minkowski's inequality)*

$$\|f + g\|_p \leq \|f\|_p + \|g\|_p \quad (1 < p < \infty).$$

8.2 YOUNG'S INEQUALITY AND ORLICZ SPACES

(8.2.17) When $f \geq 0$, $g \geq 0$, equality occurs in (8.2.16) if and only if for constants $a \geq 0$, $b \geq 0$, with $a + b > 0$, $af(t) = bg(t)$ h-almost everywhere.

Proof. To show that $f + g \in L^p$ ($0 < p < \infty$), if, for some t,

$$|f| \geq |g|, \quad |f+g|^p \leq (|f|+|g|)^p \leq (2|f|)^p.$$

If
$$|f| < |g|, \quad |f+g|^p \leq 2^p|g|^p.$$

$$|f+g|^p \leq 2^p(|f|^p + |g|^p)$$

for all t. Integrating and using the monotone convergence theorem with h-measurability, $f + g \in L^p$ ($0 < p < \infty$). Next, from Theorem 2.7.7 with $t = p^{-1}$ ($1 < p < \infty$) and $(p-1)q = p$, $|f|(|f|+|g|)^{p-1}h$ and $|g|(|f|+|g|)^{p-1}h$ are integrable on E. Using Hölder's inequality,

$$\int_E (|f|+|g|)^p \, dh = \int_E |f|(|f|+|g|)^{p-1} \, dh + \int_E |g|(|f|+|g|)^{p-1} \, dh$$

$$\leq \left(\left(\int_E |f|^p \, dh\right)^{1/p} + \left(\int_E |g|^p \, dh\right)^{1/p}\right)$$

$$\times \left(\int_E (|f|+|g|)^p \, dh\right)^{1/q}.$$

The inequality follows on dividing by the last factor when positive. When zero, then, h-almost everywhere, $|f| + |g| = 0$, $f = 0 = g$, and (8.2.16) is true. Omitting this last case, of equality, equality holds in (8.2.16) if and only if both Hölder inequalities are equalities and for some constants a, b, and h-almost everywhere,

$$|f|^p = b(|f|+|g|)^p, \quad |g|^p = a(|f|+|g|)^p, \quad a|f| = b|g|. \quad \square$$

Theorem 8.2.6.

(8.2.18) $$\lim_{q \to \infty} \left(\int_E |g|^q \, dh\right)^{1/q} = \|g\|_\infty \equiv \operatorname{ess\,sup}\{|g(t)|: t \in E^*\},$$

the supremum of all constants b for which the set X_b of t where $|g(t)| > b$, has $V(h; \mathbf{A}; E; X_b) > 0$.

Proof. $b \cdot V(h; \mathbf{A}; E; X_b)^{1/q} \leq \left(\int_E |g|^q \, dh\right)^{1/q} \leq \|g\|_\infty V(h; \mathbf{A}; E)^{1/q}$,

since the set of t with $|g(t)| > \|g\|_\infty$ has h-variation zero. (Take $|g(t)| > \|g\|_\infty + 1/m$, $m = 1, 2, \ldots$. We cannot replace $1/m$ by 0 since (a priori) the supremum could be attained.) Hence the result by letting $q \to \infty$ first. $\square$

Denoting by L^∞ the space of all h-measurable functions g with $\|g\|_\infty$ finite, Hölder's inequality becomes the obvious inequality

(8.2.19) $$\left|\int_E fg\, dh\right| \leq \int_E |fg|\, dh \leq \int_E |f|\, dh\, \|g\|_\infty = \|f\|_1 \|g\|_\infty,$$

which of course is useless unless the right side is finite, and so $\|g\|_\infty = 0$ (i.e. $g = 0$ h-almost everywhere), or $\|f\|_1 < \infty$ and $0 < \|g\|_\infty < \infty$ (i.e. g is bounded h-almost everywhere).

Theorem 8.2.7.

(8.2.20) *If f, g are in L_∞, so is $f + g$, and $\|f + g\|_\infty \leq \|f\|_\infty + \|g\|_\infty$.*

(8.2.21) *Equality holds in (8.2.20) if and only if, for each $\varepsilon > 0$, the sets X_ε and Y_ε where $|f| > \|f\|_\infty - \varepsilon$, $|g| > \|g\|_\infty - \varepsilon$, respectively, have $\bar{V}(h; \mathbf{A}; E; X_\varepsilon \cap Y_\varepsilon) > 0$. with $fg \geq 0$ h-almost everywhere.*

Proof. $|f + g| \leq |f| + |g| \leq \|f\|_\infty + \|g\|_\infty$ h-almost everywhere, for (8.2.20); the last two inequalities have to be equalities in the case of equality, and the proof of (8.2.21) is straightforward.

Theorem 8.2.8. *For fixed $f \in L^p$ and all $g \in L^q$ with $\|g\|_q = 1$,*

(8.2.22) $$\sup \int_E fg\, dh = \sup \int_E |fg|\, dh = \|f\|_p \|g\|_q \quad (1 \leq p \leq \infty).$$

provided that when $p = \infty$ each h-measurable set of infinite h-variation contains an h-measurable set of finite non-zero h-variation.

Proof. For $\text{sgn}(x) = x/|x|$ ($x \neq 0$), $\text{sgn}(0) = 1$, we put $g\,\text{sgn}^{-1}(fg)$ for g so that fg becomes $|fg|$ without altering $\|g\|_q$. Next, if $\|f\|_p = 0$, $f = 0$ h-almost everywhere, in all cases, and (8.2.22) is trivial. Then when $p = 1$ we take $g(t) = 1$ everywhere, the integral of fg being $\|f\|_1$, and the result is true. When $1 < p < \infty$, we take $\|f\|_p > 0$, $g = |f|^{p-1}/\|f\|_p^{p-1}$, $|fg| = |f|^p/\|f\|_p^{p-1}$, $\|g\|_q = 1$ since $q(p-1) = p$. Thus the integral of $|fg|$ is $\|f\|_p$. When $p = \infty$, $q = 1$, $\|f\|_\infty > 0$, let $\varepsilon > 0$. Then the set where $|f| > \|f\|_\infty - \varepsilon$ is or contains a subset X with $0 < \bar{V}(h; \mathbf{A}; E; X) = V(X) < \infty$ and $\chi(X; \cdot)h$ integrable to $V(X)$. Taking $g(t) = V(X)\chi(X; \cdot)$, $\|g\|_1 = 1$,

$$\int_E |fg|\, dh = V(X)^{-1} \int_E |f|\chi(X;\cdot)\, dh \geq \|f\|_\infty - \varepsilon,$$

and $\varepsilon > 0$ is arbitrary. □

8.2 YOUNG'S INEQUALITY AND ORLICZ SPACES

For most of the general Young functions Φ, Ψ, we cannot prove a result like Theorem 8.2.4(8.2.11), so that it is usual to take the conclusion of Theorem 8.2.8 as a definition, proceeding as follows:

Let L_Φ^* be the set of all real-valued h-measurable point functions f with $\Phi(|f|)h$ integrable over E. L_Φ^* is not always linear, for see Exercise 8.2.2. We define L_Φ to be the set of all h-measurable f for which $\|f\|_\Phi$ is finite, where for all h-measurable g with

$$\int_E \Psi(|g|)\,dh \leq 1, \quad \|f\|_\Phi = \sup \int_E |fg|\,dh.$$

Theorem 8.2.9. *The space L_Φ is linear, and if f_1, f_2 are in L_Φ,*

(8.2.23) $$\|f_1 + f_2\|_\Phi \leq \|f_1\|_\Phi + \|f_2\|_\Phi.$$

(8.2.24) *If $f \in L_\Phi^*$ then $\|f\|_\Phi \leq \int_E \Phi(|f|)\,dh + 1$ and $L_\Phi^* \subseteq L_\Phi$.*

(8.2.25) *In the definition of $\|f\|_\Phi$, $\int_E fg\,dh$ can replace $\int_E |fg|\,dh$.*

(8.2.26) *Let every h-measurable set of non-zero h-variation contain an h-measurable set of finite arbitrarily small non-zero h-variation, and let ψ be not identically zero. Then $\|f\|_\Phi = 0$ if and only if $f = 0$ h-almost everywhere.*

Proof. The scalar multiplication is clear, and addition of f_1, f_2 in L_Φ follows from

$$\int_E \Psi(|g|)\,dh \leq 1$$

and

$$\int_E |(f_1 + f_2)g|\,dh \leq \int_E |f_1 g|\,dh + \int_E |f_2 g|\,dh \leq \|f_1\|_\Phi + \|f_2\|_\Phi.$$

Hence (8.2.23). For f, g, h-measurable, Theorem 8.2.4(8.2.9) gives Young's inequality, the last integral being not greater than 1 in the definition of $\|f\|_\Phi$. Hence (8.2.24). We have already multiplied g by $\operatorname{sgn}^{-1}(fg)$ to prove (8.2.25). Then in (8.2.26), if $\|f\|_\Phi = 0$, $\int_E |fg|\,dh = 0$ for all h-measurable g with $\int_E \Psi(|g|)\,dh \leq 1$. Thus $fg = 0$, h-almost everywhere, for these g. If $f \neq 0$ on an h-measurable set X of non-zero h-variation, then X is or contains a set X_1, also h-measurable, with non-zero h-variation as small as we please. Taking a $k > 0$ for which $\Psi(k) > 0$, let

$$0 < \Psi(k).\bar{V}(h;\mathbf{A};E;X_1) \leq 1, \quad g(t) = k\chi(X_1;t), \quad \|g\|_\Psi^* \leq 1, \quad fg \neq 0$$

in X_1, contradicting that $fg = 0$ h-almost everywhere. Hence (8.2.26). $\square$

Theorem 8.2.10.

(8.2.27) *If $f \in L_\Phi$ and $\|f\|_\Phi \neq 0$, then $f/\|f\|_\Phi$ lies in L_Φ^*.*

(8.2.28) If also there is a constant $M > 0$ such that $\Phi(2t) \leq M\Phi(t)$ $(t \geq t_0 \geq 0)$, then $L_\Phi = L_\Phi^*$.

(8.2.29) *(Hölder's inequality for Orlicz spaces)* If $f \in L_\Phi$, $g \in L_\Psi$, and f, g are h-measurable, with the conditions of (8.2.26), then fgh and $|fg|h$ are integrable and

$$\left| \int_E fg\, dh \right| \leq \int_E |fg|\, dh \leq \|f\|_\Phi \|g\|_\Psi.$$

Proof. For $q > 1$, $v > 0$, ψ being monotone increasing,

$$\Psi(qv) - \Psi(v) = \int_{[v, qv)} \psi(t)\, dt \geq (qv - v)\psi(v)$$

$$\geq (q - 1)\int_{[0, v)} \psi(t)\, dt = (q - 1)\Psi(v),$$

(8.2.30) $$\Psi(qv) \geq q\Psi(v) \quad (q \geq 1, v \geq 0),$$

the extra cases being trivial. If $1 < \|g\|_\Psi^* \equiv \int_E \Psi(|g|)\, dh < \infty$, (8.2.30) gives

$$\Psi(|g|) \geq \|g\|_\Psi^* \Psi(|g|/\|g\|_\Psi^*),$$

$$\int_E \Psi(|g|/\|g\|_\Psi^*)\, dh \leq \int_E \Psi(|g|)\, dh/\|g\|_\Psi^* = 1.$$

It follows by definition that

$$\int_E |fg/\|g\|_\Psi^*|\, dh \leq \|f\|_\Phi, \quad \int_E |fg|\, dh \leq \|f\|_\Phi \|g\|_\Psi^* \quad (\|g\|_\Psi^* > 1),$$

$$\int_E |fg|\, dh \leq \|f\|_\Phi k(g), \quad k(g) \equiv \max(\|g\|_\Psi^*, 1) \qquad \text{(all } g \in L_\Psi^*\text{)}.$$

As f is h-measurable, so are $|f|/\|f\|_\Phi$ and $\Phi(|f|/\|f\|_\Phi)$. Taking $g = \varphi(|f|/\|f\|_\Phi)$, h-measurable, there is equality in Young's inequality (Theorem 8.2.4(8.2.10)) at every point, the integrals below existing. Thus

$$k(g) \geq \int_E |(f/\|f\|_\Phi)g|\, dh = \int_E \Phi(|f|/\|f\|_\Phi)\, dh + \|g\|_\Psi^*.$$

$$\int_E \Phi(|f|/\|f\|_\Phi)\, dh = 0 \quad \text{if} \quad k(g) = \|g\|_\Psi^*.$$

If $k(g) > \|g\|_\Psi^*$, then $k(g) = 1$ and so

$$\int_E \Phi(|f|/\|f\|_\Phi)\, dh \leq 1,$$

and in either case (8.2.27) is true. For (8.2.28), $L_\Phi^* \subseteq L_\Phi$ by

8.2 YOUNG'S INEQUALITY AND ORLICZ SPACES

Theorem 8.2.9(8.2.24). If $f \in L_\Phi$ and $\|f\|_\Phi \neq 0$ then $f/\|f\|_\Phi \in L_\Phi^*$ while for some p, $\|f\|_\Phi \leq 2^p$,

$$\Phi(|f|) \leq M^p \Phi(|f|/2^p) \leq M^p \Phi(|f|/\|f\|_\Phi),$$

$$\int_E \Phi(|f|) \, dh \leq M^p \int_E \Phi(|f|/\|f\|_\Phi) \, dh \leq M^p$$

and $f \in L_\Phi^*$ (assuming $t_0 = 0$). If $t_0 > 0$ we write $f = f_1 + f_2$ where $f_1 = f$ on the set where $|f|/\|f\|_\Phi < t_0$ and $f_1 = 0$ elsewhere. Then $\Phi(|f_1|)$ is bounded by $\Phi(t_0 \|f\|_\Phi)$ and is h-measurable, and so is integrable with respect to h. We prove the integrability of $\Phi(|f_2|)h$ as for $t_0 = 0$, and $\Phi(|f|) = \Phi(|f_1|) + \Phi(|f_2|)$, giving (8.2.28). For (8.2.29), f, g are h-measurable, so fgh and $|fg|h$ are integrable. For $\|g\|_\Psi = 0$ implies $g = 0$ h-almost everywhere (Theorem 8.2.9(8.2.26)) and the result is trivial. Otherwise, in the proof of (8.2.27) we replace f, Φ by g, Ψ and have $\|g/\|g\|_\Psi\|_\Psi^* \leq 1$, so that by definition

$$\int_E |fg| \, dh = \int_E |f(g/\|g\|_\Psi)| \, dh \, \|g\|_\Psi \leq \|f\|_\Phi \|g\|_\Psi. \quad \square$$

In the statement of (8.2.28) we can replace 2 by any number $N > 1$. But the condition is still very restrictive and is not satisfied by the Ψ of Example 8.2.1.

We have seen that $\|f\|_\Phi$ is a seminorm. In order to apply Baire's density theorem we look at the completeness of L_Φ, first the L_p spaces, where the only condition on h is $h \geq 0$. Then we use it to prove the result for L_Φ, with h restricted.

Theorem 8.2.11. *Let $h \geq 0$, let $(f_j(t))$ be a sequence of point functions, let p be fixed in $(0, \infty)$, and for*

$$N(f, g)_p \equiv \int_E |f - g|^p \, dh,$$

let $N(f_j, f_k)_p$ exist for every pair j, k of positive integers. Given $\varepsilon > 0$, let there be a $p(\varepsilon)$ for which

(8.2.31) $N(f_j, f_k)_p < \varepsilon$ (all $j, k \geq p(\varepsilon)$).

Then there are a point function f and a subsequence $(j(r))$ of positive integers tending to infinity, such that

(8.2.32) $f(t) = \lim_{r \to \infty} f_{j(r)}(t)$, *finite, h-almost everywhere in E^*, and for each positive integer $j \geq p(\varepsilon)$, the integral below exists and*

(8.2.33) $N(f_j, f)_p \leq \varepsilon.$

(8.2.34) *If g satisfies the conditions for f in (8.2.33) then $g = f$ h-almost everywhere, showing the essential uniqueness of f.*

A sequence satisfying (8.2.31) is called *fundamental* or *Cauchy*, one satisfying (8.2.33) is called *convergent to f*, and the theorem shows that every fundamental sequence is convergent, and the function space is complete.

Proof of theorem. For $(j(r))$ a strictly increasing sequence and X_r a set with

(8.2.35) $\quad N(f_j, f_k)_p < 2^{-2rp}(j, k \geqslant j(r)), \quad |f_{j(r)}(t) - f_{j(r+1)}(t)| \geqslant 2^{-r}$,

respectively, by Theorem 2.7.4(2.7.7),

$$2^{-rp}\bar{V}(h; \mathbf{A}; E; X_r) \leqslant \bar{V}(|f_{j(r)}(t) - f_{j(r+1)}(t)|^p h(I, t); \mathbf{A}; E)$$
$$= N(f_{j(r)}, f_{j(r+1)})_p < 2^{-2rp},$$
$$\bar{V}(h; \mathbf{A}; E; X_r) < 2^{-rp}.$$

$$\sum_{r=M}^{\infty} |f_{j(r)}(t) - f_{j(r+1)}(t)| < \sum_{r=M}^{\infty} 2^{-r} = 2^{-M+1} \quad \left(t \notin Y_M \equiv \bigcup_{r=M}^{\infty} X_r\right),$$

$$\lim_{r \to \infty} f_{j(r)}(t) = f_{j(1)}(t) + \sum_{r=1}^{\infty} (f_{j(r+1)}(t) - f_{j(r)}(t))$$

exists as a function $f(t)$ in $\setminus X$ where

$$X \equiv \bigcap_{M=1}^{\infty} Y_M, \quad \bar{V}(h; \mathbf{A}; E; X) \leqslant \bar{V}(h; \mathbf{A}; E; Y_M) \leqslant 2^{-Mp}/(1 - 2^{-p}),$$
$$\bar{V}(h; \mathbf{A}; E; X) = 0.$$

Thus f exists h-almost everywhere. By Fatou's lemma (Theorem 3.3.4),

$$\int_E |f_j(t) - f(t)|^p dh \leqslant \liminf_{r \to \infty} \int_E |f_j(t) - f_{j(r)}(t)|^p dh \leqslant \varepsilon \quad (j \geqslant p(\varepsilon)),$$

and (8.2.32), (8.2.33) are proved. If also, for every $\varepsilon > 0$,

$$\int_E |f_j(t) - g(t)|^p dh \leqslant \varepsilon \quad (j \geqslant q(\varepsilon)),$$

then by Fatou's lemma again,

$$\int_E |f(t) - g(t)|^p dh \leqslant \liminf_{r \to \infty} \int_E |f_{j(r)}(t) - g(t)|^p dh \leqslant \varepsilon,$$

$$\int_E |f(t) - g(t)|^p dh = 0.$$

By Theorem 2.7.4(2.7.7) again,

$$V(|f(t) - g(t)|^p h(I, t); \mathbf{A}; E) = 0, \quad f(t) = g(t),$$

h-almost everywhere, and we have the essential uniqueness of f. □

To extend the completeness to Orlicz spaces L_Φ, L_Ψ, we restrict h.

8.2 YOUNG'S INEQUALITY AND ORLICZ SPACES

Theorem 8.2.12. *Let $h \geq 0$ be VBG* relative to a sequence (X_j) of mutually disjoint h-measurable sets with union T. Then L_Φ is complete.*

Proof. Given $\varepsilon > 0$, there is a $p = p(\varepsilon)$ such that for all g with $0 < \|g\|_\Psi^* \leq 1$,

(8.2.36) $$\int_E |f_j - f_k| \cdot |g| \, dh < \varepsilon \quad (j, k \geq p(\varepsilon)).$$

If $\varphi = 0$ everywhere, so is Φ, L_Φ contains all h-measurable functions, and we have only to prove that the limit function of a sequence of h-measurable functions is h-measurable, the h-measurability of f being defined by the integrability of $f_m^* h$ where $f_m^* = \min\{\max(m-1, f), m\}$, and $m = 0, \pm 1, \pm 2, \ldots$. As every f_m^* is bounded by $m-1$ and m we use the Arzelà–Lebesgue test.

If φ is not identically 0, there are points $r > 0$, $s > 0$, with

$$s = \varphi(r), \quad \psi(s) \leq r, \quad \psi(u) \leq r \ (u \leq s), \quad \Psi(u) \leq ru \ (u \leq s), \quad \lim_{u \to 0+} \Psi(u) = 0,$$

and Ψ is smaller than any positive number, for u small enough. But if $\varphi(0+) > \varphi(0)$, $\Psi = 0$ in $[\varphi(0), \varphi(0+))$. Thus, given $0 < \bar{V}(h; A; E; X) < \infty$, there is a constant $v > 0$ with $\Psi(v) \leq \bar{V}(h; A; E; X)^{-1}$. If $g = v\chi(X; \cdot)$,

$$\int_E \Psi(|g|) \, dh = \Psi(v) \bar{V}(h; A; E; X) \leq 1.$$

Taking $X = X_l$ in turn, for $l = 1, 2, \ldots$, and amalgamating all sets X_l of h-variation zero with the first X_l with finite non-zero h-variation, and renumbering, (8.2.36) gives

$$\int_E |f_j - f_k| \chi(X_l; \cdot) \, dh : v_l < \varepsilon \quad (j, k \geq p(\varepsilon)),$$

and Theorem 8.2.12 with $p = 1$ ensures that there is a subsequence $(r_j(s))$ tending to infinity, of the positive integers, such that $f_{r_j(s)}(t)$ tends h-almost everywhere in X_l to a function $f(t)$. By the proof we can arrange that $(r_{j+1}(s))$ is a subsequence of $(r_j(s))$, and then the diagonal sequence $(r(j))$ with $r(j) = r_j(j)$, is a subsequence of $(r_j(s))$ from the jth term onwards. Hence $f_{r(j)}(t)$ tends to $f(t)$ h-almost everywhere in the union T of the X_l, and applying Fatou's lemma to (8.2.36) with $k = r(j)$, then uniformly in g with $\|g\|_\Psi^* \leq 1$,

$$\int_E |f_j - f| \cdot |g| \, dh \leq \varepsilon \quad (j \geq p(\varepsilon)), \quad \|f_j - f\|_\Phi \to 0 \quad (j \to \infty),$$

and L_Φ is complete. Uniqueness is proved as in Theorem 8.2.11. □

Next we need a condition to circumvent Exercise 8.2.2.

Theorem 8.2.13. *Let $h \geq 0$ be VBG* relative to a sequence (X_m) of mutually disjoint h-measurable sets with union T, with $(f_j) \subseteq L_\Phi$, $(g_j) \subseteq L_\Psi$, and let f, g satisfy $\|f_j - f\|_\Phi \to 0$, $\|g_j - g\|_\Psi \to 0$. Also, for some constant $M > 0$ let*

(8.2.37) $$\|f_j\|_\Phi \leq M, \quad \|g_j\| \leq M \quad (j = 1, 2, \ldots).$$

Then $f \in L_\Phi$, $g \in L_\Psi$, and $f_j g_j h$ and fgh are integrable with

(8.2.38) $$\int_E f_j g_j \, dh \to \int_E fg \, dh.$$

Further, if the conclusions of the Theorem hold, then (8.2.37) is true.

Proof. As in the proof of Theorem 8.2.12, a subsequence $(r(j))$ of the positive integers and tending to infinity satisfies $f_{r(j)}(t) \to f(t)$ h-almost everywhere convergent, and so for a subsequence $(s(j))$ of $(r(j))$, $g_{s(j)}(t) \to g(t)$ h-almost everywhere, and $f_{s(j)}(t) \to f(t)$ h-almost everywhere. Thus if g is an h-measurable function with $\|g\|_\Psi^* \leq 1$, and by Fatou's lemma,

$$\int_E |fg|\,dh = \int_E \liminf_{j\to\infty} |f_{s(j)} g|\,dh \leq \liminf_{j\to\infty} \int_E |f_{s(j)} g|\,dh \leq \liminf_{j\to\infty} \|f_{s(j)}\|_\Phi \leq M.$$

Hence $f \in L_\Phi$. Similarly $g \in L_\Psi$, $\|g/\|g\|_\Psi\|_\Psi^* \leq 1$, and $f(g/\|g\|_\Psi)h$ is integrable. Similarly $f_j g_j h, f_j g h, f g_j h$ are all integrable for each j. Using Hölder's inequality, Theorem 8.2.10(8.2.29),

$$\left| \int_E f_j g_j \, dh - \int_E fg \, dh \right| = \left| \int_E (f_j - f)g + f(g_j - g) + (f_j - f)(g_j - g) \, dh \right|$$

$$\leq \|f_j - f\|_\Phi \|g\|_\Psi + \|f\|_\Phi \|g_j - g\|_\Psi + \|f_j - f\|_\Phi \|g_j - g\|_\Psi \to 0 \quad (j \to \infty).$$

To prove (8.2.37) from the conclusions, by Theorem 8.2.9(8.2.23) and $f \in L_\Phi$,

$$\|f_j\|_\Phi \leq \|f_j - f\|_\Phi + \|f\|_\Phi,$$

and similarly for $(\|g_j\|_\Psi)$, giving (8.2.37). □

The next result has an affinity with Theorem 2.9.6.

Theorem 8.2.14. *For $(T, \mathcal{T}, \mathbf{A})$ an infinitely divisible, decomposable additive division space, $h: \mathcal{U}^1 \to \mathbb{R}^+$ satisfying, for every partial set P of E,*

(8.2.39) $$\bar{V}(h; \mathbf{A}; E; \mathcal{F}(E; P)) = 0,$$

and $f: T \to \mathbb{C}$, h-measurable and with $\Phi(|f|)h$ and h integrable on the elementary set E, then f is the limit h-almost everywhere of a sequence of step-functions and, given $\varepsilon > 0$, there is a step function f_1 with $\|f - f_1\|_\Phi < \varepsilon$.

8.2 YOUNG'S INEQUALITY AND ORLICZ SPACES

Proof. By Theorem 2.7.4(2.7.7), as $h \geq 0$ and $\Phi(|f|) \geq 0$,

$$\int_I dh = V(h; \mathbf{A}; I) \equiv V(I), \quad \int_I \Phi(|f|) dh = V(\Phi(|f|)h; \mathbf{A}; I) \equiv V_1(I),$$

and by Ex. 3.2.4, V_1 is absolutely continuous with respect to V. Thus we use the first part of the proof of Theorem 3.2.4 with $\Phi(|f|)$ replacing f and noting that $\Phi(t)$ is continuous, and strictly increasing in $[S, +\infty)$, where $S = \sup\{t : \varphi(t) = 0\}$. Let t_{mn} be the unique solution of

$$\Phi(t_{mn}) = m \cdot 2^{-n} \quad (m, n > 0)$$

and let f_{nN} be defined by

$$f_{nN}(t) = t_{m-2,n}(t \in P_m, 3 \leq m \leq N), \quad 0 (t \in E \setminus (P_3 \cup \ldots \cup P_N)).$$

Clearly $f_{nN}(t)$ is monotone increasing and tends to the finite limit as $N \to \infty$

$$f_n(t) = t_{m-2,n} \quad (t \in P_m, m \geq 3)$$

with $\Phi(f_n) h$ integrable, as shown in the proof of Theorem 3.2.4. Then $\Phi(f_n)$ is shown to tend to a limit h-almost everywhere. As $2^{-n} > 0$, then $t_{mn} > S$ $(m, n = 1, 2, \ldots)$ and Φ is strictly increasing and continuous in $t \geq S$. Hence f_n itself tends to a limit h-almost everywhere as $n \to \infty$. Next, taking $f \geq 0$, $(\int_E \Psi(|g|) dh \leq 1)$

$$|f - f_n||g| \leq (f + f_n)|g| \leq \Phi(f) + \Phi(f_n) + 2\Psi(|g|) \leq \Phi(f) + \Phi(f_1)$$
$$+ \sum_{j=2}^\infty |\Phi(f_j) - \Phi(f_{j-1})| + 2\Psi(|g|).$$

$$\equiv A, \text{ say}, \quad \int_E A(t) dh \leq \int_E \Phi(f) dh + \int_E \Phi(f_1) dh + (V(E) + 2)2 + 2,$$

again using Theorem 2.7.4(2.7.7), and (3.2.16).

Thus we can use the Arzelà–Lebesgue test to prove that $\|f - f_n\|_\Phi \to 0$, and so we can choose a suitable n for the last result, when $f \geq 0$.

For $f \in \mathbb{C}$ we can write $f = k_1 - k_2 + i(k_3 - k_4)$ with $k_j \geq 0$ $(j = 1, 2, 3, 4)$ and $k_j \in L_\Phi^*$, for which there is a sequence (k_{jn}) of the above type with $\|k_j - k_{jn}\|_\Phi \to 0$, $\|f - k_{1n} + k_{2n} - ik_{3n} + ik_{4n}\|_\Phi \leq \sum_{j=1}^4 \|k_j - k_{jn}\|_\Phi \to 0$, giving the theorem. □

Example 8.2.1 Let $\varphi(t) = 0$, $\Phi(t) = 0$ $(0 \leq t \leq e^{-1})$, and

$$\varphi(t) = \log_e t + 1, \quad \Phi(t) = t \log_e t + e^{-1} (t \geq e^{-1}).$$

Then $\Psi(y) = 0$ $(y = 0)$, e^{y-1} $(y > 0)$, $\Psi(0+) = e^{-1}$, $\Psi(y) = e^{y-1} - e^{-1}$, and we have the inequality

(8.2.40) $$a \log_e a + e^{b-1} \geq ab \quad (a \geq e^{-1}, b \geq 0),$$

with equality for $b = \log_e a + 1$. If $0 < a < e^{-1}$, $b \geq 0$, then

$$e^{b-1} - e^{-1} = e^{-1}(e^b - 1) \geq e^{-1} b \geq ab,$$

and $a \log_e a$ has a minimum at $a = e^{-1}$, giving (8.2.40) again, with strict inequality, and (8.2.40) is true for $a > 0$, $b \geq 0$.

Example 8.2.2 Let (f_j) be convergent in mean to f with index p and let g be any function. Prove that $(f_j + g)$ is convergent in mean to $f + g$ with index p.

Example 8.2.3 A sequence $(f_j(t))$ can converge in mean to $f(t)$ without being pointwise convergent anywhere in T. For if $T = [0, 1]$ with $f_j(t) = 0$ there, except for the two sloping sides of a triangle, the third side being on $[0, 1]$, then as $j \to \infty$ let the area of the triangle tend to 0 while the triangle moves repeatedly from 0 to 1 with height tending to infinity.

Exercise 8.2.4 Prove that if a sequence $(f_j(t))$ is pointwise convergent h-almost everywhere, and convergent in mean, then the limits are the same h-almost everywhere. (*Hint*: Use Fatou's lemma.)

Exercise 8.2.5 Show that if $\lim_{t \to \infty} \psi(t) = c < \infty$, then $L_\Phi \subseteq L_\infty$, $L_\Psi \supseteq L_1$. Prove that if also h is integrable then $L_\Phi = L_\infty$, $L_\Psi = L_1$.

Exercise 8.2.6 (Zaanen, using Lebesgue integration) Let $h \geq 0$ with $\bar{V}(h; \mathbf{A}; E)$ finite. Prove that if $\Phi_1(t) \leq \Phi(t)$ $(t \geq t_0 \geq 0)$, then an h-measurable $f \in L_\Phi^*$ is also in $L_{\Phi_1}^*$. In particular, if $f \in L^p$ then $f \in L^q$ for $1 \leq q \leq p$.

Exercise 8.2.7 (Zaanen) Prove that if $f \in L^p \cap L^q$ $(p < q)$ and if $r \in (p, q)$, then $f \in L^r$, even when $\bar{V}(h: \mathbf{A}; E) = +\infty$. Note that one or both of p, q could be in $(0, 1)$. (*Hint*: Consider sets where $|f| \leq 1$ and $|f| > 1$.)

Example 8.2.8 (Zaanen) For $[0, \frac{1}{2}]$ on the real line and $h([u, v), t) = v - u$, then $(t \log^2 t)^{-1}$ is in L^p for $p = 1$ but for no $p > 1$.

Example 8.2.9 Let $(\varphi_n(t))$ be an orthonormal sequence for the elementary set E and its space L^2, i.e. for certain h, E,

$$\int_E \varphi_j \varphi_k \, dh = \begin{cases} 0 & (j \neq k), \\ 1 & (j = k). \end{cases}$$

If, for real a_j, $\sum_{j=1}^\infty a_j^2 < \infty$, then $f_m \equiv \sum_{j=1}^\infty a_j \varphi_j$ converges in mean with index 2, as $m \to \infty$.

8.3 THE FUNCTIONAL ANALYSIS OF CONTINUOUS GENERALIZED RIEMANN INTEGRALS

Following the part of the introduction in Section 8.1 that deals with non-absolute integrals, we take a decomposable division space $(T, \mathcal{T}, \mathbf{A})$ with a topology $\mathcal{G}$, and first study the space S of functions f generalized Riemann integrable relative to a continuous integrator h over a $\mathcal{G}$-compact elementary set E. The continuity of h implies the continuity of $f(t)h(I, t)$ for a fixed $t \in E^*$, and so by Theorem 2.5.4(2.5.13) the continuity of the integral of fh for the same fixed t, as $(I, t) \in \mathcal{U}$ and $\mathcal{U}$ shrinks in $\mathbf{A} \mid E$. Since E is $\mathcal{G}$-compact, this implies that the integral of fh has values in a compact set of K, the set of values. Thus the integral is bounded in E and for seminorm we can use (8.1.2), i.e.

(8.3.1) $\qquad \|f\| = \sup_I |F(I)|, \quad \|f\| = \|F\|_\infty, \quad F(I) \equiv \int_I f \, dh.$

with the supremum over all partial intervals I of E.

If we drop some conditions of this section we may have an infinite norm, and would then need another norm.

$\|f\| = 0$ if and only if $F(I) = 0$ for every partial interval I of E, hence, if and only if $f = 0$ h-almost everywhere, by Theorem 2.4.1(2.4.4), (2.4.5). Thus as usual we regard $\|f\|$ as a norm. A linear functional $L(f)$ of the $f \in S$ is a linear functional $L_1(F)$ of the integrals F if two functions with the same integral have the same value of L, so that the two functions are equal h-almost everywhere. This holds when L is also continuous, for then $\|f - g\| = 0$ implies $L(f - g) = 0$ by continuity. Thus to study continuous linear functionals $L(f)$, it is enough to study continuous linear functionals $L_1(F)$. The F lie in the space $C(E)$ of continuous finitely additive interval functions on partial intervals of the $\mathcal{G}$-compact elementary set E, with supremum norm $\|F\|_\infty$. As h is continuous it satisfies Theorem 2.9.6(2.9.25), that the frontier star-sets $\mathcal{F}(E; P)$ have h-variation zero, so that we can use step functions f based on divisions of E (see Exercise 2.9.1), the integrals of which f are everywhere dense in $C(E)$. These integrals are absolute integrals if $h \geq 0$. Following Alexiewicz (1948), by the continuity of L_1 we can extend L_1 to the whole of $C(E)$ and then use an analogue of the Riesz representation theorem, an analogue since $C(E)$ is not a space of continuous point functions. This theorem is best proved by going a little way outside $C(E)$, as will be seen.

Theorem 8.3.1. *Let $(T, \mathcal{T}, \mathbf{A})$ be a stable infinitely divisible decomposable additive division space, let $h: \mathcal{U}^1 \to \mathbb{R}^+$ be integrable in E and*

(8.3.2) $\qquad \nabla(h; \mathbf{A}; E; \mathcal{F}(E; P)) = 0$

for all partial sets P of E, and let $L(f) = L_1(F)$ be a continuous linear functional on S using $\|F\|_\infty$, assumed finite for all $f \in S$. Then an h-measurable

$u(x)$ integrable relative to h in E exists with $|u(x)| \le \|F\|$, *the norm of F, and*

(8.3.3) $\quad L(f) = L_1(F) = (A)\int_E u(x)\,dF(f;I) = (A)\int_E u(x)f(x)\,dh,$

for every f with fh and $|f|h$ integrable in E.

Proof. $c(J) \equiv L(\chi(J^*;t)) = L_1\left((A)\int_I \chi(J^*;t)\,dh\right)$

$\qquad = L_1(H(I \cap J)), \quad H(J) \equiv \int_J dh,$

for J a partial interval of E. As H is finitely additive, so is c, and also as $h \ge 0$,

$|c(J)| \le \|L\|\,\|\chi(I \cap J^*;t)\| = \|L\|\,\|H(I \cap J)\|_\infty$

$\qquad = \|L\| \sup_I |H(I \cap J)| = \|L\| H(J),$

$\bar V(c;\mathbf{A};E;P^*) = \bar V(c;\mathbf{A};P) \le \|L\| H(P) = \|L\| \bar V(h;\mathbf{A};E;P^*),$

(8.3.4) $\qquad V(c;\mathbf{A};E;X) \le \|L\| \cdot V(h;\mathbf{A};E;X)$

for $X = P^*$, and so for X a union of a countable number of disjoint P^*. Now let X be the intersection of a monotone decreasing sequence of such sets, (X_j). Then

$\bar V(c;\mathbf{A};E;X) \le \bar V(c;\mathbf{A};E;X_j) \le \|L\| \bar V(h;\mathbf{A};E;X_j),$

$\bar V(h;\mathbf{A};E;X_j) = \int_E \chi(X_j;t)\,dh = \int_E dh - \bar V(h;\mathbf{A};E;X_j)$

$\qquad \to \int_E dh - \int_E \chi(\setminus X_j;t)\,dh$

$\qquad \to \bar V(h;\mathbf{A};E;E^*) - \bar V(h;\mathbf{A};E;\setminus X)$

$\qquad \le \bar V(h;\mathbf{A};E;X).$

Hence c is absolutely continuous relative to h in E when only countable intersections of countable unions of P^* are used. They are sufficient to prove Theorem 3.2.4 with c integrable as finitely additive. Thus, for $|u(t)| \le \|L\|$,

$$c(J) = \int_J u(t)\,dh.$$

For g a step-function based on a division $\mathscr{E}$ of E,

$g(t) \in (\mathscr{E}) \sum g(I^*)\chi(I^*;t), \quad L(g) = (\mathscr{E})\sum g(I^*)c(I) = \int_E u(t)g(t)\,dh$

Using limits of step functions we prove (8.3.3). □

8.3 CONTINUOUS GENERALIZED RIEMANN INTEGRALS

It is an interesting exercise to find which earlier theorems are used at each stage of the proof.

W.L.C. Sargent introduced α- and β-spaces, to use functional analysis methods to find properties of Denjoy–Perron integrals, see Sargent (1953). Since then, Bourbaki introduced barrel spaces and B. S. Thomson (1970b) proved directly that the spaces in question are barrelled.

Extending slightly the $F \in C(E)$ to include $F(P)$ for partial sets P of E, with F (empty set) $= 0$, we have the following.

Theorem 8.3.2. *Let* $(\mathbb{R}, \mathcal{T}, \mathbf{A})$ *be a stable, infinitely divisible decomposable, additive division space, let* $h: \mathcal{U}^1 \to \mathbb{R}^+$ *be continuous so that (8.3.2) is true, and let* $f: T \to \mathbb{C}$ *be such that* fh *is integrable in* E. *Let* X *be the subspace of* $C(E)$ *consisting of the integrals of such functions* fh. *Then* X *is barrelled.*

Proof. As fh is integrable over E, it is integrable over each partial set of E, so that

(8.3.5) If $F(\cdot) \in X$ then $F(P \cap \cdot) \in X$.

(8.3.6) If $c \in \bar{E}$, $F \in C(E)$, and $F(P \cap \cdot) \in X$ for every partial set P with $c \in \bar{P}$, then $F \in X$.

This is the property that if F is a continuous finitely additive interval function over E that is the integral of fh over P when $c \notin \bar{P}$, then F is also the integral of fh over intervals with c in their closure, which is the Cauchy limit property, see Theorem 2.10.6.

Let B be a barrel in X that is nowhere dense in X. For P a partial set of E let

$$X(P) = \{F(P \cap \cdot): F \in X\}, \quad B(P) = B \cap X(P).$$

Then clearly $B(P)$ is a barrel in $X(P)$. Now let $\mathcal{P}$ be a partition of E. If for each $I \in \mathcal{P}$, the barrel $B(I)$ is dense in some portion of $X(I)$, then as $B(I)$ is closed it contains that portion. As $B(I)$ is balanced, if $b \in B(I)$ and $|a| \leq 1$ then $ab \in B(I)$ (e.g. take $a = -1$). As $B(I)$ is also convex it contains a neighbourhood in $X(I)$ of the origin. Thus B is a neighbourhood of zero in X, contrary to hypothesis. Hence for some $I \in \mathcal{P}$, $B(I)$ is nowhere dense in $X(I)$.

Repeating this argument we can take a sequence $(\mathcal{P}_n)$ of partitions of E and a nested sequence of partial intervals (I_n) tending to a point $c \in \bar{E}$, such that each $B(I_n)$ is nowhere dense in $X(I_n)$ ($n = 1, 2, \ldots$). For each n, $c \in \bar{I}_n$. If $c \in I_n^\circ$ for all n, we divide I_n at c so that either the left or the right interval has the property, and we can keep to the same side of c, as $n \to \infty$. On the other hand, if c, is one of the two end-points after a certain stage, then after that stage c is either a right end-point throughout the rest of the sequence, or a left end-point throughout the rest of the sequence. By symmetry we need only consider the case when the interval is $[c, k_n)$ and $k_n \to c+$. As $B(c, k_1)$ is nowhere dense in $X(c, k_1)$ we can choose $F_1 \notin B$, $F_1 \in X(c, k_2)$, and $\|F_1\| < \frac{1}{2}$.

By (8.3.5), $F_1(k_2, k_1) \to F_1$ as $k_2 \to c+$ and $\setminus B$ is open. Hence there is a k_2 in $(0, k_1)$ with $H_1 \equiv F_1(k_2, k_1) \in X(k_2, k_1)$, $H_1 \in B$, $\|H_1\| < \frac{1}{2}$. Repeating these arguments we have a strictly decreasing sequence (k_n) tending to $c+$ and a sequence (H_n) with

$$H_n \in X(k_{n+1}, k_n), \quad H_n \notin nB, \quad \|H_n\| \leq 2^{-n}.$$

Now let A be the absolutely convex envelope of the set $\{H_1, H_2, \ldots\}$ with closure $\bar{A}$ in $C(E)$. Then every element of A is of the form $H = \sum_{n=1}^{\infty} \lambda_n H_n$ for some sequence of scalars (λ_n) with $\sum_{n=1}^{\infty} \lambda_n \leq 1$. Each such series obviously converges in $C(E)$, and also $\|H\| \leq 1$ for all $H \in \bar{A}$. If $k_1 > k > c$ then

$$H((k, k_1) \cap \cdot) = \sum_{n=1}^{\infty} \lambda_n H_n((k, k_1) \cap \cdot)$$

where only a finite number of terms are non-zero. Hence for all $k_1 > k > c$, $H((k, k_1) \cap \cdot) \in X$, and by (8.3.6), $H \in X$. Hence $\bar{A} \subseteq X$, and in particular, $\bar{A}$ is a complete, bounded, and absolutely convex set in X. But a barrel B absorbs every such set, so that B absorbs $\bar{A}$. But B does not absorb $(H_n) \subseteq \bar{A}$. This contradiction proves the result. □

8.4 DENSITY INTEGRATION

The numerical evaluation of Lebesgue integrals often depends on the calculus. It is usually straightforward to integrate over an interval, but it can be very difficult to integrate over other kinds of sets. The density integration of Henstock (1951, 1982) is sometimes of great help here, a simple sequence is given for evaluating:

(a) the measure of the intersection of two sets X and Y when the measures of $X \cap I$ and $Y \cap I$ are known for special intervals I;
(b) the integral of a function f over such a set X when the integral of f over the special intervals I is known (say, by calculus methods) and $|f|$ is integrable;
(c) the integral of suitable products fg when the integrals of f and g are separately known over the special intervals I.

We begin with the function of intervals

(8.4.1) $\quad K(I) \equiv K(g; X; I) \equiv g(I) D(X; I), \quad D(X; I) \equiv m(X \cap I)/mI$

for some set $X \in \bar{E}$, and m denoting Lebesgue measure. The ratio is a mean or average density over I, giving the reason for the name; it behaves as a convergence factor, and we can ask the question: when is K Burkill integrable over E for all Lebesgue measurable sets $X \subseteq \bar{E}$? The answer is that

(8.4.2) $g(I)$ is Burkill integrable and absolutely continuous.

8.4 DENSITY INTEGRATION

The proof involves some interesting functional analysis.

(8.4.3) When g is the Lebesgue integral of a function f over I, the Burkill integral of (8.4.1) is the value of the Lebesgue integral over X.

The integral over I is often evaluated by calculus methods. When we know that the integral of (8.4.1) exists as a Burkill integral, then its value can be found using any sequence of divisions of E with norm tending to 0 (e.g. we can use continued bisection). Result (a) is obtained on putting $g(I) = m(Y \cap I)$; (b) follows on taking

$$g(I) = \int_I f(x)\,dm$$

with $|f|$ integrable. K itself can be generalized to

$$g(I) \cdot \int_I f\,dm/mI,$$

and if

$$g(I) = \int_I s(x)\,dm,$$

the limit we obtain is

$$\int_E fs\,dm.$$

For example, an integral giving Euler's constant γ or C is

$$\gamma = \int_0^\infty e^{-t} \log\frac{1}{t}\,dt.$$

$$\int_u^v e^{-t}\,dt = e^{-u} - e^{-v}, \quad \int_u^v \log\frac{1}{t}\,dt = v - u + u\log u - v\log v.$$

Thus we integrate over $[0, \infty)$ the interval function

$$h(u, v) = (e^{-u} - e^{-v})(v - u + u\log u - v\log v)/(v - u)$$

obtaining

$$\gamma = \lim_{a \to 1+} \sum_{n=-\infty}^{\infty} h(a^n, a^{n+1}).$$

We have been integrating over the whole or part of the real line. We now imbed the theory in the work of this book. Let $(T, \mathcal{T}, \mathbf{A})$ be a stable decomposable additive division space with $h: \mathcal{U}^1 \to \mathbb{R}^+$ integrable over a fixed elementary set E, such that for every partial set P of E,

$$H(P) \equiv \int_P dh > 0.$$

Let $g: \mathcal{U}^1 \to \mathbb{C}$. For some $X \subseteq T$ and each such P let

$$M(X; P) \equiv \int_P \chi(X; \cdot) \, dh$$

exist. Or for some $f: E^* \to \mathbb{C}$ let

$$\int_E f \, dh$$

exist. We then consider the integrability of

$$J(I, t) \equiv J(g; X; I, t) \equiv g(I, t) M(X; I)/H(I), \quad L(I, t) \equiv L(g; f; I, t) \equiv \frac{g(I, t)}{H(I)} \int_I f \, dh.$$

Of course, $J(g; X; I, t) = L(g; \chi(X; \cdot); I, t)$.

Theorem 8.4.1. *The space $\mathscr{V}$ of all $X \subseteq T$ for which $M(X; E)$ exists, is complete for the pseudometric on sets of T,*

$$\rho(X; Y) \equiv V(d(X, Y)), \quad V(Z) \equiv \bar{V}(h; \mathbf{A}; E; Z), \quad d(X, Y) \equiv (X \setminus Y) \cup (Y \setminus X).$$

Proof. Let (X_n) be a fundamental sequence of sets of T, i.e. given $\varepsilon > 0$, there is an integer N depending on ε, for which

$$V(d(X_m, X_n)) < \varepsilon \quad (m, n \geq N).$$

When $\varepsilon = 2^{-j}$ let $N = N(j)$ and put

$$X = \limsup_{j \to \infty} X_{N(j)} = \bigcap_{k=1}^{\infty} \bigcup_{j=k}^{\infty} X_{N(j)}.$$

$$X \setminus X_{N(j)} \subseteq \bigcup_{k=j+1}^{\infty} X_{N(k)} \setminus X_{N(j)} \subseteq \bigcup_{k=j+1}^{\infty} X_{N(k)} \setminus X_{N(k-1)},$$

$$X_{N(j)} \setminus X \subseteq \bigcup_{k=j+1}^{\infty} X_{N(k-1)} \setminus X_{N(k)},$$

$$V(X \setminus X_{N(j)}) \leq \sum_{k=j}^{\infty} 2^{-k} = 2^{1-j}, \quad V(X_{N(j)} \setminus X) \leq 2^{1-j},$$

$$V(d(X, X_{N(j)})) \leq 2^{2-j}.$$

The proof that $V(d(X, Y))$ is a pseudometric needs the following steps.

$$(X \setminus Y) \cup (Y \setminus Z) \supseteq X \setminus Z, \quad d(X, Y) \cup d(Y, Z) \supseteq d(X, Z).$$

Then for $N(j) > n \geq N$,

$$V(d(X, X_n)) \leq V(d(X, X_{N(j)})) + V(d(X_{N(j)}, X_n)) < 2^{2-j} + \varepsilon,$$

and X is the limit of (X_n) for the given pseudometric, and $\mathscr{V}$ is complete. □

8.4 DENSITY INTEGRATION 205

Theorem 8.4.2. *Let $(T, \mathcal{T}, \mathbf{A}^+)$ be a division space. Given $\delta > 0$, let there be a partition $\mathcal{P}$ of P from a $\mathcal{U} \in \mathbf{A}^+ | P$ such that $V(I) \leq \delta$ for each $I \in \mathcal{P}$, and let $V(\mathcal{P}(E; P)) = 0$, for every partial set P of E. Let*

(8.4.4) $$(\mathcal{E}) \sum J(g; X; I, t)$$

be bounded for each fixed $X \in \mathcal{V}$ and all divisions $\mathcal{E}$ of E from $\mathcal{U}$. Then g is of bounded variation.

Proof. Let $\mathcal{V}_N$ be the set of those $X \in \mathcal{V}$ for which, for all divisions $\mathcal{E}$ of E from $\mathcal{U}$,

$$|(\mathcal{E}) \sum J(g; X; I, t)| \leq N.$$

Then clearly $\mathcal{V}$ is the union of the $\mathcal{V}_N$, which we now prove closed for the pseudometric of Theorem 8.4.1. For if X is in the closure and $\varepsilon > 0$, there is a $Y \in \mathcal{V}_N$ satisfying $V(d(X, Y)) < \varepsilon$,

$$|(\mathcal{E}) \sum g M(X; I)/H(I) - (\mathcal{E}) \sum g M(Y; I)/H(I)|$$
$$\leq (\mathcal{E}) \sum |g| \{M(X \setminus Y; I) + M(X \cap Y; I) - M(Y \setminus X; I)$$
$$- M(X \cap Y; I)\}/H(I) \leq (\mathcal{E}) \sum |g| M(d(X, Y); I/H(I)$$
$$\leq (\mathcal{E}) \sum |g| \varepsilon / H(I),$$
$$|(\mathcal{E}) \sum g M(X; I)/H(I)| \leq N + \varepsilon (\mathcal{E}) \sum |g|/H(I).$$

As $\varepsilon > 0$ is independent of $\mathcal{E}$ we can let $\varepsilon \to 0+$ and so have $X \in \mathcal{V}_N$. By Theorems 8.4.1, 0.4.4 (Baire's category theorem), some $\mathcal{V}_N$ contains a closed sphere S, centre some X_0 and radius $\delta > 0$, so that each $X \in \mathcal{V}$ with $d(X, X_0) \leq \delta$ lies in $\mathcal{V}_N$.

If $Y \in \mathcal{V}$ has $V(Y) \leq \delta$ then

$$V(d(X_0 \cup Y, X_0)) = V((X_0 \cup Y) \setminus X_0) \leq V(Y) \leq \delta,$$
$$V(d(X_0 \setminus Y, X_0)) = V(X_0 \setminus (X_0 \setminus Y)) \leq V(Y) \leq \delta,$$
$$X_0 \cup Y \in \mathcal{V}_N, \quad X_0 \setminus Y \in \mathcal{V}_N, \quad (X_0 \cup Y) \setminus (X_0 \setminus Y) = Y,$$
$$|(\mathcal{E}) \sum J(g; X_0 \cup Y; I, t)| \leq N, \quad |(\mathcal{E}) \sum J(g; X_0 \setminus Y; I, t)| \leq N,$$
$$(\mathcal{E}) \sum J(g; Y; I, t) \leq 2N \quad (Y \in \mathcal{V}, V(Y) \leq \delta)$$

for all divisions $\mathcal{E}$ of E from $\mathcal{U}$. Using the special partition $\mathcal{P}$ of P containing r intervals, we can replace Y by P on putting $2rN$ on the right, while $M(P; I) = H(I)$ when $I \subseteq P$, $M(P; I) = 0$ when $I \subseteq \setminus P$. Taking an $\mathcal{E}$ that refines $\mathcal{P}$, we obtain

$$|(\mathcal{E} \cdot P) \sum g(I, t)| \leq 2rN$$

and by the usual argument we have

$$(\mathscr{E})\sum |g(I, t)| \leq 2qrN$$

where $q = 2$ for g real-valued, $q = 4$ for g complex-valued, and so g is of bounded variation. (The converse is easy.) □

Theorem 8.4.3. *Let $(T, \mathscr{T}, \mathbf{A}^+)$ satisfy the conditions of Theorem 8.4.2. Then J is $\mathbf{A}^+$-integrable over E for all $X \in \mathscr{V}$ if and only if g is $\mathbf{A}^+$-integrable and h-absolutely continuous in E, and also of bounded variation there.*

Proof. If $X = T$, $M(T, E)$ exists and is $H(E)$. Hence $T \in \mathscr{V}$ and $J = g(I, t)$. If J is $\mathbf{A}^+$-integrable over E then g is $\mathbf{A}^+$-integrable. Next, given $\varepsilon > 0$, let $\mathscr{W}_N$ be the set of all $X \in \mathscr{V}$ for which, for all $\mathscr{E}, \mathscr{E}'$ divisions over E from $\mathscr{U}$,

$$|(\mathscr{E})\sum J(g; X; I, t) - (\mathscr{E}')\sum J(g; X; I, t)| \leq \varepsilon.$$

Then $\mathscr{W}_N$ is proved closed just as $\mathscr{V}_N$ has been proved closed, and

$$|(\mathscr{E})\sum J(g; Y; I, t) - (\mathscr{E}')\sum J(g; Y; I, t)| \leq 2\varepsilon \quad (M(Y; E) \leq \delta)$$

for some $\delta > 0$. We fix $\mathscr{E}'$ and take δ' in $0 < \delta' \leq \delta$ such that

$$|(\mathscr{E}')\sum J(g; Y; I, t)| \leq M(Y, E)(\mathscr{E}')\sum |g(I, t)|/H(I) < \varepsilon$$

when $M(Y, E) \leq \delta'$, and then by the usual argument

$$|(\mathscr{E})\sum J(g; Y; I, t)| \leq 3\varepsilon, \quad (\mathscr{E}_1)\sum |g(I, t)| \leq 12\varepsilon$$

for $\mathscr{E}_1 \subseteq \mathscr{E}$ and

(8.4.5) $$(\mathscr{E}_1)\sum H(I) \leq \delta'.$$

To show the h-absolute continuity of g let $X \subseteq T$ have

$$V(h; \mathbf{A}; E; X) = \bar{V}(H; \mathbf{A}; E; X) < \delta'.$$

Then for some $\mathscr{U} \in \mathbf{A}^+|E$, (8.4.5) is true for all $\mathscr{E}_1 \subseteq \mathscr{E}$ with $t \in \backslash X$ for all $(I, t) \in \mathscr{E} \backslash \mathscr{E}_1$, and all divisions $\mathscr{E}$ of E from $\mathscr{U}$. Hence

$$V(g; \mathbf{A}; E; X) \leq V(g; \mathscr{U}; E; X) = \sup_{\mathscr{E}} (\mathscr{E})\sum |g(I, t)|\chi(X; t)$$

$$= \sup_{\mathscr{E}} (\mathscr{E}_1)\Sigma |g(I, t)| \leq 12\varepsilon,$$

finishing the proof one way.

If g is $\mathbf{A}^+$-integrable to G over the partial sets of E and h-absolutely continuous, then for all partial intervals I of E and some j with j, $|j|$

integrable,
$$G(I) = \int_I j(t)\, dh,$$
$$|(\mathscr{E})\sum J(g; X; I, t) - (\mathscr{E})\sum J(G; X; I, t)| \leqslant (\mathscr{E})\sum |g - G|,$$
$$(\mathscr{E})\sum J(G; X; I, t) = (\mathscr{E})\sum \int_I j(t)\frac{M(X; I)}{H(I)}\, dh,$$
$$0 \leqslant M(X; I)/H(I) \leqslant 1.$$

Thus the Arzelà–Lebesgue limit theorem shows that J is integrable to
$$\int_E j(t)\chi(X; t)\, dh,$$
which finishes the proof of the converse. □

REFERENCES

The bulk of these references has been collected over more than forty years of research into integration theory, and I have greatly benefited from a bibliography of non-absolute integration by P. S. Bullen. His and my collections overlap considerably, as is natural, and are given here. But this collection forms only a small sample, but a significant sample, of all research on integration theory, as can be seen from a perusal of *Mathematical Reviews* (*MR*) and *Zentralblatt für Mathematik* (*Zbl*).

Ahmed, S. I. and Pfeffer, W. F. (1986). A Riemann integral in a locally compact Hausdorff space *J. Austral. Math. Soc. Ser. A*, **41**, 115–37. *MR* 87g: 26015.

Albeverio, S. A. and Hoegh-Krohn, R. J. (1976). Mathematical theory of Feynman path integrals. *Lec. Notes Math.* **523**, *MR* 58 #14535.

Aleksandrov, P. S. (1924a). Über die Äquivalenz des Perronschen und des Denjoyschen Integralbegriffes. *Math. Zeit.* **20**, 213–22.

Aleksandrov, P. S. (1924b). L'intégration au sens de M. Denjoy considérée comme recherche des fonctions primitives. *Rec. Math. Soc. Math. Moscou* **31**, 465–76.

Alexiewicz, A. (1948). Linear functionals on Denjoy-integrable functions. *Colloq. Math.* **1**, 289–93. *MR* 10–717.

Alexiewicz, A. (1950). On Denjoy integrals of abstract functions. *Soc. Sci. Lett. Varsovie C.R.Cl. III Sci. Math. Phys.* **41**, 97–129. *MR* 14–A27.

Amemiya, I. and Ando, T. (1961). Measure-theoretic singular integrals. (Japanese with English summary). *Bull. Res. Inst. App. El., Hokkaido University* **13**, 33–50.

Appling, W. D. L. (1962a). Concerning nonnegative valued interval functions. *Proc. Am. Math. Soc.* **13**, 784–8. *MR* 25 #5150.

Appling, W. D. L. (1962b). Interval functions and the Hellinger integral. *Duke Math. J.* **29**, 515–20. *MR* 25 #4075.

Appling, W. D. L. (1962c). Interval functions and real Hilbert spaces. *Rend. Circ. Mat. Palermo* **11**(2), 154–6. *MR* 27 #4040.

Appling W. D. L. (1963a). Infinite series and non-negative valued interval functions. *Duke Math. J.* **30**, 107–11. *MR* 26 #2576.

Appling, W. D. L. (1963b). Interval functions and nondecreasing functions. *Can. J. Math.* **15**, 752–4. *MR* 27 #4903.

Arley, N. and Borchsenius, V. (1945). On the theory of infinite systems of differential equations and their application to the theory of stochastic processes and the perturbation theory of quantum mechanics. *Acta Math.* **76**, 261–322. *MR* 7–161.

Armstrong, G. M. (1971). A classical approach to the Denjoy integral by parametric derivatives. *J. Lond. Math. Soc.* **3**(2), 346–9. *MR* 43 #4971.

Armstrong, G. M. (1986). Properties of a general integral equivalent to the Denjoy–Perron integral. *Tamkang J. Math.* **17**, 29–39 *MR* 88k: 26007.

Artstein, Z. (1976). Lyapounov convexity theorem and Riemann-type integrals. *Indiana Univ. Math. J.* **25**, 717–24. *MR* 54 #5429.

Artstein, Z. (1977). Topological dynamics of ordinary differential equations and Kurzweil equations. *J. Diff. Eq.* **23**, 224–43. *MR* 55 #5964.

Artstein, Z. and Burns, J. A. (1975). Integration of compact set-valued functions. *Pacific J. Math.* **58**, 297–307. *MR* 52 #5931.

Arzelà, C. (1885). Sulla integrazione per serie. *Atti Acc. Lincei Rend., Roma* **1**(4), 532–7, 596–9.

Arzelà, C. (1900). Sulle serie di funzioni. *Mem. Inst. Bologna* **8**(5), 131–86, 701–44.
Babcock, B. S. (1975). On properties of the approximate Peano derivatives. *Trans. Am. Math. Soc.* **212**, 279–94. *MR* 54 #2895.
Baidya, S. and Bose, M. K. (1985). On the proximal Denjoy–Stieltjes integral. *Indian J. Pure Appl. Math.* **16**, 975–93. *MR* 86m: 26005.
Banach, S. (1923). Sur le problème de mesure. *Fundamenta Math.* **4**, 1–34.
Banach, S. (1932). Théorie des opérations lineaires. Warsaw. *Zbl*.5.209.
Banach, S. (1937). The Lebesgue integral in abstract spaces. In S. Saks (1937), *Theory of the Integral* (2nd English edn), pp. 320–30, Warsaw, *Zbl*.17, 300.
Banach, S. and Steinhaus, H. (1927). Sur le principe de la condensation de singularités *Fundamenta Math.* **9**, 50–61.
Bartle, R. G. (1956). A general bilinear vector integral. *Studia Math.* **15**, 337–52. *MR* 18–289.
Bartle, R. G. Dunford, N. and Schwartz, J. (1955). Weak compactness and vector measures. *Can. J. Math.* **7**, 289–305. *MR* 16–1123.
Bauer, H. (1915). Der Perronsche Integralbegriff und seine Beziehung zum Lebesgueschen. *Monatsh. Math. Phys.*, **26**, 153–98.
Bellamy, O. S. and Ellis, H. W. (1975). On extensions of the Riemann and Lebesgue integrals by nets. *Canadian Math. Bull.* **18**, 7–17. *MR* 51 #13152.
Benedicks, M. and Pfeffer, W. F. (1985). The Dirichlet problem with Denjoy–Perron integrable boundary condition. *Canadian Math. Bull.* **28**, 113–19. *MR* 86h: 31003.
Benninghofen, B. (1984). Superinfinitesimals and the calculus of the generalized Riemann integral. Models and Sets. *Lecture Notes Math.* **1103**, 9–52. *MR* 86g: 26008.
Bergin, J. A. (1977). A new characterization of the Çesàro-Perron integrals using Peano derivatives. *Trans. Am. Math. Soc.* **228**, 287–305. *MR* 55 #8272.
Bhakta, P. C. (1965a). On functions of bounded ω-variation. *Riv. Mat. Univ. Parma* **6**(2), 55–64. *MR* 36 #327.
Bhakta, P. C. (1965b). On functions of bounded ω-variation, II. *J. Australian Math. Soc.* Ser. A. **5**, 380–87. *MR* 33 #2771.
Bhattacharaya, S. and Das, A. G. (1985). The LS_k integrals. *Bull. Inst. Math. Acad. Sinica* **13**, 385–401. *MR* 88a: 26012.
Bhattacharyya, P. (1973). On modified Perron integral. *Indian J. Math.* **15**, 103–17. *MR* 51 #824.
Bhattacharyya, P. and Lahiri, B. K. (1978a). On the modified Perron integral, II. *Indian J. Math.* **20**, 233–41. *MR* 82h: 26012.
Bhattacharyya, P. and Lahiri, B. K. (1978b). On upper and lower integral functions. *Bull. Calcutta Math. Soc.* **70**, 331–5. *MR* 81i: 26008.
Birkhoff, G. (1935). Integration of functions with values in a Banach space. *Trans. Am. Math. Soc.* **38**, 357–78. *Zbl*.13.008.
Birkhoff, G. (1937a). Moore–Smith convergence in general topology. *Ann. Math.* **38**, 39–56. *Zbl*.16.085.
Birkhoff, G. (1937b). On product integration. *J. Math. Phys.* [Massachusetts Inst. Technol.] **16**, 104–32. *Zbl*.18.134.
Bliss, G. A. (1917a). Existence of a Stieltjes integral. *Proc. Nat. Acad. Sci.* **3**, 633–7.
Bliss, G. A. (1917b). Integrals of Lebesgue. *Bull. Am. Math. Soc.* **24**, 1–47.
Bochner, S. (1933). Integration von Funktionen deren Werte die Elemente eines Vektorraumes sind. *Fundamenta Math.* **20**, 262–76. *Zbl*.7.109.
Bochner, S. and Fan, K. (1947). Distributive order preserving operations in partially ordered vector sets. *Ann. Math.* **48**(2), 168–79. *MR* 8–387.

Bochner, S. and Taylor, A. E. (1938). Linear functionals on certain spaces of abstractly valued functions *Ann. Math.* **39**, 913–44. *Zbl.* 20.371.

Bogdanowicz, W. (now Bogdan) (1965). An approach to the theory of L_p-spaces of Lebesgue–Bochner summable functions and generalized Lebesgue–Bochner–Stieltjes integral. *Bull. de l'Académie Polonaise des Sciences Sér. Sci. Math.* **13**, 793–800. *MR* 33 #7488a.

Bois-Reymond, P. du (1868). Über die allgemeinen Eigenschaften der Klasse von Doppelintegralen zu welcher das Fouriersche Doppelintegral gehört. *J. Reine Angew. Math.* **69**, 65–108.

Bois-Reymond, P. du (1875). *Beyer Akad. Wiss. Math.-Natur. Kl. Abh.* **12**, 129.

Bongiorno, B. and Preiss, D. (1985). Unusual descriptive definition of integral. *Contemp. Math.* **42**, 13–32. *MR* 87a: 26006.

Borel, E. (1910a). Sur la définition de l'intégrale définie, *Comptes Rendus* **150**, 375–8.

Borel, E. (1910b). Sur une condition générale de l'intégrabilite. *Comptes Rendus* **150**, 508–10.

Bosanquet, L. S. (1930a). On Abel's integral equation and fractional integrals. *Proc. Lond. Math. Soc.* **31**(2), 134–43.

Bosanquet, L. S. (1930b). On the summability of Fourier series. *Proc. Lond. Math. Soc.* **31**(2), 144–64.

Bosanquet, L. S. (1940). A property of Çesàro–Perron integrals. *Proc. Edinburgh Math. Soc.* **6**(2), 160–5. *MR* 2–131.

Bosanquet, L. S. (1945). Some properties of the Çesàro–Lebesgue integrals. *Proc. Lond. Math. Soc.* **49**(2), 40–62. *MR* 7–280.

Bosanquet, L. S. and Kestelman, H. (1939). The absolute convergence of series of integrals. *Proc. Lond. Math. Soc.* **45**(2), 88–97. *Zbl.*20.354.

Bose, M. K. (1977). Results on AC*-ω (C-sense) and ACG*-ω (C-sense) functions. *Comment Math. Prace Mat.* **20**, 7–28. *MR* 58 #6097.

Bose, M. K. (1979). On Perron–Stieltjes and special Denjoy–Stieltjes integral. *Rev. Roumaine Math. Pures Appl.* **24**, 3–26, *MR* 80i: 26010.

Bose, M. K. (1980a). On special Çesàro–Denjoy–Stieltjes integral. *Comment Math. Prace Mat.* **22**, 17–31. *MR* 82i: 26003.

Bose, M. K. (1980b). On the Denjoy–Stieltjes integral. *Glasnik Mat. Ser. III*, **15**(35), 41–50, *MR* 82a: 26005.

Bose, M. K. (1985). Two properties of the Çesàro–Perron–Stieltjes integral. *Glasnik Mat. Ser. III* **20**(40), 15–33, *MR* 87e: 26007.

Bose, M. K. and Baidya, S. (1986). On (ω) proximal extreme limits and (ω) proximal semicontinuity. *Bull. Calcutta Math. Soc.* **78**, 155–9. *MR* 87h: 26006.

Bose, M. K. and Nath, R. K. (1985). The proximal Çesàro–Perron integral. *Math. Japon.* **30**, 357–69, *MR* 86m: 26006.

Botsko, M. W. (1986). An easy generalization of the Riemann integral. *Am. Math. Monthly* **93**, 728–32. *MR* 87m: 26007.

Botsko, M. W. (1987). A unified treatment of various theorems in elementary analysis. *Am. Math. Monthly* **94**, 450–2.

Bourbaki, N. (1935). Sur un théorème de Carathéodory et la mesure dans les espaces topologiques. *Comptes Rendus Acad. Sci. Paris* **201**, 1309–11. *Zbl.*13.155.

Bourbaki, N. (1959). Livre VI, *Intégration*, Chap. 6 (Intégration vectorielle). *MR* 23 #A2033.

Bourbaki, N. (1964). *Espaces vectoriels topologiques*, Chap. III–V, Herman, Paris.

Bray, H. E. (1918/1919). Elementary properties of Stieltjes integrals. *Ann. Math.* **20**, 177–86.

Brille, J. (1931). Sur une propriété des fonctions présentent un certain caractère complexe de résolubilité. *Comptes. Rendus Acad. Sci. Paris* **192**, 1191–3. Zbl.1.329.

Bromwich, T. J. I'a and Hardy, G. H. (1908). The definition of an infinite integral as the limit of a finite or infinite series. *Quart. J. Math.* **39**, 222–40. (*Collected Works of G. H. Hardy*, 5, pp. 414–33, Clarendon Press, Oxford, 1972).

Brooks, J. K. (1969a) Representations of weak and strong integrals in Banach spaces. *Proc. Nat. Acad. Sci. USA* **63**, 266–70. MR 43 #459.

Brooks, J. K. (1969b). Transforming bilinear vector integrals. *Studia Math.* **33**, 165–71. MR 39 #7060

Brooks, J. K. (1969c). On absolute continuity in transformation theory. *Monatsch. Math.* **73**, 1–6. MR 39 #1627

Brooks, J. K. (1969d). On the Gelfand–Pettis integral and unconditionally convergent series. *Bull. Acad. Polon. Sci. Ser. Sci. Math. Astronom. Phys.* **17**, 809–13. MR 41 #8624.

Bruckner, A. M. (1978). *Differentiation of Real Functions*, Lecture Notes in Mathematics, 659, Springer–Verlag. MR 80h: 26002.

Bruckner, A. M., Fleissner, R. J. and Foran, J. (1986). The minimal integral which includes Lebesgue integrable functions and derivatives. *Colloq. Math.* **50**, 289–93. MR 87k: 26008.

Bruneau, M. (1974). *Variation totale d'une fonction*. Lecture Notes in Mathematics, 413, Springer–Verlag. MR 52 #14190.

Buczolich, Z. (1987/1988). Nearly upper semicontinuous gauge functions in $\mathbb{R}^m$. *Real Analysis Exchange* **13**, 436–41. MR 89f; 26013.

Buczolich, Z. A general Riemann complete integral in the plane. *Acta Math. Hung.* (details not known).

Bullen, P. S. (1961). Construction of primitives of generalized derivatives with applications to trigonometric series. *Can. J. Math.* **13**, 48–58. MR 22 #12186.

Bullen, P. S. (1965). A general Perron integral. *Can. J. Math.* **17**, 17–30. MR 30 #4959.

Bullen, P. S. (1966). *A General Perron Integral III*. Math. Res. Center U.S. Army, University of Wisconsin Tech. Summary Rep. 696.

Bullen, P. S. (1967a). A general Perron integral II. *Can. J. Math.* **19**, 457–73. MR 36 #421.

Bullen P. S. (1967b). On a theorem of Privaloff. *Can. Math. Bull.* **10**, 353–9. MR 36 #422.

Bullen, P. S. (1970). A constructive definition of an integral. *Tohoku Math. J.* **22**(2), 597–603. MR 43 #6368.

Bullen P. S. (1971). A criterion for n-convexity. *Pacific J. Math.* **36**, 81–98. MR 43 #443.

Bullen, P. S. (1972). The P^n-integral. *J. Australian Math. Soc.* Ser. A **14**, 219–36. MR 47 #8783.

Bullen, P. S. (1979). Axiomatisations of various non-absolute integrals. *South East Asia Bull. Math.*, Special Issue b, 173–89. MR 81i: 26009.

Bullen, P. S. (1979/1980). Non-absolute integrals: a survey. *Real Analysis Exchange* **5**, 195–259. MR 82a: 26006.

Bullen, P. S. (1981). Approximate continuous integrals: *Res. Rep.* **18**, Department of Mathematics, University of Melbourne.

Bullen, P. S. (1982a). Lecture notes on approximately continuous integrals I, II. *Lecture Notes* 11, 13, Department of Mathematics National University of Singapore.

Bullen, P. S. (1982b). Integration by parts for the Perron integral. *Res. Rep.* **39**, Department of Mathematics, National University of Singapore.

Bullen, P. S. (1982c). Denjoy and porosity. *Res. Rep.* **40**, Department of Mathematics, National University of Singapore.

Bullen, P. S. (1982d). Applications of Denjoy's index I, II. *Res. Rep.* **43, 62**. Department of Mathematics, National University of Singapore.

Bullen, P. S. (1983a). The Burkill approximately continuous integral. *J. Australian Math. Soc. Ser. A.* **35**, 236–53. *MR* 84m: 26010.

Bullen, P. S. (1983b). The Burkill approximately continuous integral II. *Math. Chron.* **12**, 93–8. *MR* 85e: 26007.

Bullen, P. S. (1983/1984a). A simple proof of integration by parts for the Perron integral. *Real Analysis Exchange* **9**, 366–8.

Bullen, P. S. (1983/1984b). Some applications of partitioning covers. *Real Analysis Exchange* **9**, 539–57. *MR* 86a: 26014.

Bullen, P. S. (1984/1985). Denjoy's index and porosity. *Real Analysis Exchange* **10**, 85–144. *MR* 87c: 26012.

Bullen, P. S. (1985). A simple proof of integration by parts for the Perron integral. *Can. Math. Bull.* **28**, 195–9. *MR* 86d: 26016.

Bullen, P. S. (1986). A survey of integration by parts for Perron integrals. *J. Australian Math. Soc. Ser. A* **40**, 343–63. *MR* 87g: 26011.

Bullen, P. S. (1986/1987). Query 178, *Real Analysis Exchange* **12**, 393.

Bullen, P. S. (1988). An unconvincing counterexample. *Int. J. Math. Educ. Sci. Technol.* **19**, 455–9.

Bullen, P. S. (1988/1989). Some applications of a theorem of Marcinkiewicz. *Real Analysis Exchange* **14**, 12–13.

Bullen, P. S. and Lee, C. M. (1973a). On the integrals of Perron type. *Trans. Am. Math. Soc.* **182**, 481–501. *MR* 49 #3057.

Bullen, P. S. and Lee, C. M. (1973b). The SC_nP-integral and the P^{n+1}-integral. *Can. J. Math.* **25**, 1274–84. *MR* 53 #13489.

Bullen, P. S. and Mukhopadhyay, S. N. (1973a). On the Peano derivatives. *Can. J. Math.* **25**, 127–40. *MR* 47 #407.

Bullen, P. S. and Mukhopadhyay, S. N. (1973b). Peano derivatives and general integrals. *Pacific J. Math.* **47**, 43–58. *MR* 48 #6332.

Bullen, P. S. and Mukhopadhyay, S. N. (1974). Integration by parts formulae for some trigonometric integrals. *Proc. London Math. Soc.* **29**(3), 159–73. *MR* 51 #825.

Bullen, P. S. and Sarkhel, D. N. (1987). On the solution of $(dy/dx)_{ap} = f(x, y)$. *J. Math. anal. Appl.* **127**, 365–76. *MR* 88j: 34007.

Bullen, P. S., Sarkhel, D. N. and Vyborny, R. (1988). A Riemann view of the Lebesgue integral. *South East Asian Bull. Math. Soc.* **12**, 39–51.

Burkill, H. (1951). Almost periodicity and non-absolutely integrable functions. *Proc. Lond. Math. Soc.* **53**(2), 32–42. *MR* 13–230.

Burkill, H. (1952). Cesàro–Perron almost periodic functions. *Proc. Lond. Math. Soc.* **2**(3), 150–74. *MR* 14–162.

Burkill, H. (1957). The Cesàro–Perron scale of almost periodicity. *Proc. Lond. Math. Soc.* **7**(3), 481–97. *MR* 20 #1167.

Burkill, H. (1962). Sums of trigonometric series. *Proc. Lond. Math. Soc.* **12**(3), 690–706. *MR* 26 #6692.

Burkill, H. (1972). A note on trigonometric series. *J. Math. Anal. Appl.* **40**, 39–44. *MR* 47 #2250.

Burkill, H. (1977). Fourier series of SCP integrable functions. *J. Math. Anal. Appl.* **57**, 587–609. *MR* 55 #10939.

Burkill, J. C. (1923). The fundamental theorem of Denjoy integration. *Proc. Camb. Phil. Soc.* **21**, 659–63.

Burkill, J. C. (1924a). Functions of intervals. *Proc. Lond. Math. Soc.* **22**(2), 275–310.

Burkill, J. C. (1924b). The expression of area as an integral. *Proc. Lond. Math. Soc.* **22**(2), 311–36.

Burkill, J. C. (1924c). The derivates of functions of intervals. *Fundamenta Math.* **5**, 321–7.

Burkill, J. C. (1929). On Hobson's convergence theorem for Denjoy integrals. *J. Lond. Math. Soc.* **4**, 127–32.

Burkill, J. C. (1931). The approximately continuous Perron integral. *Math. Zeit.* **34**, 270–8. *Zbl*.2.386

Burkill, J. C. (1932). The Çesàro–Perron integral. *Proc. Lond. Math. Soc.* **34**(2), 314–22. *Zbl*.5.392

Burkill, J. C. (1935). The Çesàro–Perron scale of integration. *Proc. Lond. Math. Soc.* **39**(2), 541–52. *Zbl*.12.204

Burkill, J. C. (1936a). The expression of trigonometrical series in Fourier form. *J. Lond. Math. Soc.* **11**, 43–8. *Zbl*.13.260

Burkill, J. C. (1936b). Fractional orders of integrability. *J. Lond. Math. Soc.* **11**, 220–6. *Zbl*.14.258.

Burkill, J. C. (1948). Differential properties of Young–Stieltjes integrals. *J. Lond. Math. Soc.* **23**, 22–8. *MR* 10–185.

Burkill, J. C. (1951a). Integrals and trigonometric series. *Proc. Lond. Math. Soc.* **1**(3), 46–57. *Corrigendum ibid.* **47** (1983), 192. *MR* 13–126.

Burkill, J. C. (1951b) Uniqueness theorems for trigonometric series and integrals. *Proc. Lond. Math. Soc.* **1**(3), 163–9. *MR* 13–935.

Burkill, J. C. and Gehring, F. W. (1953). A scale of integrals from Lebesgue's to Denjoy's. *Quart. J. Math.* (*Oxford*) **4**(2), 210–20. *MR* 15–204.

Burkill, J. C. and Haslam-Jones, U. S. (1931). The derivatives and approximate derivatives of measurable functions. *Proc. Lond. Math. Soc.* **32**(2), 346–55. *Zbl*.2.20.

Burry, J. H. W. and Ellis, H. W. (1970). On measures determined by continuous functions that are not of bounded variation. *Can. Math. Bull.* **13**, 121–4. *MR* 41 #5566.

Butković, D. (1970). On Denjoy integrals which depend on a parameter. (Macedonian) *Proc. 5th Congress Math. Phys. Astron. Yugoslavia*, Ohrid, **1**, 77–81. *MR* 56 #8765.

Caccioppoli, R. (1955). L'integrazione e la ricerca delle primitive rispetto ad una funzione continua qualunque. *Ann. Mat. Pura Appl.* **40**(4), 15–34. *MR* 17–954.

Cameron, R. H. (1960/1961). A family of integrals serving to connect the Wiener and Feyman integrals. *J. Math. and Phys.* **39**, 126–40. *MR* 23 #8821.

Cameron, R. H. (1962/1963). The Ilstow and Feynman integrals. *J. d'Analyse Math.* **10**, 287–361. *MR* 27 #831.

Cameron, R. H. and Donsker, M. D. (1959). Inversion formulae for characteristic functionals of stochastic processes. *Ann. Math.* **69**(2), 15–36. *MR* 21 #7562.

Cameron, R. H. and Martin, W. T. (1941). An unsymmetric Fubini theorem. *Bull. Am. Math. Soc.* **47**, 121–5. *MR* 2–257.

Cameron, R. H. and Martin, W. T. (1944a). Transformations of Wiener integrals under translations. *Ann. Math.* **45**(2), 386–96, *MR* 6–5.

Cameron, R. H. and Martin, W. T. (1944b). An expression for the solution of a class of nonlinear integral equations. *Am. J. Math.* **66**, 281–98. *MR* 5–243.

Cameron, R. H. and Martin, W. T. (1945). Evaluation of various Wiener integrals by use of certain Sturm–Liouville differential equations. *Bull. Am. Math. Soc.* **51**, 73–90. *MR* 6–160.

Cameron, R. H. and Storvick, D. A. (1966a). A translation theorem for analytic Feynman integrals. *Trans. Am. Math. Soc.* **125**, 1–6. *MR* 34 #872.

Cameron, R. H. and Storvick, D. A. (1966b). Analytic continuation for functions of several complex variables. *Trans. Am. Math. Soc.* **125**, 7–12. *MR* 34 #2939.

Cameron, R. H. and Storvick, D. A. (1976). An L_2 analytic Fourier–Feynman transform. *Michigan Math. J.* **23**, 1–30. *MR* 53 #8371.

Cameron, R. H. and Storvick, D. A. (1979). *Some Banach algebras of analytic Feynman integrable functionals*. Lecture Notes in Mathematics 798, 18–67, Springer-Verlag. *MR* 83f: 46059

Carathéodory, C. (1927). Vorlesungen über reelle Funktionen. [Leipzig–Berlin] 2 Aufl.

Carathéodory, C. (1938). Entwurf fuer eine Algebraisierung des Integralbegriffs. S.-B. *Math. Nat. Abt Bayer. Akad. Wiss.* 27–69. *Zbl*.20.297

Carmichael, R. D. (1919). Existence of a Stieltjes integral. *Proc. Nat. Acad. Sci.* **5**, 551–5.

Carrington, D. C. and Pacquement, A. (1972). Sur une extension du procède d'integration de M. R. Henstock, *C.R. Acad. Sci. Paris*, Ser. A-B, **274**, A1901–A1904. *MR* 46 #3712.

Cauchy, A. L. (1821). Cours d' analyse de l'École Royale Polytechnique. Analyse Algebrique. *Works* **3**(2), [Gauthier-Villars] Paris (1900).

Cauchy, A. L. (1823). Résumé des leçons données à l'École Royale Polytechnique sur le calcul infinitésimal. *Works* **4**(2), 5–261 (Paris).

Cauchy, A. L. (1827). De la différentiation sous le signe intégrale. *Works* **7**(2), 160–76.

Cauchy, A. L. (1882–99) Oeuvres complètes (Paris).

Čelidze, V. G. (Chelidze) (1937). Über derivierte Zahlen einer Funktion zweier variablen. *C. R. Acad. Sci. URSS* **15**, 13–15. *Zbl*.16.297.

Čelidze, V. G. (1947). Double Denjoy integrals (Russian). *Akad. Nauk Gruzin. SSR. Trudy Tbiliss. Mat. Inst. Razmadze* **15**, 155–242. *MR* 14–735.

Čelidze, V. G. and Dzvarseisvili, A. G. (Dzhvarsheishvili) (1978). Theory of the Denjoy integral and some of its applications (Russian). *Tbilis. Gos. Univ.*, Tbilisi. *MR* 81e: 26002.

Çesari, L. (1962). Quasi additive set functions and the concept of integral over a variety. *Trans. Am. Math. Soc.* **102**, 94–113. *MR* 26 #292.

Chakrabarti, P. S. (1983). *Studies on Approximate Peano Derivatives and Approximate Çesàro–Perron Integrals*. Thesis, University of Burdwan.

Chakrabarti, P. S. and Mukhopadhyay, S. N. (1981). Integration by parts for certain approximate CP-integrals. *Bull. Inst. Math. Acad. Sinica* **9**, 493–507. *MR* 83g: 26013.

Chakrabarti, P. S. and Mukhopadhyay, S. N. (1982). A scale of approximate Çesàro–Perron integrals. *Bull. Inst. Math. Acad. Sinica* **10**, 323–346. *MR* 84g: 26007.

Chakrabarty, M. C. (1969a). Some results on ω-derivatives and BV-ω functions. *J. Australian Math. Soc.* Ser. A, **9**, 345–60. *MR* 40 #1547.

Chakrabarty, M. C. (1969b). Some results on AC-ω functions. *Fundamenta Math.* **64**, 219–30. *MR* 41 #5567.

Chakrabarty, M. C. (1969/1970). On functions of generalized bounded ω-variation. *Fundamenta Math.* **66**, 293–300. *MR* 41 #5568.

REFERENCES

Chakrabarty, M. C. (1971). On the space of BV-ω functions. *Fundamenta Math.* **70**, 13–23. *MR* 44 #4162.

Chakrabarty, M. C. On modified Perron integral. *Indian J. Math.* (details not known).

Chan Kai-Meng, (1967). The equivalence of Henstock's two definitions of the Riemann complete integral. *J. Lond. Math. Soc.* **42**, 349–50. *MR* 34 #7761.

Chandrasekhar, S. (1943). Stochastic problems in physics and astronomy. *Rev. Modern Phys.* **15**, 1–89. Selected papers on noise and stochastic processes (Dover, New York, 1954). *MR* 4.248.

Chatfield, J. A. (1973). Equivalence of integrals. *Proc. Am. Math. Soc.* **38**, 279–85. *MR* 47 #409.

Chatfield, J. A. (1979). Solution for an integral equation with continuous interval functions. *Pacific J. Math.* **80**, 47–57. *MR* 80h: 45022.

Chatterji, S. D. (1974). Sur l'intégrabilité de Pettis. *Math. Zeit.* **136**, 53–58. *MR* 49 #7767.

Chew Tuan-Seng (1985). *Orthogonally Additive Functionals.* Thesis, National University of Singapore.

Chew Tuan-Seng (1986). *The Superposition of Operators in the Denjoy Space.* Res. Report, 256. Department of Mathematics, National University of Singapore.

Chew Tuan-Seng (1988a). On the generalized dominated convergence theorem. *Bull Australian Math. Soc.* **37**, 165–171. *MR* 89a: 26012.

Chew Tuan-Seng (1988b). Nonlinear Henstock–Kurzweil integrals and representation theorems. *Southeast Asian Bull. Math.* **12**, 97–108.

Chew Tuan-Seng (1988/1989). The superposition of operators in the space of Henstock–Kurzweil integrable functions. *Real Analysis Exchange* **14**, 62–63. *MR*.

Choksi, J. R. (1958). On compact contents. *J. Lond. Math. Soc.* **33**, 387–98. *MR* 20 #7088.

Choquet, G. (1944). Primitive d'une fonction par rapport à une fonction à variation non bornée. *C. R. Acad. Sci. Paris* **218**, 495–97. *MR* 6–204.

Choquet, G. (1948). Application des propriétés descriptives de la fonction contingent à la théorie des fonctions de variable réelle et à la géométrie différentielle de variétés cartesiennes. *J. Math. Pures Appl.* **26**(9), 115–226. *MR* 9–419.

Copeland, A. H. (1937). A new definition of a Stieltjes integral. *Bull. Am. Math. Soc.* **43**, 581–8. *Zbl*.17.107.

Coppin, C. A. and Vance, J. F. (1972), On a generalized Riemann–Stieltjes integral. *Rivista Mat. Univ. Parma* **1**(3), 73–8. *MR* 51 #827.

Cousin, P. (1895). Sur les fonctions de n variables complexes. *Acta Math.* **19**, 1–62.

Cross, G. E. (1960a). The relation between two definite integrals. *Proc. Am. Math. Soc.* **11**, 578–9. *MR* 22 #8094.

Cross, G. E. (1960b). The expression of trigonometrical series in Fourier form. *Can. J. Math.* **12**, 694–8. *MR* 22 #8275.

Cross, G. E. (1963). A relation between two symmetric integrals. *Proc. Am. Math. Soc.* **14**, 185–90. *MR* 26 #281.

Cross, G. E. (1967). An integral for Çesàro summable series. *Can. Math. Bull.* **10**, 85–97. *MR* 35 #4669.

Cross, G. E. (1971). On the generality of the AP-integral. *Can. J. Math.* **23**, 557–561. *MR* 44 #378.

Cross, G. E. (1975). The P^n-integral. *Can. Math. Bull.* **18**, 493–7. *MR* 53 #3224.

Cross, G. E. (1978a). The representation of (C, k) summable series in Fourier form. *Can. Math. Bull.* **21**, 149–58. *MR* 58 #11270.

Cross, G. E. (1978b). Additivity of the P^n-integral. *Can. J. Math.* **30**, 783–96. *MR* 58 #6108.

REFERENCES

Cross, G. E. (1978c). The SC_{k+1} P-integral and trigonometric series. *Proc. Am. Math. Soc.* **69**, 297–302. MR 57 #16494.

Cross, G. E. (1981). Some conditions for n-convex functions. *Proc. Am. Math. Soc.* **82**, 587–92. MR 82f: 26010.

Cross, G. E. (1982a). The exceptional sets in the definition of the P^n-integral. *Can. Math. Bull.* **25**, 385–91. MR 84a: 26004.

Cross, G. E. (1982b). Additivity of the P^n-integral (II). *Can. J. Math.* **34**, 506–12. MR 83m: 26009.

Cross, G. E. (1985/1986). Higher order Riemann complete integrals. *Real Analysis Exchange* **11**, 347–64. MR 87g: 26012.

Cross, G. E. (1986). The integration of exact Peano derivatives. *Can. Math. Bull.* **29**, 334–40. MR 87g: 26006.

Cross, G. E. (1987/1988). Generalized integrals as limits of Riemann-like sums. *Real Analysis Exchange* **13**, 390–403. MR 89e: 26018.

Cross, G. E. and Shisha, O. (1985/1986). A new approach to integration. *Real Analysis Exchange* **11**, 85–6. *J. Math. Anal. Appl.* **114** (1986), 289–94. MR 87g: 26013.

Daniell, P. J. (1917/1918). A general form of integral. *Ann. Math.* **19**(2), 279–94. *Jbuch* **47**, 911.

Daniell, P. J. (1919/1920). Further properties of the general integral. *Ann. Math.* **22**(2), 203–20. *Jbuch* **47**, 911.

Darboux, J. G. (1875). Mémoire sur les fonctions discontinues. *Ann. Sci. Ec. Norm. Sup.* **4**(2), 57–112.

Das, A. G. (1984). Convergence in u-second variation and RS_u-integrals. *Studia Sci. Math. Hungar.* **19**, 177–85. MR 88h: 26003.

Das, A. G. (1987). A new improper Riemann integral and the Dirac delta function. *Indian J. Pure Appl. Math.* **18**, 997–1001. MR 89a: 26014.

Das, A. G. and Lahiri, B. K. (1977). On RS_u-integral. *Studia Sci. Math. Hungar.* **12**, 117–24. MR 81d: 26004.

Das, A. G. and Lahiri, B. K. (1979/1980). On absolutely kth continuous functions. *Fundamenta Math.* **105**, 159–69. MR 81h: 26008.

Das, A. G. and Lahiri, B. K. (1980). On RS_k integrals. *Comment Math. Prace Mat.* **22**, 33–41. MR 82g: 26013.

Das, U. (1986). *Generalized Derivatives and Generalized Integrals*. Thesis, University of Kalyani.

Das, U. and Das, A. G. (1981). Convergence in kth variation and RS_k integrals. *J. Australian Math. Soc. Ser. A.* **31**, 163–74. MR 83a: 26015.

Das, U. and Das, A. G. (1986). Approximate extensions of P^k- and C_kD-integrals. *Indian J. Math.* **26**, 183–94. MR 88b: 26011.

Das, U. and Das, A. G. (1988). Integration by parts for some general integrals. *Bull. Australian Math. Soc.* **37**, 1–15. MR 89b: 26008.

Davies, R. O. and Schuss, Z. (1970). A proof that Henstock's integral includes Lebesgue's. *J. Lond. Math. Soc.* **2**(2), 561–2. MR 42 #435.

Deheuvels, P. (1986). *L'intégrale*. Presses Universitaires de France, Paris. MR 87j: 26001.

Demuth, O. (Demut) (1978). Constructive analogues of generalized absolutely continuous functions of generalized bounded variation (Russian). *Comment Math. Univ. Carolinae* **19**, 471–87. MR 80b: 03100.

Demuth, O. (1979). Constructive Denjoy integrals (Russian). *Comment Math. Univ. Carolinae* **20**, 213–27. MR 80k: 03065.

Demuth, O. (1980). On the constructive Perron integral (Russian). *Acta Univ. Carolinae-Math. Phys.* **21**, 3–57. MR 82h: 26013.

Denjoy, A. (1912a). Une extension de l'intégrale de M. Lebesgue. *C.R. Acad. Sci. Paris* **154**, 859–62.

Denjoy, A. (1912b). Calcul de la primitive de la fonction dérivée la plus générale. *C.R. Acad. Sci. Paris* **154**, 1075–8.

Denjoy, A. (1914a). Sur une propriété des fonctions à nombres dérivés finis. *C.R. Acad. Sci. Paris* **158**, 99–101.

Denjoy, A. (1914b). Exemples des fonctions dérivées. *C.R. Acad. Sci. Paris* **158**, 1003.

Denjoy, A. (1915a). Sur la théorie descriptive des nombres dérivés d'une fonction continue. *C.R. Acad. Sci. Paris* **160**, 707–9.

Denjoy, A. (1915b). Sur les nombres dérivées. *C.R. Acad. Sci. Paris* **160**, 763–5.

Denjoy, A. (1915c). Les quatre cas fundamentaux des nombres dérivés. *C.R. Acad. Sci. Paris* **161**, 124–6.

Denjoy, A. (1915d). Mémoire sur les nombres dérivés des fonctions continues. *J. Math. Pures Appl.* **1**(7), 105–240.

Denjoy, A. (1915e). Sur les fonctions dérivées sommables. *Bull. Soc. Math. France* **43**, 161–248.

Denjoy, A. (1916a). Sur la dérivation et son calcul inverse. *C.R. Acad. Sci. Paris* **162**, 377–80.

Denjoy, A. (1916b). Mémoire sur la totalisation des nombres dérivés non-sommables. *Ann. Sci. Ecole Norm. Sup.* **33**(3), 127–222.

Denjoy, A. (1917). Mémoire sur la totalisation des nombres dérivés la plus generale. *Ann. Sci. Ecole Norm. Sup.* **34**(3), 181–236.

Denjoy, A. (1919). Sur l'intégration riemanienne. *C.R.Acad. Sci. Paris* **169**, 219–20.

Denjoy, A. (1920). Sur les ensembles clairsemés. *Nederl. Akad. Wetensch. Proc.* Ser. A. **22**, 419–28.

Denjoy, A. (1921a). Sur une classe de fonctions admettant une dérivée seconde généralisée. *Nederl. Akad. Wetensch. Proc.* Ser A **23**, 363–74.

Denjoy, A. (1921b). Sur une propriété des séries trigonométriques. *Nederl. Akad. Wetensch. Proc.* Ser. A **23**, 375–87.

Denjoy, A. (1921c). Sur un calcul de totalisation à deux degrés. *C.R. Acad. Sci. Paris* **172**, 653–5.

Denjoy, A. (1921d). Sur la détermination des fonctions intégrables présentant un certain caractère complexe de résolubilité. *C.R. Acad. Sci. Paris* **172**, 833–5.

Denjoy, A. (1921e). Caractères de certaines fonctions intégrables et opérations correspondantes, *C.R. Acad. Sci. Paris* **172**, 903–6.

Denjoy, A. (1921f). Calcul des coefficients d'une série trigonométrique convérgente quelconque dont la somme est donné. *C.R. Acad. Sci. Paris* **172**, 1218–21.

Denjoy, A. (1921g). Sur une mode d'intégration progressif et les caractères d'intégrabilité correspondants. *C.R. Acad. Sci. Paris* **173**, 127–9.

Denjoy, A. (1931). Sur la définition riemannienne de l'intégrale de Lebesgue. *C.R. Acad. Sci. Paris* **193**, 695–8. *Zbl*.3.106.

Denjoy, A. (1933). Sur le calcul des coefficients de séries trigonométriques. *C.R. Acad. Sci. Paris* **196**, 237–9. *Zbl*.6.302.

Denjoy, A. (1935). Sur l'intégration des coefficients différentiels d'ordre supérieur. *Fundamenta Math.* **25**, 273–326. *Zbl*.12.346.

Denjoy, A. (1940a). Totalisation simple des fonctions ramenée à celle des séries. *C.R. Acad. Sci. Paris* **210**, 73–6. *MR* 1–208.

Denjoy, A. (1940b). Exemples de séries trigonométriques nonsommables. *C.R. Acad. Sci. Paris* **210**, 94–7. *MR* 1–225.

Denjoy, A. (1941). *Leçons sur le calcul des coefficients d'une série trigonométrique*, I–III, pp. xiv–326, Gauthier-Villars, Paris. *MR* 8–260.

Denjoy, A. (1949). *Leçons sur le calcul des coefficients d'une série trigonométrique*, IV, pp. 327–481, 483–715, Gauthier-Villars, Paris. *MR* 11–99.

Denjoy, A. (1954). Mémoire sur la dérivation et son calcul inverse. (A collection of 1915d, e, 1916b, 1917), Gauthier-Villars, Paris. *MR* 16–22.

Denjoy, A. (1955a). Totalisation des dérivées premières généralisées, I. *C.R. Acad. Sci. Paris* **241**, 617–620. *MR* 17–353.

Denjoy, A. (1955b). Totalisation des dérivées premières symétriques, II, *C. R. Acad Sci Paris* **241**, 829–32. *MR* 17–353.

Denjoy, A. (1957). *Une demi-siècle (1907–1956) de notes communiquées aux Académies, II Le champ réel*. (A collection of 1916a, 1919 to 1933, 1940 and 1955), Gauthier-Villars, Paris. *Zbl* 44,58.

De Sarkar, S. (1984). *Fouriers of bounded kth variation, related absolute continuity and general integrals*. Thesis, University of Kalyani.

De Sarkar, S. and Das, A. G. (1983). On functions of bounded kth variation. *J. Indian Inst. Sci.* **64**, 299–309. *MR* 86c: 26012.

De Sarkar, S. and Das, A. G. (1985). A convergence theorem in essential kth variation. *Soochow J. Math.* **11**, 109–15. *MR* 87h: 26009.

De Sarkar, S. and Das, A. G. (1987). Riemann derivatives and general integrals. *Bull. Australian. Math. Soc.* **35**, 187–211. *MR* 88c: 26009.

De Sarkar, S., Das, A. G. and Lahiri, B. K. (1985). Approximate Riemann derivative and approximate P^k-, D^k-integrals. *Indian J. Math.* **27**, 1–32. *MR* 87j: 26011.

Dienes, P. (1947). Sur l'intégrale de Riemann–Stieltjes. *Revue Sci.* **85**, 259–74. *MR* 9–275.

Diestel, J. and Uhl, J. J. (1977). *Vector Measures*. Am. Math. Soc. Math. Surveys, 15. *MR* 56 #12216.

Dieudonné, J. (1941). Sur le théorème de Lebesgue Nikodym. *Ann. Math.* **42**(2), 547–55. *MR* 3–50.

Dieudonné, J. (1944). Sur le théorème de Lebesgue Nikodym, II. *Bull. Soc. Math. France* **72**, 193–239. *MR* 7–305.

Dieudonné, J. (1948). Sur le théorème de Lebesgue Nikodym, III. *Ann. de l'Université de Grenoble, Sect. Sci. Math. Phys. N. S.* **23**, 25–53. *MR* 10–519.

Dinculeanu, N. (1966). *Vector Measures*. Berlin (Pergamon, New York, 1967). *MR* 34 #6011a,b.

Ding Chuan-Song and Lee Peng-Yee (1986/1987). On absolutely Henstock integrable functions. *Real Analysis Exchange* **12**, 524–9. *MR* 88f: 26007.

Dobrakov, I. (1970a). On integration in Banach spaces, I. *Czech. Math. J.* **20**(95), 511–36. *MR* 51 #1391.

Dobrakov, I. (1970b). On integration in Banach spaces, II, *Czech. Math. J.* **20**(95), 680–95. *MR* 51 #1392.

Dobrakov, I. (1971). On representation of linear operators on C_0 (T, X). *Czech. Math. J.* **21**(96), 13–30. *MR* 43 #2544.

Dobrakov, I. (1979). On integration in Banach spaces, III. *Czech. Math. J.* **29**(104), 478–99. *MR* 81b: 28008.

Dobrakov, I. (1980a). On integration in Banach spaces IV. *Czech. Math. J.* **30**(105), 259–79. *MR* 81m: 28006.

Dobrakov, I. (1980b). On integration in Banach spaces V. *Czech. Math. J.* **30**(105), 610–28. *MR* 81m: 28007.

Dollard, J. D. and Friedman, C. N. (1978a). On strong product integration. *J. Functional Anal.* **28**, 309–54. *MR* 58 #11742a.
Dollard, J. D. and Friedman, C. N. (1978b). Product integrals II: Contour integrals. *J. Functional Anal.* **28**, 355–68. *MR* 58 #11742b.
Dollard, J. D. and Friedman, C. N. (1979). Product integration with applications to differential equation. *Encyclopedia of Math.* **10**, Addison-Wesley. *MR* 81e: 34003.
Dubois, J. and M'-Khalfi, A. (1988a). Une intégral de type Riemann généralisée contenant l'intégral de Birkhoff. *Ann. Soc. Math. Québec* **12**, 31–53. *MR* 89h: 26008.
Dubois, J. and M'Khalfi, A. (1988b). Propriétés et théorèmes de convergence pour une intégral de type Riemann généralisée. *Ann. Soc. Math. Québec* **12**, 189–210. *MR*.
Dubuc, S. (1969). La dérivée de la primitive de l'intégrale de Perron. *Can. Math. Bull.* **12**, 521–22. *MR* 40 #1550.
Dunford, N. (1935a). Integration in general analysis. *Trans. Am. Math. Soc.* **37**, 441–53. *Zbl*.11.341.
Dunford, N. (1935b). Corrections to the paper 'Integration in general analysis'. *Trans. Am. Math. Soc.* **38**, 600–1. *Zbl*.13.155.
Dunford, N. (1936a). Integration of abstract functions. *Bull. Am. Math. Soc.* **41**, 178.
Dunford, N. (1936b). Integration and linear operations. *Trans. Am. Math. Soc.* **40**, 474–94.
Dunford, N. (1936c). A particular sequence of step functions. *Duke Math. J.* **2**, 166–70.
Dunford, N. (1937). Integration of vector valued functions (abstract). *Bull. Am. Math. Soc.* **42**, 24.
Dunford, N. (1938). Uniformity in linear spaces. *Trans. Am. Math. Soc.* **44**, 305–56. *Zbl*.19.416.
Dunford, N. and Schwartz, J. (1958). *Linear Operators, Part I: General Theory*, Interscience, New York. *MR* 22 #8302.
Dunford, N. and Schwartz, J. (1963). *Linear Operators, Part II: Spectral Theory. Self-adjoint Operators in Hilbert space*. Interscience, New York. *MR* 32 #6181.
Dunford, N. and Schwartz, J. (1971). *Linear Operators, Part III: Spectral Operators*. Interscience, New York. *MR* 54 #1009.
Dunford, N. and Tamarkin, J. D. (1941). A principle of Jessen and general Fubini theorems. *Duke Math. J.* **8**, 743–9. *MR* 3–207.
Dushnik, B. (1931). *On the Stieltjes Integral.* Dissertation, University of Michigan.
Dutta, D. K. (1973). Some results on BV-ω points. *Riv. Mat. Univ. Parma* **2**(3), 233–9. *MR* 52 #14191.
Dutta, D. K. (1975). Upper and lower Lebesgue–Stieltjes integrals. *Fund. Math.* **87**, 121–40. *MR* 51 #13153.
Dutta, D. K. (1978/1979). Çesàro–Perron–Stieltjes integral. *Colloq. Math.* **40**, 291–304. *MR* 80j: 26005.
Dutta, D. K. (1981). On the (ω) C-derivates of a function. *Indian. J. Math.* **23**, 95–104. *MR* 85b: 26006.
Dutta, D. K. (1984). A note on the definition of the CPS-integral. *Bull. Calcutta Math. Soc.* **76**, 180–3. *MR* 87e: 26008.
Džvaršeĭšvili, A. G. (Dzhvarsheĭshvili) (1950). On the representation by singular integrals of functions integrable in the sense of Denjoy–Perron (Russian). *Soobšč. Akad. Nauk Gruzin. SSR*, **11**, 473–8. *MR* 14–635.
Dzvarseisvili, A. G. (1951a). On a double integral of Denjoy–Čelidze (Russian). *Soobšč. Akad. Nauk. Gruzin. SSR*, **12**, 193–9. *MR* 14–28.

Dzvarseisvili, A. G. (1951b). On a sequence of integrals in the sense of Denjoy (Russian). *Akad. Nauk. Gruzin. SSR, Trudy Mat. Inst. Razmadze* **18**, 221–36. *MR* 14–628.

Dzvarseisvili, A. G. (1951c). On integration and differentiation under the Denjoy integral sign (Russian). *Soobšč. Akad. Nauk Gruzin. SSR.* **12**, 385–92. *MR* 14–628.

Dzvarseisvili, A. G. (1953a). On Fubini's theorem for double Denjoy integrals (Russian). *Soobšč. Akad. Nauk. Gruzin. SSR*, **14**, 393–8. *MR* 16–345.

Dzvarseisvili, A. G. (1953b). On the normed space of D^*-integrable functions. *Akad. Nauk. Gruzin SSR, Trudy Tbiliss Mat. Inst. Razmadze* **19**, 153–62 *MR* 16–490.

Dzvarseisvili, A. G. (1954). On the generalized absolutely continuous functions of two variables. *Soobšč. Akad. Nauk Gruzin. SSR*, **15**, 129–33. *MR* 16–1092.

Dzvarseisvili, A. G. (1955). On generalized absolutely continuous functions of two variables (Russian). *Akad. Nauk Gruzin. SSR, Trudy Tbiliss. Mat. Inst. Razmadze* **21**, 77–110. *MR* 17–954.

Dzvarseisvili, A. G. (1956a). On a sequence of integrals (Russian). *Akad. Nauk Gruzin. SSR*, **17**, 297–302. *MR* 18–297.

Dzvarseisvili, A. G. (1956b). On the summation of conjugate series and series of Fourier–Denjoy (Russian), *Akad. Nauk Gruzin. SSR, Trudy Tbiliss. Mat. Inst. Razmadze* **22**, 203–25. *MR* 18–393.

Dzvarseisvili, A. G. (1957). On integration of the product of two functions (Russian). *Akad. Nauk Gruzin SSR, Trudy Tbiliss. Mat. Inst. Razmadze* **24**, 35–51. *MR* 20 #5265.

Dzvarseisvili, A. G. (1958a). The Denjoy integral and some questions of analysis (Russian). *Akad. Nauk Gruzin. SSR, Trudy Tbiliss Mat. Inst. Razmadze* **25**, 273–372 *MR* 21 #5711a.

Dzvarseisvili, A. G. (1958b). Analytic functions in the interior of the unit circle (Russian). *Akad. Nauk Gruzin. SSR, Trudy Tbiliss. Mat. Inst. Razmadze* **25**, 373–410. *MR* 21 #5711b.

Dzvarseisvili, A. G. (1959). On functions conjugate to D^*-integrable functions (Russian). *Soobšč. Akad. Nauk Gruzin. SSR*, **23**, 385–9. *MR* 24 #A2195.

Dzvarseisvili, A. G. (1970). The Cauchy–Denjoy type integrals (Russian). *Sakharth. SSR Mechn. Akad. Math. Inst. Srom.* **38**, 52–64. *MR* 44 #417.

Egorov, D.-Th. (Egoroff) (1912). Sur l'intégration des fonctions mesurables. *C.R. Akad. Sci. Paris* **155**, 1474–5.

Einstein, A. (1906). Zur Theorie der Brownschen Bewegung. *Ann. d. Physik* **19**, 371–81.

Ellis, H. W. (1949). Mean-continuous integrals. *Can. J. Math.* **1**, 113–24. *MR* 10–520.

Ellis, H. W. (1950). Examples of integrals that are discontinuous in sets of positive measure. *Trans. Roy. Soc. Canada* Sect. III **44**(3), 37–42. *MR* 12–399.

Ellis, H. W. (1951a). On the compatibility of the approximate Perron and Çesàro–Perron integrals. *Proc. Am. Math. Soc.* **2**, 396–7. *MR* 13–331.

Ellis, H. W. (1951b). Darboux properties and applications to non-absolutely convergent integrals. *Can. J. Math.* **3**, 471–85. *MR* 13–332.

Ellis, H. W. (1952). On the relation between the P^2-integral and the Çesàro–Perron scale of integrals. *Trans. Roy. Soc. Canada*, Sect. III **46**(3), 29–32. *MR* 14–628.

Ellis, H. W. and Jeffery, R. L. (1967a). On measures determined by functions with finite right and left limits everywhere. *Can. Math. Bull.* **10**, 207–25. *MR* 35 #4356.

Ellis, H. W. and Jeffery, R. L. (1967b). Derivatives and integrals with respect to a base function of generalized bounded variation. *Can. J. Math.* **19**, 225–41. *MR* 35 #1725.

REFERENCES

Ellis, R. (1957). A note on the continuity of the inverse. *Proc. Am. Math. Soc.* **8**, 372–3. MR 18–745.

Ene, G. and Ene, V. (1985/1986). Nonabsolutely convergent integrals. *Real Analysis Exchange* **11**, 121–33. MR 87c: 26013.

Ene, V. (1988/1989). Integrals of Lusin and Perron type. *Real Analysis Exchange* **14**, 115–39.

Enomoto, S. (1954a). Notes sur l'intégration I Quelques propriétés des fonctions d'intervalle. *Proc. Japan Acad.* **30**, 176–9. MR 16–344.

Enomoto, S. (1954b). Notes sur l'intégration II. Une propriété du recouvrement fermé de l'intervalle. *Proc. Japan Acad.* **30**, 289–90. MR 16–344.

Enomoto, S. (1954c). Notes sur l'intégration III Théorème de Fubini. *Proc. Japan Acad.* **30**, 437–42. MR 16–345.

Enomoto, S. (1955a). Sur une totalisation dans les espaces de plusieurs dimensions, I. *Osaka Math. J.* **7**, 69–102. MR 17–246.

Enomoto, S. (1955b). Sur une totalisation dans les espaces de plusieurs dimensions, II. *Osaka Math. J.* **7**, 157–78. MR 19–399. (Change of name to S. Nakanishi)

Evans, G. C. (1927). The logarithmic potential. *Am. Math. Soc. Coll. Publication* 61, New York.

Fatou, P. (1906). Séries trigonométriques et séries de Taylor. *Acta Math.* **30**, 335–400 (see p. 375).

Feynman, R. P. (1948). Space–time approach to non-relativistic quantum mechanics. *Rev. Modern Phys.* **20**, 367–87. MR 10–224.

Fichtenholz, G. (1936). Sur une généralisation de l'intégrale de Stieltjes. *C.R. Acad. Sci. URSS*, NS **3**, 95–100. Zbl.15.106.

Fichtenholz, G. and Kantorovic, L. (1935). Sur les operations lineaire dans l'espace des fonctions bornées. *Studia Math.* **5**, 69–98. Zbl.13.065.

Finetti, B. de and Jacob, M. (1935). Sull' integrale di Stieltjes–Riemann. *Giornale Istituto Italiano degli Attuari* **6**, 303–19. Zbl.13.106.

Fischer, E. (1907). Sur la' convergence en moyenne, *C.R. Acad. Sci. Paris.* **144**, 1022–4.

Fleissner, R. and O'Malley, R. (1979). Conditions implying the summability of approximate derivatives. *Coll. Math.* **41**, 257–63. MR 81m: 26002.

Foglio, S. (1963). Absolute N-integration. *J. Lond. Math. Soc.* **38**, 87–8. MR 26 #3868.

Foglio, S. (1968). The N-variational integral and the Schwartz distributions. *Proc. Lond. Math. Soc.* **18**(3), 337–48. MR 37 #3353.

Foglio, S. (1970).The N-variational integral and the Schwartz distributions II. *J. Lond. Math. Soc.* **2**(2), 14–18. MR 40 #6257.

Foglio, S. (1985). Arzelà dominated convergence for generalized integrals. *Atas do decimo quarto colòquio Brasileiro di Matemàtica* 1 (Rio de Janeiro) 205–221, MR.

Foglio, S. (1988/1989). New and old results connected with Henstock's integrals. *Real Analysis Exchange* **14**, 70–1.

Foglio, S. and Henstock, R. (1973). The N-variational integral and the Schwartz distributions III. *J. Lond. Math. Soc.* **6**(2), 693–700. MR 48 #2755.

Foran, J. (1975). An extension of the Denjoy integral. *Proc. Am. Math. Soc.* **49**, 359–65. MR 51 #3370.

Foran, J. (1981). On extending the Lebesgue integral. *Proc. Am. Math. Soc.* **81**, 85–8. MR 81j: 26007.

Foran, J. (1981/1982). A note on Denjoy integrable functions. *Real Analysis Exchange* **7**, 255–8. MR 83g: 26014.

Foran, J. (1982/1983). A chain rule for the approximate derivative and change of variables for the D-integral. *Real Analysis Exchange* **8**, 443–54. *MR* 84g: 26005.

Foran, J. (1983/1984). The chain rule and change of variable for the Denjoy integral. *Real Analysis Exchange* **9**, 138–40.

Foran, J. and Meinershagen, S. (1987/1988). Some answers to a question of P. Bullen. *Real Analysis Exchange* **13**, 265–77 *MR* 88m: 26014.

Foran, J. and O'Malley, R. J. (1984/1985). Integrability conditions for approximate derivatives. *Real Analysis Exchange* **10**, 294–306. *MR* 86j: 26008.

Fréchet, M (1915). Sur l'intégrale d'une fonctionnelle étendue à un ensemble abstrait. *Bull. Soc. Math. France* **43**, 249–67.

Fréchet, M (1936). Sur quelques définitions possible de l'intégrale de Stieltjes. *Duke Math. J.* **2**, 383–95. *Zbl.*14.207.

Fréchet, M (1944). L'intégrale abstraite d'une fonction abstraite d'une variable abstraite et son application à la moyenne d'un élément aléatoire de nature quelconque. *Rev. Sci. (Rev. Rose Illus.)* **82**, 483–512. *MR* 8–141.

Frenkel, Y. and Cotlar, M. (1950). Non-additive majorants and minorants in the theory of the Perron–Denjoy integral, (Spanish). *Rev. Acad. Ci., Madrid* **44**, 411–26. *MR* 13–121.

Freudenthal, H. (1936). *Teilweise geordnete Moduln*. *Neder. Akad. Wetensch. Amsterdam, Proc.* **39**, 641–51. *Zbl.*14.313.

Fubini, G. (1907). Sugli integrali multipli. *Atti. Accad. Naz. Lincei Rend.* **16**, 608–14.

Gál, I. S. (1952). Sur la méthode de résonance et sur un théorème concernant les espaces de type (B). *Ann. Inst. Fourier Univ. Grenoble* **3**, 23–30. *MR* 14–288.

Gál, I. S. (1953). The principle of condensation of singularities. *Duke Math. J.* **20**, 27–35. *MR* 14–657.

Gál, I. S. (1957). On the fundamental theorems of the calculus. *Trans. Am. Math. Soc.* **86**, 309–20. *MR* 20 #86.

Gelfand, I. M. (1936). Sur un lemme de la théorie des espaces linéaires. *Commun. Inst. Sci. Math. et Mecan. de Univ. Kharkoff et soc. Math. Kharkoff* **13**(4), 35–40. *Zbl.*14.162.

Gelfand, I. M. (1938). Abstrakte Funktionen und lineare Operatoren. *Rec. Math. Moscou*, N.S. **4**, 235–84, *Zbl.*20.367.

Geöcze, Z. de (1910). Quadrature des surfaces courbes. *Math. Naturwiss. Ber. Ungarn.* **26**, 1–88.

Getchell, B. C. (1935). On the equivalence of two methods of defining Stieltjes integrals. *Bull. Am. Math. Soc.* **41**, 413–18. *Zbl.*11.395.

Ghosh, B. and Bose, M. K. (1987). On the Çesàro derivatives. *Soochow J. Math.* **13**, 149–58. *MR* 90a: 26010.

Gillespie, D. C. (1915). The Cauchy definition of integral. *Ann. Math.* **17**, 61–3.

Gleyzal, A. N. (1941). Interval-functions. *Duke Math. J.* **8**, 223–30. *MR* 3–226.

Glivenko, V. I. (1936). Stieltjes integral (Moskou U. Leningrad. verl. onti) (Russian). *Zbl.*17.061.

Goldstine, H. H. (1941). Linear functionals and integrals in abstract spaces. *Bull. Am. Math. Soc.* **47**, 615–20. *MR* 3–7, *MR* 8–207.

Gomes, A. Pereira (1946). Introduction to the notion of functional in spaces without points (Portuguese). *Portug. Mat.* **5**, 1–120. *MR* 8–275.

Gomes, R. L. (1946). Sur la notion de fonctionnelle. *Portug. Mat.* **5**, 202–6. *MR* 8–588.

Goodman, G. S. (1977). N-functions and integration by substitution. *Rend. Sem. Mat. Fis. Milano* **47**, 123–34. *MR* 80e: 26006.

Goodman, G. S. (1978). Integration by substitution. *Proc. Am. Math. Soc.* **70**, 89–91. *MR* 57 #16497.

Goodman, G. S. (1984). Zygmund's lemma and Riemann integrability. *Studia Math.* **77**, 519–22. *MR* 86a: 26010.

Gordon, L. (1966/1967). Perron's integral for derivatives in L^r. *Studia Math.* **28**, 295–316. *MR* 36 #322.

Gordon, L. I. and Lasher, S. (1967). An elementary proof of integration by parts for the Perron integral. *Proc. Am. Math. Soc.* **18**, 394–8. *MR* 35 #1726.

Gordon, R. A. (1986–1987). Equivalence of the generalized Riemann and restricted Denjoy integrals. *Real Analysis Exchange* **12**, 551–74. *MR* 88d: 26013.

Gould, G. G. (1965). Integration over vector-valued measures. *Proc. Lond. Math. Soc.* **15**(3), 193–225. *MR* 30 #4894.

Gowurin, M. (1936). Ueber die Stieltjessche integration abstrakter funktionen. *Fundamenta. Math.* **27**, 254–68. *Zbl.*16.061.

Grattan-Guinness, I. (1970). *The Development of the Foundations of Mathematical Analysis from Euler to Riemann.* The M.I.T. Press, Cambridge, Massachusetts. *MR* 58 #15948.

Graves, L. M. (1927). Riemann integration and Taylor's theorem in general analysis. *Trans. Am. Math. Soc.* **29**, 163–77.

Grimshaw, M. E. (1934). The Cauchy property of the generalised Perron integrals. *Proc. Camb. Phil. Soc.* **30**, 15–18. *Zbl.*8.150.

Guzmán, M. de (1975). *Differentiation of Integrals in $\mathbb{R}^n$*. Lecture Notes in Mathematics. Springer-Verlag 481. *MR* 56 #15866.

Haber, S. and Shisha, O. (1973). An integral related to numerical integration. *Bull. Am. Math. Soc.* **79**, 930–2. *MR* 47 #7288.

Haber, S. and Shisha, O. (1974). Improper integrals, simple integrals and numerical quadratures. *J. Approximation Theory* **11**, 1–15. *MR* 50 #5309.

Hackenbroch, W. (1968). Integration vektorwertiger funktionen nach operatorwertigen massen. *Math. Zeit.* **105**, 327–44. *MR* 40 #2813.

Hahn, H. (1914). Über annaherung der Lebesgueschen integrale durch Riemannsche summen. *Sitzber. Akad. Wiss. Wien. Abt.* **123**, 713–43.

Hahn, H. (1915). Ueber eine Verallgemeinerung der Riemannschen Integral-definition. *Monatsh. Math. Phys.* **26**, 3–18.

Hake, H. (1921). Über de la Vallée Poussins Ober- und Unterfunktionen einfacher Integrale und die Integraldefinitionen von Perron. *Math. Ann.* **83**, 119–42.

Halmos, P. R. (1950). *Measure Theory*, Van Nostrand, New York. *MR* 11–504.

Hardy, G. H. (1901). Notes on some points in the integral calculus (I). On the formula for integration by parts. *Messenger of Math.* **30**, 185–7.

Hardy, G. H. (1907). Notes on some points in the integral calculus (XIX). On Abel's Lemma and the second theorem of the mean. *Messenger of Math.* **36**, 10–13.

Hardy, G. H. (1918). Notes on some points in the integral calculus (L). On the integral of Stieltjes and the formula for integration by parts. *Messenger of Math.* **48**, 90–100.

Harnack, A. (1884). Die allgemeinen Sätze über den Zusammenhang der Funktionen einer reelen Variabeln mit ihren Ableitungen, II. *Math. Annalen* **24**, 217–52.

Haupt, O., Aumann, G. and Pauc, C. Y. (1955). *Differential-und Integralrechnung* III. Walter de Gruyter, Berlin. *MR* 17–1066; 18–1118; 19–1431.

Hayashi, Y. (1962a). A trial production on the integral, I. *Bull. Univ. Osaka Prefecture* Ser. A, **11**(1), 121–31. *MR* 27 #2600a.

Hayashi, Y. (1962b). A trial production on the integral, II. *Bull. Univ. Osaka Prefecture* Ser. A, **11**(2), 117–26. *MR* 27 #2600b.

Hayashi, Y. (1963). A trial production on the integral, III. *Bull. Univ. Osaka Prefecture* Ser. A, **12**(1), 111–26. *MR* 28 #180.

Hayashi, Y. (1963/1964). A trial production on the integral, IV. *Bull. Univ. Osaka Prefecture* Ser. A, **12**(2), 127–38. *MR* 28 #5163.

Hayes, C. A. and Pauc, C. Y. (1970). *Derivation and Martingales*. Band 49. Springer-Verlag, New York, Berlin *Zbl*.192, 406.

Hellinger, E. (1907). *Die Orthogonal invarianten quadratischer Formen von unendlichvielen Variablen*. Dissertation, Göttingen.

Hellinger, E. (1909). Neue Begrundung der Theorie quadratischer Formen von unendlichvielen Veränderlichen. *J. Reine Angew. Math.* **136**, 210–71.

Helly, E. (1912). Über lineare Funktional operationen. S.- B.K. Akad. Wiss. Wien Math. -Naturwiss. kl 121, IIa, 265–297.

Helton, B. W. (1966). Integral equations and product integrals. *Pacific J. Math.* **16**, 297–322. *MR* 32 #6167.

Helton, B. W. (1969). A product integral representation for a Gronwall inequality. *Proc. Am. Math. Soc.* **23**, 493–500. *MR* 40 #1562.

Helton, B. W. (1973). The solution of a nonlinear Gronwall inequality. *Proc. Am. Math. Soc.* **38**, 337–42. *MR* 46 #9287.

Helton, B. W. (1976). A special integral and a Gronwall inequality. *Trans. Am. Math. Soc.* **217**, 163–81. *MR* 53 #10998.

Helton, J. C. (1975a). Mutual existence of sum and product integrals. *Pacific. J. Math.* **56**, 495–516. *MR* 53 #8894.

Helton, J. C. (1975b). Product integrals and the solution of integral equations. *Pacific J. Math.* **58**, 87–103 *MR* 52 #6341.

Helton, J. C. (1978). Nonlinear operations and the solution of integral equations. *Trans. Am. Math. Soc.* **237**, 373–90. *MR* 80b: 45023.

Helton, J. C. and Stackwisch, S. (1978). An approximation technique for nonlinear integral operations. *J. Math. Anal. Appl.* **65**, 365–74. *MR* 80b: 45026.

Henstock, R. (1946). On interval functions and their integrals. *J. Lond. Math. Soc.* **21**, 204–9. *MR* 8–572.

Henstock, R. (1948). On interval functions and their integrals (II). *J. Lond. Math. Soc.* **23**, 118–28. *MR* 10–239.

Henstock, R. (1951). Density integration. *Proc. Lond. Math. Soc.* **53**(2), 192–211. *MR* 13–20.

Henstock, R. (1955a). Linear functions with domain a real countably infinite dimensional space. *Proc. Lond. Math. Soc.* **5**(3), 238–56. *MR* 17–176.

Henstock, R. (1955b). The efficiency of convergence factors for functions of a continuous real variable. *J. Lond. Math. Soc.* **30**, 273–86. *MR* 17–359.

Henstock, R. (1956). Linear and bilinear functions with domain contained in a real countably infinite dimensional space. *Proc. Lond. Math. Soc.* **6**(3), 481–500. *MR* 18–584.

Henstock, R. (1957a). On Ward's Perron–Stieltjes integral. *Can. J. Math.* **9**, 96–109. *MR* 18–880.

Henstock, R. (1957b). The summation by convergence factors of Laplace–Stieltjes integrals outside their half plane of convergence. *Math. Zeitschrift* **67**, 10–31. *MR* 18–880.

Henstock, R. (1960a). A new descriptive definition of the Ward integral. *J. Lond. Math. Soc.* **35**, 43–48. *MR* 22 #1648.

Henstock, R. (1960b). The use of convergence factors in Ward integration. *Proc. Lond. Math. Soc.* **10**(3), 107–21. *MR* 22 #12197.

Henstock, R. (1960c). The equivalence of generalized forms of the Ward, variational, Denjoy–Stieltjes, and Perron–Stieltjes integrals. *Proc. Lond. Math. Soc.* **10**(3), 281–303. *MR* 22 #12198.

Henstock, R. (1961a). N-variation and N-variational integrals of set functions. *Proc. Lond. Math. Soc.* **11**(3), 109–33. *MR* 23 #A955.

Henstock, R. (1961b). Definitions of Riemann type of the variational integrals. *Proc. Lond. Math. Soc.* **11**(3), 402–18. *MR* 24 #A1994.

Henstock, R. (1963a). Difference-sets and the Banach–Steinhaus theorem. *Proc. Lond. Math. Soc.* **13**(3), 305–21. *MR* 26 #6776.

Henstock, R. (1963b). Tauberian theorems for integrals. *Can. J. Math.* **15**, 433–9. *MR* 28 #2187.

Henstock, R. (1963c). *Theory of Integration*, Butterworths, London. *MR* 28 #1274.

Henstock, R. (1964). The integrability of functions of interval functions. *J. Lond. Math Soc.* **39**, 589–97. *MR* 29 #5975.

Henstock, R. (1966). Majorants in variational integration. *Can. J. Math.* **18**, 49–74. *MR* 32 #2545.

Henstock, R. (1968a). A Riemann-type integral of Lebesgue power. *Can. J. Math.* **20**, 79–87. *MR* 36 #2754.

Henstock, R. (1968b). *Linear Analysis*, Butterworths, London. *MR* 54 #7725, *Zbl.* 172, #390.

Henstock, R. (1969). Generalized integrals of vector-valued functions. *Proc. Lond. Math. Soc.* **19**(3), 509–36. *MR* 40 #4420.

Henstock, R. (1973a). Integration by parts. *Aequationes Math.* **9**, 1–18. *MR* 47 #3608.

Henstock, R. (1973b). Integration in product spaces, including Wiener and Feynman integration. *Proc. Lond. Math. Soc.* **27**(3), 317–44. *MR* 49 #9145.

Henstock, R. (1974). Additivity and the Lebesgue limit theorems. *Proc. Carathéodory International Symposium*, Athens, September 3–7, 1973, pp. 223–41. Greek Math. Soc. Athens. *MR* 57 #6355.

Henstock, R. (1978). Integration, variation and diferentiation in division spaces. *Proc. Roy. Irish Acad.*, Sect A, **78**, 69–85. *MR* 80d: 26011.

Henstock, R. (1979). The variation on the real line. *Proc. Roy. Irish. Acad.* Sect A, **79**(1), 1–10. *MR* 81d:26005.

Henstock, R. (1980a). Generalized Riemann integration and an intrinsic topology. *Can. J. Math.* **32**, 395–413. *MR* 82b:26010.

Henstock, R. (1980b). Division spaces, vector-valued functions and backwards martingales. *Proc. Roy. Irish Acad.*, **80**(2), 217–32. *MR* 82i: 60091.

Henstock, R. (1982). Density integration and Walsh functions. *Bull. Malaysian Math. Soc.* **5**(2), 1–19. *MR* 84i:26010.

Henstock, R. (1983). A problem in two-dimensional integration. *J. Australian Math. Soc.*, Series A, **35**, 386–404. *MR* 84k: 26010.

Henstock, R. (1983/1984). The Lebesgue syndrome. *Real Analysis Exchange* **9**, 96–110.

Henstock, R. (1986). The reversal of power and integration. *Bull. Inst. Math. Appl.* **22**, 60–1.

Henstock, R. (1988a). Lectures on the theory of integration, World Scientific, Singapore. *MR*

Henstock, R. (1988b). A short history of integration theory. *Southeast Asia Math. Bull.* **12**(2), 75–95. *MR* 90e: 01031

Henstock, R. (1988/1989). Integration over function spaces. *Real Analysis Exchange* **14**, 20–23.

Henstock, R. and Macbeath, A. M. (1953). On the measure of sum-sets (I). The theorems of Brunn, Minkowski, and Lusternik. *Proc. Lond. Math. Soc.* **3**(3), 182–94. *MR* 15–109.

Hildebrandt, T. H. (1917). On integrals related to and extensions of the Lebesgue integrals. *Bull. Am. Math. Soc.* **24**(2), 113–44.

Hildebrandt, T. H. (1918). On integrals related to and extensions of the Lebesgue integrals. *Bull. Am. Math. Soc.* **25**, 177–202.

Hildebrandt, T. H. (1927). Lebesgue integration in general analysis. *Bull. Am. Math. Soc.* **33**, 646.

Hildebrandt, T. H. (1931). On the interchange of limit and Lebesgue integral for a sequence of functions. *Trans. Am. Math. Soc.* **33**, 441–3. *Zbl.* 2.251.

Hildebrandt, T. H. (1938). Definitions of Stieltjes integrals of the Riemann type. *Am. Math. Monthly* **45**, 265–78. *Zbl.*19.056.

Hildebrandt, T. H. (1953). Integration in abstract spaces. *Bull. Am. Math. Soc.* **59**, 111–39. *MR* 14–735.

Hildebrandt, T. H. (1959). On systems of linear differentio-Stieltjes-integral equations. *Illinois J. Math.* **3**, 352–73. *MR* 21 #4339.

Hille, E. and Phillips, R. S. (1957). Functional analysis and semigroups. *Am . Math. Soc. Colloquium Publication.* **31**, *MR* 19–664.

Hinčin, A. (Khintchine) (1916). Sur une extension de l'integral de M. Denjoy. *C.R. Acad. Sci. Paris* **162**, 287–91.

Hinčin, A. (1917). Sur la dérivation asymptotique. *C.R. Acad. Sci. Paris*, **164**, 142–4.

Hinčin, A. (1918). On the process of integration of Denjoy (Russian). *Rec. Math. Soc. Math. Moscou* **30** (Mat. Sb. **20**), 543–57.

Hinčin, A. (1927). Recherches sur la structure des fonctions mesurable. (Recherches sur la generalisation de la notion de derivée). *Fundamenta Math.* **9**, 212–79.

Hobson, E. W. (1909). On the second mean value theorem of the integral calculus. *Proc. Lond. Math. Soc.* **7**(2), 14–23.

Hobson, E. W. (1920). On Hellinger's integrals. *Proc. Lond. Math. Soc.* **18**(2), 249–65.

Hobson, E. W. (1921, 1926). The Theory of Functions of a Real Variable and the Theory of Fourier's Series, Vol. 1 (1921), vol. 2 (1926), Cambridge University Press, Cambridge.

Hölder, E. (1889). Über einen Mittelwertsatz. *Nachr. Akad. Wiss. Göttingen (Math. Phys.)* 38–47.

Holec, J. and Mařík, J. (1965). Continuous additive mappings. *Czechoslovak Math. J.* **15**(90), 237–43. *MR* 31 #1354.

Huggins, F. N. (1971). A generalization of a theorem of F. Riesz. *Pacific J. Math.* **39**, 695–701. *MR* 46 #3713.

Huggins, F. N. (1973). Bounded slope variation. *Texas J. Sci.* **24**, 431–7.

Huggins, F. N. (1976). On the equivalence of certain integrals. *J. Math. Phys. Sci.* **10**, 359–64. *MR* 55 #8291.

Huggins, F. N. (1977). Bounded slope variation and generalized convexity. *Proc. Am. Math. Soc.* **65**, 65–69. *MR* 57 #6326.

Hyslop, J. (1926). A theorem on the integral of Stieltjes. *Proc. Edinburgh Math. Soc.* **44**, 79–84.

Ionescu Tulcea, C. T. (1947). Sur l'intégration des nombres dérivés. *C.R. Acad. Sci. Paris* **225**, 558–60. *MR* 9–179.

Iseki, K. (1962). On quasi-Denjoy integration. *Proc. Japan Acad.* **38**, 252–7. *MR* 26 #6348.

Iseki, K. (1981). On the summation of linearly ordered series. *Nat. Sci. Rep. Ochanomizu Univ.* **32**, 1–12. *MR* 82k: 26005.

Iseki, K. (1983). An attempt to generalize the Denjoy integration. *Natural Sci. Rep. Ochanomizu Univ.* **34**, 19–33. *MR* 85c: 26005.

Iseki, K. (1984). On the powerwise integration. *Nat. Sci. Rep. Ochanomizu Univ.* **35**, 1–46. *MR* 86i: 26009.

Iseki, K. (1985a). On the Dirichlet continuity of functions. *Nat. Sci. Rep. Ochanomizu Univ.* **36**, 1–13. *MR* 87h: 26007a.

Iseki, K. (1985b). On the powerwise integration in the wide sense. *Nat. Sci. Rep. Ochanomizu Univ.* **36**, 15–39. *MR* 87h:26007b.

Iseki, K. (1985c). On an integration called Dirichlet totalization. *Nat. Sci. Rep. Ochanomizu Univ.* **36**, 115–30. *MR* 88c:26010.

Iseki, K. (1985d). On the incremental integration. *Nat. Sci. Rep Ochanomizu Univ.* **36**, 131–45. *MR* 88c:26011.

Iseki, K. (1986a). On the normal integration. *Nat. Sci. Rep. Ochanomizu Univ.* **37**, 1–34. *MR* 88c:26012.

Iseki, K. (1986b). On the sparse integration. *Nat. Sci. Rep. Ochanomizu Univ.* **37**, 91–9. *MR*. 88k: 26008.

Iseki, K. (1987). On two theorems of Nina Bary type. *Nat. Sci. Rep. Ochanomizu Univ.* **38**, 33–98. *MR* 89k: 26006.

Iseki, K. and Maeda, M. (1971). On a generalization of Denjoy integration. *Nat. Sci. Rep. Ochanomizu Univ.* **22**, 101–10. *MR* 47 #2015.

Izumi, S. I. (1933). A new concept of integrals. *Proc. Imp. Acad. Japan* **9**, 570–3. *Zbl*.8.248.

Izumi, S. I. (1934). A new concept of integrals II. *Proc. Imp. Acad. Tokyo* **10**, 57–8. *Zbl*.8.345.

Izumi, S. I. (1935). On the Verblunsky's generalization of the Denjoy integrals. *Sci. Rep. Tohoku Univ.* **24**, 344–51. *Zbl*.12.204.

Izumi, S. I. (1937). Lebesgue integral as the inverse of the differentiation. *Tohoku Math. J.* **43**, 225–32. *Zbl*.17.203.

Izumi, S. I. (1940a). An abstract integral. *Proc. Imp. Acad. Tokyo* **16**, 21–5. *MR* 1–239.

Izumi, S. I. (1940b). An abstract integral II. *Proc. Imp. Acad. Tokyo* **16**, 87–9. *MR* 1–305.

Izumi, S. I. (1941). An abstract integral IV. *Proc. Imp. Acad. Tokyo* **17**, 1–4. *MR* 2–355.

Izumi, S. I. (1942a). An abstract integral VII. *Proc. Imp. Acad. Tokyo* **18**, 53–6. *MR* 9–19.

Izumi, S. I. (1942b). An abstract integral X. *Proc. Imp. Acad. Tokyo* **18**, 543–7. *MR* 9–19.

Izumi, S. I., Matuyama, N. and Orihara, M. (1942). An abstract integral V. *Proc. Imp. Acad. Tokyo* **18**, 45–9. *MR* 9–19.

Izumi, S. I. and Nakamura, M. (1940). An abstract integral III. *Proc. Imp. Acad. Tokyo* **16**, 518–23. *MR* 2–258.

Jacquier-Bryssine, N. and Pacquement, A. (1975). Sur une extension du procédé d'intégration de O. Perron. *C.R. Acad. Sci. Paris*. Sér. A-B, **281**, A839–A842. *MR* 55 #8273.

James, R. D. (1950). A generalized integral II. *Can. J. Math.* **2**, 297–306. *MR* 12–94.

James, R. D. (1954). Generalized nth primitives. *Trans. Am. Math. Soc.* **76**, 149–76. *MR* 15–611.

James, R. D. (1955). Integrals and summable trigonometric series. *Bull. Am. Math. Soc.* **61**, 1–15. *MR* 16–692.

James, R. D. (1956). Summable trigonometric series. *Pacific J. Math.* **6**, 99–110. *MR* 17–1198.

James, R. D. and Gage, W. H. (1946). A generalized integral. *Trans. Roy. Soc. Canada* Sect. III, **40**(3), 25–35. *MR* 9–19.

Jarnik, J. and Kurzweil, J. (1983a) Integral of multivalued mappings and its connection with differential relations (Russian). *Casopis Pěst. Mat.* **108**, 8–28. *MR* 84m: 34016.

Jarnik, J. and Kurzweil, J. (1984). Non-absolutely convergent integral which admits C^1-transformations. *Casopis Pest. Mat.* **109**, 157–67. *MR* 86b:26013.

Jarnik, J. and Kurzweil, J. (1985a). A non-absolutely convergent integral which admits C^1-transformations and which can be used for integration on manifolds. *Czech. Math. J.* **35**(110), 116–39. *MR* 86e:26011.

Jarnik, J. and Kurzweil, J. (1985b). A report on three topics in integration theory and ordinary differential equations. *Česk. Akad. Věch.* **11**. *MR*

Jarnik, J. and Kurzweil, J. (1987). A general form of the product integral and linear ordinary differential equations. *Czech. Math. J.* **37**(112), 642–59. *MR* 89i:34009.

Jarnik, J. and Kurzweil, J. (1988). A new and more powerful concept of the *PU*-integral. *Czech. Math. J.* **38**(113), 8–48. *MR* 89c:26022.

Jarnik, J., Kurzweil, J. and Schwabik, S. (1983). On Mawhin's approach to multiple nonabsolutely convergent integrals. *Casopis Pest. Mat.* **108**, 356–80. *MR* 85h:26011.

Jeffery, R. L. (1932). Non-absolutely convergent integrals with respect to a function of bounded variation. *Trans. Am. Math. Soc.* **34**, 645–75. *Zbl*.4.390.

Jeffery, R. L. (1937). Functions defined by sequences of integrals and the inversion of approximate derived numbers. *Trans. Am. Math. Soc.* **41**, 171–92. *Zbl*.16.158.

Jeffery, R. L. (1938). The equivalence of sequence integrals and non-absolutely convergent integrals. *Bull. Am. Math. Soc.* **44**, 840–5. *Zbl*.20.011.

Jeffery, R. L. (1939). Functions of bounded variation and non-absolutely convergent integrals in two or more dimensions. *Duke Math. J.* **5**, 753–74. *MR* 1–208.

Jeffery, R. L. (1940). Integration in abstract space. *Duke Math. J.* **6**, 706–18. *MR* 2–103.

Jeffery, R. L. (1942). Perron integrals. *Bull. Am. Math. Soc.* **48**, 714–17. *MR* 4–75.

Jeffery, R. L. (1951a). Non-absolutely convergent integrals. *Proc. 2nd Can. Math. Congress* 93–145. (University of Toronto Press). *MR* 13–449.

Jeffery, R. L. (1951b). The theory of functions of a real variable (2nd edn)., University of Toronto Press. *MR* 13–216.

Jeffery, R. L. (1956). Trigonometric series. *Can. Math. Congress Lecture* Ser. No. 2, University of Toronto Press, Toronto. *MR* 20 #4137.

Jeffery, R. L. (1958). Generalized integrals with respect to functions of bounded variation. *Can. J. Math.* **10**, 617–26. *MR* 21 #113.

Jeffery, R. L. (1963). Derivatives and integrals with respect to a base function. Presidential Address, *Proc. 5th Candian Math. Congress*, 25–47.

Jeffery, R. L. (1970). Generalized integrals with respect to base functions which are not of bounded variation. *Lecture Notes in Math.* **419**, 211–20. *MR* 54 #2898.

Jeffery, R. L. and Ellis, H. W. (1942). Çesàro totalization. *Trans. Roy. Soc. Canada* III, **36**, 19–44. *MR* 4–154.

Jeffery, R. L. and Macphail, M. S. (1941). Non-absolutely convergent integrals. *Trans. Roy. Soc. Canada*, III, **35**, 41–58. *MR* 3–227.

Jeffery, R. L. and Miller, D. S. (1945). Convergence factors for generalized integrals. *Duke Math J.* **12**, 127–42. *MR* 6–204.

Jessen, B. (1934). The theory of integration in a space of an infinite number of dimensions. *Acta Math.* **63**, 249–323.

Kac, M. (1959). *Probability and Related Topics in Physical Sciences*, Vol. 1, Interscience, New York. (see pp. 168–171). *MR* 21 #1635.

Kahane, J.-P. (1988). *Enseign. Math.* **34**(2), 255–68.

Kakutani, S. (1948). On equivalence of infinite product measures. *Ann. Math.* **49**(2), 214–24. *MR* 9–340.

Kaltenborn, H. S. (1934). Linear functional operations on functions having discontinuities of the first kind. *Bull. Am. Math. Soc.* **40**, 702–8. *Zbl.*10.169.

Kaltenborn, H. S. (1937). Existence conditions and a substitution theorem for Stieltjes mean integrals. *Tohoku Math. J.* **44**, 1–11. *Zbl.*18.248.

Kamke, E. (1956). *Das Lebesgue–Stieltjes Integral*, B. G. Teubner Verlagsgesellschaft Leipzig. *MR* 18–384.

Kappos, D. A. (1949). Ein Beitrag zur Carathéodoryschen Definition der Ortsfunktionen in Booleschen Algebren. *Math. Zeit.* **51**, 616–34. *MR* 10–437.

Kappos, D. A. (1969). Probability algebras and stochastic spaces, Academic Press, New York and London. *MR* 57 #4271.

Karták, K. (1955). On the theory of many dimensional integrals (Czech). *Casopis Pěst. Mat.* **80**, 400–414. *MR* 19–640.

Karták, K. (1956). A theorem on the substitution in Denjoy integrals (Czech). *Casopis Pěst. Mat.* **81**, 410–19. *MR* 20 #2416.

Karták K. (1967). A generalization of the Carathéodory theory of differential equations. *Czech Math. J.* **17**(92), 482–514. *MR* 36 #4050.

Karták, K. and Mařik, J. (1965). A non-absolutely convergent integral in E_m and the theorem of Gauss. *Czech. Math. J.* **15**(90), 253–60. *MR* 31 #1356.

Karták, K. and Mařik, J. (1969). On the representations of some Perron integrable functions. *Czech. Math. J.* **19**(94), 745–9. *MR* 40 #2797.

Kassimatis, C. (1958). Functions which have generalized Riemann derivatives. *Can. J. Math.* **10**, 413–20. *MR* 20 #1737.

Kassimatis, C. (1963). The integration of the generalized derivatives. *Duke Math. J.* **30**, 101–5. *MR* 26 #3835.

Kay, A. J. (1975). Nonlinear integral equations and product integrals. *Pacific J. Math.* **60**, 203–22. *MR* 53 #3623.

Kelley, J. L. (1955). *General Topology*, Van Nostrand. *MR* 16–1136.

Kempisty, S. (1925). Un nouveau procédé d'integration de fonctions mesurables non-sommables. *C.R. Acad. Sci. Paris* **180**, 812–15.

Kempisty, S. (1927). Sur l'intégral (A) de M. Denjoy. *C.R. Acad. Sci. Paris* **185**, 749–51.

Kempisty, (1931). Sur l'intégral (A) de M. Denjoy. *C.R. Acad. Sci. Paris* **192**, 1186–9. *Zbl.*2.020.

Kempisty, S. (1932*a*). L'integration des fonctions sommables. *Ann. Soc. Polon. Math.* **10**, 1–11. *Zbl.*6.195.

Kempisty, S. (1932*b*). Sur les dérivées des fonctions des systèmes simples d'intervalles. *Bull. Soc. Math. France* **60**, 106–26. *Zbl.*5.058.

Kempisty, S. (1934). Sur la totalisation des fonctions de deux variables. *C.R. Acad. Sci. Paris.* **198**, 2060–2. *Zbl.*9.208.

Kempisty, S. (1936*a*). Intégrale Denjoy–Stieltjes d'une fonction de deux variables. *C.R. Acad. Sci. Paris,* **202**, 1241–4. *Zbl.*13.299.

Kempisty, S. (1936*b*). Sur les fonctions absolument continues d'intervalle. *Fundamenta Math.* **27**, 10–37. *Zbl.*15.105.

Kempisty, S. (1939). *Fonctions d'Intervalle Non-Additives*. Actualités Sci. Indust. #824. Ensembles et Fonctions III, Paris. *MR* 1–207.

Kennedy, M. D. (1930). Determinate functions of intervals and their rate of increase. *Proc. Lond. Math. Soc.* **30**(2), 58–80.

Kennedy, M. D. (1931). Upper and lower Lebesgue integrals. *Proc. Lond. Math. Soc.* **32**(2), 21–50.

Kennedy, M. D. and Pollard, S. (1934). Upper and lower integrals. *Math. Zeit.* **39**, 432–54. Zbl.10.158.

Keogh, F. R. (1960). Some generalizations of the Riemann–Lebesgue theorem. *J. Lond. Math. Soc.* **35**, 283–93. MR 23 #A1981.

Kestelman, H. (1960). *Modern Theories of Integration.* (2nd edn), Dover. MR 23 #A282.

Khintchine, A. See Hinčin, A.

Kisyński, J. (1968). On the generation of tight measures. *Studia Math.* **30**, 141–51. MR 38 #290.

Kolmogorov, A. (1930). Untersuchungen über den Integralbegriffe. *Math. Annalen* **103**, 654–96.

Kolmogorov, A. (1933). *Grundbegriffe der Wahrscheinlichkeitsrechnung*, Springer, Berlin. Zbl.7.216.

Kozlov, V. Ya. (1951). Goldovskiĭ's example (Russian). *Mat. Sb.* **28**(70), 197–204. MR 12–599.

Král, J. (1985). Note on generalized Perron integral. *Casopis Pest. Mat.* **110**, 371–4, 423. MR 87c:26014.

Krause, M. (1903). Mittelovertsatze in Gebiete der Doppelsummen und Doppelintegrale. Leipziger Bericht. *Math. Phys. Klasse*, 239–63.

Krżyzański, M. (1934). Sur les fonctions absolument continues généralisées de deux variables. *C.R.Acad Sci. Paris* **198**, 2058–60. Zbl.9.207.

Krżyzański, M. (1935). O uogólnionych funkjach bezwlednie ciąglych dwoch zniennych. *Ann. Soc. Polon. Math.*

Krżyzański, M. (1939). Sur l'éxtension de l'opération intégral de Denjoy aux fonctions de deux variables. *Bull. Sém. Math. Univ. Wilno*, **2**, 41–51. MR 1–47.

Krzyzewski, K. (1961). Remarks on totalisation of series. *Colloq. Math.* **8**, 257–62. MR 23 #A3821.

Krzyzewski, K. (1962a). On change of variable in the Denjoy–Perron integral (I). *Colloq. Math.* **9**, 99–104. MR 24 #A2652.

Krzyzewski, K. (1962b). On change of variable in the Denjoy–Perron integral (II). *Colloq. Math.* **9**, 317–23. MR 26 #283.

Krzyzewski, K. (1968). A note on the Denjoy integral. *Colloq. Math.* **19**, 121–30. MR 36 #6559.

Kubota, Y. (1959). On the definition of Çesàro–Perron integrals. *Tôhoku Math. J.* **11**(2), 266–70. MR 21 #7283.

Kubota, Y. (1960). The Cauchy property of the generalized approximately continuous Perron integral. *Tôhoku Math. J.* **12**(2), 171–4. MR 22 #12199.

Kubota, Y. (1963). On the approximately continuous Denjoy integral. *Tôhoku Math. J.* **15**(2), 253–64. MR 27 #3772.

Kubota, Y. (1964a). The mean continuous Perron integral. *Proc. Japan Acad.* **40**, 171–5. MR 29 #2355.

Kubota, Y. (1964b). An integral of the Denjoy type. *Proc. Japan Acad.* **40**, 713–17. MR 31 #2371.

Kubota, Y. (1965). A generalized derivative and integrals of the Perron type. *Proc. Japan Acad.* **41**, 443–8. MR 32 #1321.

Kubota, Y. (1966a). The Çesàro–Perron–Stieltjes integral I. *Proc. Japan. Acad.* **42**, 605–10. MR 34 #7762.

Kubota, Y. (1966b). An integral of the Denjoy type II. *Proc. Japan. Acad.* **42**, 737–42. MR 34 #7763.

Kubota, Y. (1967a). An integral of the Denjoy type III. *Proc. Japan. Acad.* **43**, 441–4. MR 36 #5276.

Kubota, Y. (1967b). An abstract integral. *Proc. Japan Acad.* **43**, 949–52. MR 37 #4217.

Kubota, Y. (1968a). On a characterization of the CP-integral. *J. Lond. Math. Soc.* **43**, 607–11. MR 37 #4228.

Kubota, Y. (1968b). On the compatibility of the AP- and the D-integrals. *Proc. Japan. Acad.* **44**, 330–3. MR 37 #2937.

Kubota, Y. (1970). A characterization of the approximately continuous Denjoy integral. *Can. J. Math.* **22**, 219–26. MR 41 #1943.

Kubota, Y. (1971). An approximately continuous Perron integral. *Can. Math. Bull.* **14**, 261–3. MR 47 #2010.

Kubota, Y. (1972). A constructive definition of the approximately continuous Denjoy integral. *Can. Math. Bull.* **15**, 103–8. MR 47 #6960.

Kubota, Y. (1973). An integral with basis and its application to trigonometric series. *Bull. Fac. Sci. Ibaraki Univ.* Ser. A no. 5, 1–8. MR 48 #8707.

Kubota, Y. (1976). On Romanovski's general Denjoy integral. *Bull. Fac. Sci. Ibaraki Univ.* Ser. A no. 8, 17–22. MR 58 #11271.

Kubota, Y. (1977). Some theorems on the approximately continuous Denjoy integral. *Math. Japon.* **22**, 289–93. MR 57 #3331.

Kubota, Y. (1978). The preponderantly continuous Denjoy integral. *Tôhoku Math. J.* **30**(2), 537–41. MR 80c:26004.

Kubota, Y. (1979/1980) An elementary theory of the special Denjoy integral. *Math. Japon.* **24**, 507–20. MR 81c:26006.

Kubota, Y. (1980). A direct proof that the RC-integral is equivalent to the D^*-integral. *Proc. Am. Math. Soc.* **80**, 293–6. MR 81h:26006.

Kubota, Y. (1981). A characterisation of the Denjoy integral. *Math. Japon.* **26**, 389–92. MR 82m:26006.

Kubota, Y. (1985). Remark on the Henstock integral of Stieltjes type. *Bull. Fac. Sci. Ibaraki Univ.* Ser. A **17**, 25–29. MR 86k:26006.

Kubota, Y. (1986). Notes on integration. *Math. Japon.* **31**, 617–21. MR 87i:26006.

Kubota, Y. (1988/1989). Extensions of the Denjoy integral. *Real Analysis Exchange* **14**, 72–3. MR

Kunisawa, K. (1943). Integrations in a Banach space. *Proc. Phys.-Math. Soc. Japan* **25**(3), 524–9. MR 7–455.

Kunugi, K. (1954). Sur les espaces complets et régulièrement complets. *Proc. Japan. Acad.* **30**, I 553–6, II 912–16.

Kunugi, K. (1956). Application de la méthode des espaces rangés à la théorie de l'intégration, I. *Proc. Japan. Acad.* **32**, 215–20.

Kurzweil, J. (1957). Generalized ordinary differential equations and continuous dependence on a parameter. *Czech. Math. J.* **7**(82), 418–49 (see 422–8). MR 22 #2735.

Kurzweil, J. (1973a). On Fubini theorem for general Perron integral. *Czech. Math. J.* **23**(98), 286–97. MR 48 #11411.

Kurzweil, J. (1973b). On multiplication of Perron-integrable functions. *Czech. Math. J.* **23**(98) 542–66. MR 49 #485.

Kurzweil, J. (1978). The Perron–Ward integral and related topics. Appendix A of Jacobs, K., *Measure and Integral*, pp. 515–33, Academic Press, New York. MR 80k:28002.

Kurzweil, J. (1980). Nichtabsolut Konvergente Integrale, Leipzig. *MR* 82m:26007.

Kurzweil, J. and Jarnik, J. (1986). A Perron integral, Perron product integral and ordinary linear differential equations. *Lecture Notes Math.* **1192**, 149–54. *MR* 88e:34004a.

Kurzweil, J. and Jarnik, J. (1987). On some extensions of the Perron integral on one-dimensional intervals. An approach by integral sums fulfilling a symmetry condition. *Funct. Approx. Comment. Math.* **17**, 49–55. *MR* 88j:26010.

Kurzweil, J. and Jarnik, J. (1988). The PU-integral and its properties. *Cesk. Akad. Ved. Math. Ustav.* **33**, 1–10. *MR*

Kurzweil, J. and Jarnik, J. (1988/1989). The PU-integral and its properties. *Real Analysis Exchange* **14**, 34–43. *MR*

Lagare, E. M. (1987). Improper Riemann integrals and uniformly regular matrices. *Southeast Asian Bull. Math.* **11**, 23–26. *MR* 88k: 26010.

Leader, S. (1986). What is a differential? A new answer from the generalized Riemann integral. *American Math. Monthly* **93**, 348–56. *MR* 87e:26002.

Leader, S. (1986/1987). A concept of differential based on variational equivalence under generalized Riemann integration. *Real Analysis Exchange* **12**, 144–75. *MR* 88a:26007.

Leader, S. (1988/1989). 1-diferentials on 1-cells: a further study. *Real Analysis Exchange* **14**, 74. *MR*

Lebesgue, H. (1902). Integrale, longueur, aire. *Annali Mat. Pura. Appl.* **7**(3), 231–359. *Jbuch* **33**, 307.

Lebesgue, H. (1904). *Leçons sur l'Intégration et la Recherche des Fonctions Primitives* (2nd edn, 1928), Gauthiers-Villars, Paris.

Lebesgue, H. (1909a). Sur les intégrales singulières. *Ann. Fac. Sci. Univ. Toulouse* **1**(3), 25–117.

Lebesgue, H. (1909b). Remarques sur un énoncé dû à Stieltjes et concernant les intégrales singulières. *Ann. Fac. Sci. Univ. Toulouse* **1**(3), 119–28.

Lebesgue, H. (1910). Sur l'intégration des fonctions discontinues. *Ann. École Norm.* **27**(3), 361–450.

Lebesgue, H. (1926). Sur la recherche de fonctions primitives. *Acta Math.* **49**, 245–62.

Lebesgue, H. (1971). A propos de quelques travaux mathématiques récents, article inédit écrit en 1905, avec notes de G. Choquet. *Enseign. Math.* **17**, 1–48. *MR* 48 #3675.

Lee C. M. (1975/1976). On approximate Peano derivatives. *J. Lond. Math. Soc.* **12**(2), 475–8. *MR* 53 #3222.

Lee, C. M. (1976a). An approximate extension of Çesàro–Perron integrals. *Bull. Inst. Math. Acad. Sinica* **4**, 73–82. *MR* 54 #484.

Lee, C. M. (1976b). On functions with summable approximate Peano derivative. *Proc. Am. Math. Soc.* **57**, 53–57. *MR* 53 #3223.

Lee, C. M. (1976/1977). Monotonicity theorems for approximate Peano derivatives and integrals. *Real Analysis Exchange* **1**, 52–62. *MR*

Lee, C. M. (1978). An analogue of the theorem of Hake–Alexandroff–Looman. *Fund. Math.* **100**, 69–74. *MR* 58 #6109.

Lee, C. M. (1981a). On integrals and summable trigonometric series. *Can. Math. Bull.* **24**, 433–40. *MR* 83b:26009.

Lee, C. M. (1981b). Generalizations of Çesàro continuous functions and integrals of Perron type. *Trans Am. Math. Soc.* **266**, 461–81. *MR* 83b:26010.

REFERENCES

Lee, C. M. (1982). Regular approximate Peano derivatives are the ordinary ones. *Bull. Inst. Math. Acad. Sinica* **10**, 401–4. *MR* 84c:26014.

Lee, C. M. (1982/1983). On absolute Peano derivatives. *Real Analysis Exchange* **8**, 228–43. *MR* 84m:26007.

Lee, C. M. (1983). On generalized Peano derivatives. *Trans. Am. Math. Soc.* **275**, 381–96. *MR* 84c:26008.

Lee, C. M. and O'Malley, R. J. (1975). The second approximate derivative and the second approximate Peano derivative. *Bull. Inst. Math. Acad. Sinica* **3**, 193–7. *MR* 52 #3446.

Lee, P. Y. (1964). An alternative proof for an example involving variation. *Bull. Math. Soc. Nanyang Univ.*, 63–5. *MR* 32 #1309.

Lee, P. Y. (1965a). An equality for variational integrals. *Bull. Math. Soc. Nanyang Univ.*, 45–7. *MR* 34 #2824.

Lee, P. Y. (1965b). Integrals involving parameters. *J. Lond. Math. Soc.* **40**, 338–44. *MR* 30 #3956.

Lee, P. Y. (1966a). A note on some generalizations of the Riemann–Lebesgue theorem. *J. Lond. Math. Soc.* **41**, 313–17. *MR* 34 #3157.

Lee, P. Y. (1966b). Integrals involving parameters (II). *J. Lond. Math. Soc.* **41**, 680–4. *MR* 34 #2825.

Lee, P. Y. (1967a). A scale of spaces of interval functions. *J. Lond. Math. Soc.* **42**, 443–6. *MR* 35 #3426.

Lee, P. Y. (1967b). A note on an equality for variational integrals. *J. Nanyang Univ.* **1**, 263–4. *MR* 38 #1232.

Lee, P. Y. (1968). Integration by completion of normed linear spaces. *Nanta Math.* **2**, 16–20. *MR* 40 #1572.

Lee, P. Y. (1970a). A non-standard example of a distribution. *Am. Math. Monthly* **77**, 984–7. *MR* 42 #2298.

Lee, P. Y. (1970b). A nonstandard convergence theorem for distributions. *Math. Chronicle* **1**(2) 81–4. *MR* 42 #8273.

Lee, P. Y. (1970c). The uniqueness of a Riesz-type deinition of the Lebesgue integral. *Math. Chronicle* **1**(2), 85–7. *MR* 42 #6163.

Lee, P. Y. (1972/1973). Some problems in integration theory. *Math. Chronicle* **2**, 105–16. *MR* 51 #826.

Lee, P. Y. (1976). *Uniqueness and Convergence Theorems in Integration Theory*. Occ. Paper 4, Lee Kong Chian Inst. Math. Comp. Sci., Nanyang Univ.

Lee, P. Y. (1985). *Convergence Theorems in Integration Theory*. Res. Rep. 226, Nat. Univ. Singapore.

Lee, P. Y. (1985–86). *Notes on Classical Integration Theory* (I), (II). Res. Rep. 275, 277. Nat. Univ. Singapore.

Lee, P. Y. (1986a). *Some Further Problems in Integration Theory*. Res. Rep. 243, Nat. Univ. Singapore.

Lee, P. Y. (1986b). *Two Proofs of the Generalized Mean Covergence Theorems* Res. Rep. 244, Nat. Univ. Singapore.

Lee, P. Y. (1986c). *Lanzhou Lectures on Classical Integration Theory*. Lecture Notes, Nat. Univ. Singapore. *MR*

Lee, P. Y. (1986d). Riesz representation theorems. *Southeast Asia Bull. Math.* **10**, 96–101. *MR* 88e:26005.

Lee, P. Y. (1988/1989). Generalized convergence theorems for Denjoy–Perron integrals. *Real Analysis Exchange* **14**, 48–9. *MR*

Lee, P. Y. and Chew, T. S. (1984). *Non Linear Integrals*. Res. Rep. 173, Nat. Univ. Singapore.

Lee, P. Y. and Chew, T. S. (1985). A better convergence theorem for Henstock integrals. *Bull. Lond. Math. Soc.* **17**, 557–64. MR 87b: 26010.

Lee, P. Y. and Chew T. S. (1985/1986). A Riesz-type definition of the Denjoy integral. *Real Analysis Exchange* **11**, 221–7. MR 87e:26009.

Lee, P. Y. and Chew, T. S. (1986). On convergence theorems for the nonabsolute integrals. *Bull. Australian Math. Soc.* **34**, 133–40. MR 87f:26007.

Lee, P. Y. and Chew, T. S. (1987). A short proof of the controlled convergence theorem for Henstock integrals. *Bull. Lond. Math. Soc.* **19**, 60–2. MR 87k:26009.

Lee, P. Y. and Lu, S. (1988). *Notes on Classical Integration Theory (VIII)*. Res. Rep. 324, Nat. Univ. Singapore.

Lee, P. Y. and Soeparno, D. (1988). The controlled convergence theorem for the approximately continuous integral of Burkill. *Proc. Analysis Conf. Singapore 1986*, 63–8. MR 89c:26013.

Lee, P. Y. and Wittaya, N-I. (1982). A direct proof that the Henstock and Den joy integrals are equivalent. *Bull. Malaysian Math. Soc.* **5**(2), 43–7. (Corrigendum: Lee, P. Y. et al. (1986). MR 84f:26011.

Lee, P. Y., Lu, S. and Xu, D. F. (1986). Corrigendum: A direct proof that Henstock and Denjoy integrals are equivalent. *Bull. Malaysian Math. Soc.* **9**(2), 69–70. MR 88i:26016.

Lee, T. W. (1971/1972). On an extension of Fubini's theorem. *J. Lond. Math. Soc.* **4**(2), 519–22. MR 46 #315.

Lee, T. W. (1972). On the connections between Henstock's and Ridder's approaches to generalized Riemann integrals. *J. Lond. Math. Soc.* **5**(2), 337–46. MR 47 #410.

Lee, T. W. (1976). On the generalized Riemann integral and stochastic integral. *J. Australian Math. Soc. Ser. A* **21**, 64–71. MR 55 #8294.

Levi, B. (1906) Ricerche sulle funzioni derivate. *Atti Accad. Naz Lincei, Rend.* **15**, 433–8.

Levi, B. (1941). A theory of the Lebesgue integral independent of the notion of measure (Spanish). *Publ. Inst. Mat. Univ. Nac. Litoral.* **3**, 65–116. MR 3–227.

Lewis, D. R. (1970) Integration with respect to vector measures. *Pacific J. Math.* **33**, 157–65. MR 41 #3706.

Lewis, J. T. and Shisha, O. (1983). The generalized Riemann, simple, dominated and improper integrals, *J. Approximation Theory* **38**, 192–9. MR 84h:26014.

Lewis, J. T., Osgood, C. F. and Shisha, O. (1978). Infinite Riemann sums, the simple integral, and the dominated integral. General Inequalities. *Proc. First Intern Conf. Math. Res. Inst., Oberwolfach*, **1**, 233–42. MR 58 #11272.

Liao, K. C. (1987). A refinement of the controlled convergence theorem for Henstock integrals. *Southeast Asian Bull. Math.* **11**, 49–51. MR 88m:26016.

Liu, G. Q. (1987/1988). The measurability of δ in Henstock integration. *Real Analysis Exchange* **13**, 446–50. MR 89e:26019.

Lojásiewicz, S. (1957). Sur la valeur et la limite d'une distribution en un point. *Studia Math.* **16**, 1–36. MR 19–433.

Looman, H. (1923). Sur la totalisation des dérivées des fonctions continues des plusiers variables indépendantes. *Fundamenta Math.* **4**, 246–85.

Looman, H. (1925). Über die Perronsche Integraldefinition. *Math. Ann.* **93**, 153–6.

Lukashenko, T. P. (1971). The functions that are associated with Denjoy–Hinčin integrable functions (Russian). *Izv. Akad. Nauk SSSR Ser. Mat.* **35**, 381–407. English translation *Math. USSR Izv.* **5**, 383–420. MR 44 #379.

Lukashenko, T. P. (1972). Conjugate functions and integrals (Russian). *Izv. Akad. Nauk. SSSR Ser. Mat.* **36**, 435–49. English translation *Math. USSR Izv.* **6**, 429–44. MR 46 #5542.

Lukashenko, T. P. (1973). D-integrable conjugate functions (Russian). *Izv. Akad. Nauk SSSR Ser. Mat.* **37**, 946–58. English translation *Math. USSR Izv.* **7**, 949–58. MR 48 #11885.

Lukashenko, T. P. (1977a). Conjugate functions and the restricted Denjoy integral (Russian). *Mat. Sb.* **104**(146), 89–139. English translation *Math. USSR Sbornik* **33**, 81–124. MR 57 #10345.

Lukashenko, T. P. (1977b). Denjoy–Hinčin integrable functions and their conjugate functions (Russian). *Izv. Akad. Nauk SSSR Ser. Mat.* **41**, 663–702, 718. English translation *Math. USSR. Izv.* **11**, 635–63. MR 56 #6265.

Lukashenko, T. P. (1979a). Majorants of Denjoy-integrable functions (Russian). *Mat. Sb.* **110**(152), 440–53, 72. English translation *Math. USSR Sbornik* **38**(1981), 407–20. MR 81b:42069.

Lukashenko, T. P. (1979b). Integrals of conjugate functions (Russian). *Izv. Akad. Nauk SSSR Ser. Mat.* **43**, 795–830, 966. English translation *Math. USSR Izv.* **15**(1980), 53–86. MR 81a:42021.

Lukashenko, T. P. (1982a). The Perron–Stieltjes integral (Russian). *Vesthik Moskov. Univ. Ser. I. Mat. Mekh.* **37**, 39–47, 110. MR 83m:26010.

Lukashenko, T. P. (1982b). Functions of generalized bounded variation and generalized absolutely continuous functions. *Dokl. Akad. Nauk. SSSR*, **263**, 537–40. English translation *Soviet Math. Dokl.* **25**, 379–82. MR 83g:26015.

Lukashenko, T. P. (1982c). Functions of generalized bounded variation (Russian). *Izv. Akad. Nauk SSSR Ser. Mat.* **46**, 276–313, 431. English translation *Math. USSR Izv.* **20**(1983), 267–301. MR 84m:26011.

Luxemburg, W. A. J. (1971). Arzelà's dominated convergence theorem for the Riemann integral. *Am. Math. Monthly* **78**, 970–9. MR 45 #6992.

Luzin, N. N. (Lusin) (1912a). Sur les propriétés des fonctions mesurables. *C.R. Acad. Sci. Paris* **154**, 1688–90.

Luzin, N. N. (1912b). Kosnovnoĭ teorem' integral 'nago iscisleniya. *Mat. Sb.* **12**(28), 266–94.

Luzin, N. N. (1912c). Sur les propriétés de l'intégrale de M. Denjoy. *C.R. Acad. Sci. Paris* **155**, 1475–8.

Luzin, N. N. (1915). *Integrals and Trigonometric Series* (Russian), Moscow.

McGill, P. (1973). Conditions for the Henstock integral to include the Lewis integral. *Proc. Roy. Irish Academy*, Sec. A. **73**, 275–8. MR 48 #11438.

McGill, P. (1974/1975). An extension of a concept of Henstock. *J. Lond. Math. Soc.* **9**(2), 49–53. MR 51 #10579.

McGill, P. (1975a). Properties of the variation. *Proc. Roy. Irish Acad.* Sec. A, **75**, 73–7. MR 52 #3448.

McGill, P. (1975b). Comparisons of vector integrals. *Quart. J. Math. Oxford* **26**(2), 315–22. MR 52 #5936.

McGill, P. (1975c). Integration in vector lattices. *J. Lond. Math. Soc.* **11**(2), 347–60. MR 52 #14224.

McGill, P. (1976). Vector measures and convergence theorems. *J. Lond. Math. Soc.* **14**(2), 563–9. MR 55 #8305.

McGill, P. (1977). Constructing smooth measures on certain classes of paved sets. *Proc. Roy. Irish Acad.* Sec. A, **77**, 31–43. MR 57 #6347.

McGill, P. (1980). An elementary integral. *Lecture Notes in Math.* **794**, 310–16. MR 82b:28017.

McGill, P. (1981). Measure extensions and decompositions: a topological approach. *Proc. Roy. Irish Acad.* Sec. A, **81**, 157–66. MR 83g:28021.

McGrotty, J. J. (1962). A theorem on complete sets. *J. Lond. Math. Soc.* **37**, 338–40. MR 25 #4052.

McLeod, R. M. (1980). *The Generalized Riemann Integral*, Carus Math. Monographs 20, Math. Association of America. MR 82h:26015.

MacNeille, H. M. (1938). Extension of measure. *Proc. Nat. Acad. Sci. USA* **24**, 188–93. Zbl.18.349.

MacNeille, H. M. (1941). A unified theory of integration. *Proc. Nat. Acad. Sci. USA* **27**, 71–6. MR 2–258.

MacNerney, J. S. (1960). Hellinger integrals in inner product spaces. *J. Elisha Mitchell Sci. Soc.* **76**, 251–73. MR 23 #A4002.

MacNerney J. S. (1963). An integration-by-parts formula. *Bull. Am. Math. Soc.* **69**, 803–5. MR 27 #3770.

MacNerney J. S. (1964). A nonlinear integral operation. *Illinois J. Math.* **8**, 621–38. MR 29 #5082.

MacPhail, M. S. (1945). Integration of functions in a Banach space. *Nat. Math. Mag.* **20**, 69–78. MR 7–455.

McShane, E. J. (1942). On Perron integration. *Bull. Am. Math. Soc.* **48**, 718–26. MR 4–75.

McShane, E. J. (1944). *Integration*, Princeton University Press, Princeton, NJ. MR 6–43

McShane, E. J. (1949). Remark concerning integration. *Proc. Nat. Acad. Sci. USA* **35**, 46–9. MR 10–360.

McShane, E. J. (1969). A Riemann type integral that includes Lebesgue–Stieltjes, Bochner and stochastic integrals. *Mem. Am. Math. Soc.* **88**. MR 42 #436.

McShane, E. J. (1973). A unified theory of integration. *Am. Math. Monthly* **80**, 349–59. MR 47 #6981.

McShane, E. J. (1974). *Stochastic Calculus and Stochastic Models*, Academic Press, New York. MR 56 #1457.

McShane, E. J. (1983). Unified integration, Academic Press, New York. MR 86c:28002.

Malliavin, P. (1949). Majorantes et minorantes des fonctions simplement totalisables. *C.R. Acad. Sci. Paris* **229**, 286–7. MR 11–90.

Manougian, M. N. (1969). On the convergence of a sequence of Perron integrals. *Proc. Am. Math. Soc.* **23**, 320–2. MR 40 #278.

Marcinkiewicz, J. (1936). Sur les séries de Fourier. *Fundamenta Math.* **27**, 38–69. Zbl.14.215.

Marcinkiewicz, J. and Zygmund, A. (1936). On the differentiability of functions and summability of trigonometrical series. *Fundamenta Math.* **26**, 1–43. Zbl.14.111.

Mařík, J. (1952). Foundations of the theory of the integral in Euclidean spaces (Czech). *Casopis Pest. Mat.* **77**, 1–51, 125–45, 267–301. MR 15.691. Abstract (Russian). *Czechoslovak Math. Journal* **2**(77), (1952), 273–7. Unpublished English translation by L. I. Trudzik.

Mařík, J. (1956). The surface integral. *Czechoslovak Math. J.* **6**, 522–58.

Mařík, J. (1965). Extensions of additive mappings. *Czechoslovak Math. J.* **15**(90), 244–52. *MR* 31 #1355.

Mařík, J. and Matyska, J. (1965). On a generalization of the Lebesgue integral in E_m. *Czechoslovak Math. J.* **15**(90), 261–9. *MR* 31 #1357.

Martin, R. H. (1973). Product integral approximations of solutions to linear operator equations. *Proc. Am. Math. Soc.* **41**, 506–12. *MR* 52 #1363.

Masani, P. R. (1947). Multiplicative Riemann integration in normed rings. *Trans. Am. Math. Soc.* **61**, 147–92. *MR* 8–321.

Masani, P. R. (1981). Multiplicative partial integration and the Trotter product formula. *Adv. Math.* **40**, 1–9, 308. *MR* 82m:47030a, b.

Matsuyama, N. (1942). An abstract integral IX. *Proc. Imp. Acad. Tokyo* **18**, 539–42. *MR* 9–19.

Matyska, J. (1968). On β-integration in E_1. *Czechoslovak Math. J.* **18**(93), 523–6. *MR* 37 #6413.

Mawhin, J. (1981a). Generalized multiple Perron integrals and the Green–Goursat theorem for differentiable vector fields. *Czechoslovak Math. J.* **31**(106), 614–32. *MR* 82m: 26010.

Mawhin, J. (1981b). Generalized Riemann integrals and the divergence theorem for differentiable vector fields. *E. B. Christoffel*, 704–14. (Birkhauser, Basel). *MR* 83m:26016.

Mawhin, J. (1983a). Présences des sommes de Riemann dans l'évolution du calcul intégral. *Cahiers Sém. Hist. Math.* **4**, 117–47. *MR* 84j:01047.

Mawhin, J. (1983b). Introduction a l'Analyse. Cabay, Louvain-la-Neuve.

Mawhin, J. (1986). Nonstandard analysis and generalized Riemann integrals. *Casopis Pěst. Mat.* **111**, 34–47, 89. *MR* 87h:26033.

Mawhin, J. (1986/1987). Classical problems in analysis and new integrals. *Real Analysis Exchange* **12**, 69–84. *MR*

Mawhin, J. (1988/1989). Multiple generalised Riemann integrals on compact intervals. *Real Analysis Exchange* **14**, 51–5. *MR*

Meinershagen, S. (1986/1987) D^* derivation basis and the Lebesgue–Stieltjes integral. *Real Analysis Exchange* **12**, 265–81. *MR* 88a:26013.

Menšov, D. E. (Menchoff) (1916). Vzaimootnošenie meždy opred'leniyami integrala Borle'ya i Denjoy. *Mat. Sb.* **30**, 188–95.

Meyer, P. A. (1966). Probability and potentials, Blaisdell, Waltham, Mass. *MR* 34 #5118, 5119.

Mikusiński, J. G. (1964a). Sur une définition de l'intégrale de Lebesgue. *Bull. Acad. Polon. Sci. Sér. Sci. Math. Astron. Phys.* **12**, 203–4. *MR* 29 #4857.

Mikusiński, J. G. (1964b). *An Introduction to the Theory of the Lebesgue and Bochner Integrals*, University of Florida, Gainesville.

Mikusiński, P. and Ostaszewski, K. (1988/1989). Embedding Henstock integrable functions into the space of Schwartz distributions. *Real Analysis Exchange*, **14**, 24–29. *MR*

M'Khalfi, A. (1988). On a generalized multiple integral and the divergence theorem. *Bull. Soc. Math. Belgique* **40B**, 111–30.

Minkowski, H. (1896). Geometrie der Zahlen, I, pp. 115–17, Leipzig.

Moore, E. H. (1900). *Trans. Am. Math. Soc.* **1**, 499–506; Jbuch 31, 398.

Moore, E. H. (1915). Definition of limit in general integral analysis. *Proc. Nat. Acad. Sci. USA* **1**, 628–32.

Moore, E. H. (1939). *General Analysis I, Part II. The fundamental notions of general analysis.* Mem. American Philos. Soc. Philadelphia. 1. Zbl.20.366.

Moore, E. H. and Smith, H. L. (1922). A general theory of limits. *Am. J. Math.* **44**, 102–21.

Morrison, T. J. (1976). A note on the Denjoy integrability of abstractly-valued functions. *Proc. Am. Math. Soc.* **61**, 385–6. *MR* 58 #6149.

Mugalov, A. G. (1966). Integration and differentiation under the integral sign in the sense of Çesàro–Denjoy (Russian). *Izv. Akad. Nauk Azerbaĭdžan SSR. Ser. Fiz.-Tehn. Mat. Nauk.* no. 6, 22–9. *MR* 35 #313.

Mugalov, A. G. (1967). On passage to the limit under the Çesàro–Denjoy integral sign (Russian). *Izv. Akad. Nauk Azerbaĭdžan SSR. Ser. Fiz.-Tehn. Mat. Nauk.* no. 1 33–41. *MR* 35 #5557.

Mugalov, A. G. (1970). Iterated integrals (Russian). Special Questions on Differential Equations and Function Theory. 70–76, Izdat 'Elm'–Baku. *MR* 43 #2171.

Mugalov, A. G. (1976). The curvilinear integral in the sense of Çesàro–Denjoy (Russian). *Akad. Nauk Azerbaidzan SSR* **32**, 3–8. 55 #5820.

Mukhopadhyay, S. N. (1974). On the regularity of the P^n-integral and its application to summable trigonometric series. *Pacific J. Math.* **55**, 233–47. *MR* 51 #10546.

Mukhopadhyay, S. N. (1975). On the approximate Peano derivatives. *Fundamenta Math.* **88**, 133–43. *MR* 51 #13149.

Mukhopadhyay, S. N. (1978). Summable trigonometric series. *J. Math. Anal. Appl.* **66**, 427–32. *MR* 80d:42004.

Muldowney, P. (1987). *A General Theory of Integration in Function Spaces*, including Wiener and Feynman integration. Pitman Research Notes in Mathematics, 153, Longman, Harlow. *MR* 89i:28006.

Muldowney, P. (1988/1989). Infinite dimensional generalised Riemann integrals. *Real Analysis Exchange* **14**, 14–15. *MR*

Musielak, J. and Orlicz, W. (1967a). Notes on the theory of integral. I. *Bull. Acad. Polon. Sci. Sér. Sci. Math. Astron. Phys.* **15**, 329–37. *MR* 36 #3945.

Musielak, J. and Orlicz, W. (1967b). Notes on the theory of integral II. *Bull Acad. Polon. Sci. Sér. Sci. Math. Astron. Phys.* **15**, 723–30. *MR* 37 #6426.

Musielak, J. and Orlicz, W. (1968). Notes on the theory of integral III. *Bull. Acad. Polon. Sci. Sér. Sci. Math. Astron. Phys.* **16**, 317–26. *MR* 38 #287.

Nakamura, M. (1942). An abstract integral VI. *Oroc. Imp. Acad. Tokyo.* **18**, 50–52. *MR* 9–19.

Nakanishi, S. (formerly Enomoto) (1956). L'intégrale de Denjoy et l'intégration au moyen des espaces rangés I. *Proc. Japan Acad.* **32**, 678–83. *MR* 19–256.

Nakanishi, S. (1957a). L'intégrale de Denjoy et l'intégration au moyen des espaces rangés II. *Proc. Japan Acad.* **33**, 13–18. *MR* 19–1167.

Nakanishi, S. (1957b). L'intégrale de Denjoy et l'intégration au moyen des espaces rangés III. *Proc. Japan Acad.* **33**, 265–70. *MR* 19–1167.

Nakanishi, S. (1958). L'intégrale de Denjoy et l'intégration au moyen des espaces rangés IV. *Proc. Japan Acad.* **34**, 96–101. *MR* 21 #1375.

Nakanishi, S. (1968a). On generalized integrals I. *Proc. Japan Acad.* **44**, 133–8. *MR* 38 #279.

Nakanishi, S. (1968b). On generalized integrals II. *Proc. Japan Acad.* **44**, 225–30. *MR* 38 #279.

Nakanishi, S. (1968c). On generalized integrals III. *Proc. Japan Acad.* **44**, 904–9. *MR* 39 #378.

Nakanishi, S. (1969a). On generalized integrals IV. *Proc. Japan Acad.* **45**, 86–91. *MR* 40 #289.

Nakanishi, S. (1969b). On generalized integrals V. *Proc. Japan Acad.* **45**, 374–9. *MR* 40 #7415.

Nakanishi, S. (1974). On the strict union of ranked metric spaces. *Proc. Japan Acad.* **50**, 603–7. *MR* 52 #1599.

Nakanishi, S. (1978a). Integration of functions with values in a convex ranked space. *Math. Japon.* **23**, 85–103. *MR* 58 #11278.

Nakanishi, S. (1978b). The method of ranked spaces proposed by Professor Kinjiro Kunugi. *Math. Japon.* **23**, 291–323. *MR.* 80f:54002.

Nakanishi, S. (1978/1979). On ranked union spaces. *Math. Japon.* **23**, 249–57. *MR* 80i:46006.

Nakanishi, S. (1979). The space of distribution treated as a ranked space. *Proc. Japan Acad.* **55**, 395–8. *MR* 81k:46042.

Nakanishi, S. (1984). The Bochner integral for functions with values in certain ranked vector spaces and the Radon–Nikodym theorem. *Math. Japon.* **29**, 797–813. *MR* 86c:28022.

Nakanishi, S. and Fujita, K. (1970). On generalized integrals VI. Restrictions of (E.R.) integral I. *Proc. Japan Acad.* **46**, 41–6. *MR* 41 #8622.

Nakano, H. (1942). Über Erweiterungen von allgemein teilweise geordneten Moduln I. *Proc. Imp. Acad. Tokyo* **18**, 626–30. *MR* 8–387.

Nakano, H. (1943). Über Erweiterungen von allgemein teilweise geordneten Moduln. II. *Proc. Imp. Acad. Tokyo* **19**, 138–43. *MR* 8–387.

Nalli, P. (1914) *Esposizione e confronto critico delle diverse defizioni proposte per l'intégrale definita di una funzione limita o no*, Palermo.

Nalli, P. (1915). Sulle serie di Fourier delle funzione non assolutamente integrabili. *Rend. Circ. Math. Palermo* **40**, 33–7.

Nasibov, M. H. (1964). Reconstruction of a function of two variables from its Schwarzian derivative (Russian). *Izv. Akad. Nauk Azerbaĭdžan. SSR Ser. Fiz.-Tehn. Mat. Nauk*, no. 1, 25–34. *MR* 29 #2339.

Natanson, I. P. (1950). *Theory of Functions of a Real Variable* (Russian). Vols. I, II, GITTL, Moscow. English translation (1st edn. 1955) (2nd edn revised), Ungar, New York (1961). 16–804. *MR* 26 #6309.

Natanson, I. P. and Natanson, G. I. (1957). On the mutual relation between the integrals of Denjoy in the narrow and in the wide sense (Russian). *Uspehi Mat. Nauk* **12**, 161–8. *MR* 20 #949.

Nath, R. K. and Bose, M. K. (1984). On the proximal Çesàro–Denjoy integral. *Soochow J. Math.* **10**, 99–110. *MR* 86k:26007.

Neuberger, J. W. (1958). Continuous products and nonlinear integral equations. *Pacific J. Math.* **8**, 529–49. *MR* 21 #1509.

Nicolescu, L.-J. (1954). Intégrales Perron–Stieltjes (Romanian). *Acad. R.P.Romîne Bull. Şti. Sect. Şti. Mat. Fiz.*, **6**, 755–70. *MR* 17–353

Nikodym, O. (1930). Sur une généralization des intégrales de M. J. Radon. *Fundamenta Math.* **15**, 131–79.

Okano, H. (1957). Some operations on the ranked spaces I. *Proc. Japan Acad.* **33**, 172–6.

Okano, H. (1958a). (ER)-integral of Radon–Stieltjes type. *Proc. Japan Acad. Tokyo* **34**, 580–4. *MR* 21 #2032.

Okano, H. (1958b). Multiplication of (ER)-integrable functions. *Proc . Japan Acad.* **34**, 585–6. *MR* 21 #2033.

Okano, H. (1959a). L'intégration des fonctions à valeurs vectorielles d'après la méthode des espaces rangés. *Proc. Japan Acad.* **35**, 77–82. *MR* 21 #7422.

Okano, H. (1959b). Une généralisation d'un théorème de Fatou concernant l'intégrale de Poisson *Proc. Japan Acad.* **35**, 461–4. *MR* 26 #3869.

Okano, H. (1959c). Sur les intégrales (*ER*) et ses applications. *Osaka Math. J.* **11**, 187–212. *MR* 25 #4067.

Okano, H. (1960). Les intégrales E.R. genéralisées sous une forme de Radon–Stieltjes. *Proc. Japan. Acad.* **36**, 324–6. *MR* 25 #4068.

Okano, H. (1962a). Sur une généralisation de l'intégrale (*E.R.*) et un théorème général de l'intégration par parties. *J. Math. Soc. Japan* **14**, 430–42. *MR* 26 #2579.

Okano, H. (1962b). Une nouvelle méthode pour considérer la série comme une intégrale, I. *Proc. Japan Acad.* **38**, 213–16. *MR* 26 #511.

Oliver, H. W. (1954). The exact Peano derivative. *Trans. Am. Math. Soc.* **76**, 444–56. *MR* 15–944.

Olmsted, J. M. H. (1942). Lebesgue theory on a Boolean algebra. *Trans. Am. Math. Soc.* **51**, 164–93. *MR* 4–11.

O'Malley, R. J. (1977). Selective derivatives. *Acta Math. Sci. Hungar.* **29**, 77–97. *MR* 55 #10614.

Orihara, M. and Sunouchi, G. (1942). An abstract integral VIII. *Proc. Imp. Acad. Tokyo* **18**, 535-8. *MR* 9–19.

Orlicz, W. (1968). On spaces $L^*\phi$ based on the notion of a finitely additive integral. *Prace Mat.* **12**, 99–113. *MR* 40 #4750.

Osgood, C. F. (1985). Obtaining a function of bounded coarse variation by a change of variable. *J. Approximation Theory* **44**, 14–20. *MR* 86i:26010.

Osgood, C. F. and Shisha, O. (1976a). On simple integrability and bounded coarse variation. *Approximation Theory II*, pp. 491–501, Academic Press, New York. *MR* 55 #8275.

Osgood, C. F. and Shisha, O. (1976b). The dominated integral. *J. Approximation Theory* **17**, 150–65. *MR* 54 #6467.

Osgood, C. F. and Shisha, O. (1977). Numerical quadrature of improper integrals and the dominated integral. *J. Approximation Theory* **20**, 139–52. *MR* 56 #7128.

Osgood, W. F. (1897). Non-uniform convergence and the integration of series term-by-term. *Am. J. Math.* **19**, 155–90.

Ostaszewski, K. M. (1981/1982). Continuity in the density topology. *Real Analysis Exchange* **7**, 259–70. *MR* 83e #26007.

Ostaszewski, K. M. (1982). Two modifications of the approximate derivative. *Demonstratio Math.* **15**, 899–911. *MR* 85c:26003.

Ostaszewski, K. M. (1983). Continuity in the density topology II. *Rend. Circ. Mat. Palermo* **32**(2), 398–414. *MR* 85g:26003.

Ostaszewski, K. M. (1983/1984a). A topology for the space of Denjoy-integrable functions. *Real Analysis Exchange* **9**, 79–85. *Zbl.*576.26013.

Ostaszewski, K. M. (1983/1984b). Density topology and the Luzin (*N*)-condition. *Real Analysis Exchange* **9**, 390–3. *MR* 86b:26004.

Ostaszewski, K. M. (1984/1985). Variational equivalence and generalized absolute continuity. *Real Analysis Exchange* **10**, 220–9. *MR* 86h:26008.

Ostaszewski, K. M. (1985/1986). Nonabsolute integration in the plane. *Real Analysis Exchange* **11**, 30–39. *Zbl.*614.26006.

Ostaszewski, K. M. (1986). Henstock integration in the plane. *Mem. Am. Math. Soc.* **63**, 353. *MR* 87j:26016.

Ostaszewski, K. M. (1988). The space of Henstock integrable functions. *Intern. J. Math. Math. Sci.* **11**, 15–21. *MR* 88m:46039.

Ostaszewski, K. M. and Sochacki, J. (1987). Gronwall's inequality and the Henstock integral. *J. Math. Anal. Appl.* **128**, 370–4. *MR* 88j:26016.
Ostrowski, A. M. (1976). On Cauchy–Frullani integrals. *Comment. Math. Helvetici* **51**, 57–91. *MR* 53 #8347.
Pacquement, A. (1969). Sur une définition de l'intégrale généralisant celle de M. Denjoy. *C.R. Acad. Sci. Paris*, Sér. A–B, **269**, A1128–A1131. *MR* 40 #7398.
Pacquement, A. (1970a). Sur le calcul inverse de la dérivation approximative. *C.R. Acad. Sci. Paris*, Sér. A–B, **271**, A80–A83. *MR* 42 #3236.
Pacquement, A. (1970b). Sur un cas de détermination d'une fonction numérique par la connaissance de la dérivée approximative. *Ann. Univ. Madagascar Sér. Sci. Nature Math.* **7**, 41–7. *MR* 52 #5899.
Pacquement, A. (1971). Intégration indéfinie dans $\mathbb{R}^2$. *Ann. Univ. Madagascar Sér. Sci. Nature Math.* **8**, 7–13. *MR* 52 #10972.
Pacquement, A. (1972). Intégrales indéfinies non nécessairement continues dans. $\mathbb{R}^2$. *C.R. Acad. Sci. Paris*, Ser. A–B, **274**, A451–A454. *MR* 46 #3711.
Pacquement, A. (1977). Détermination d'une fonction par sa dérivée sur un réseau binaire. *C.R. Acad. Sci. Paris*, Ser A–B, **284**, A365–A368. *MR* 55 #3199.
Pal, B. K. and Mukhopadhyay, S. N. (1983). The Çesàro–Denjoy–Bochner scale of integration. *Acta Math. Hungarica* **42**, 243–55. *MR* 85m:26011.
Pal, B. K. and Mukhopadhyay, S. N. (1985). The Çesàro–Denjoy– Pettis scale of integration. *Acta Math. Hungarica* **45**, 289–95. *MR* 86m:28005.
Perron, O. (1914) Über den Integralbegriff. *Sitzber. Heidelberg Akad. Wiss., Math.- Naturw. Klasse Abt.* A, **16**, 1–16.
Perron, O. (1950). Die Lehre von den Kettenbruchen (2nd edn.) Chelsea, New York. *MR* 12–254.
Pesin I. N. (1966). *Development of the concept of Integral*, Moscow. *MR* 34 #7764. English translation: *Classical and Modern Integration Theory*, Academic Press, New York (1970). *MR* 41 #8614.
Petrovskiĭ, I. (Petrovsky, Petrowsky) (1934). Sur l'unicité de la fonction primitive par rapport à une fonction continue arbitraire. *Mat. Sb.* **41**, 48–58. *Zbl.*9.307.
Pettineo, B. (1956). Sur la dérivabilité des fonctions. *C.R. Acad. Sci. Paris*, **243**, 553–4. *MR* 18–76.
Pettineo, B. (1956/1957). Sulla derivazione delle funzione continue. *Atti Accad. Sci. Lett. Palermo*, **7**(4), 211–38. *MR* 24 #1343.
Pettineo, B. (1965). Sulla derivabilità delle funzione. *Atti Accad. Naz. Lincei Mem. Cl. Sci. Fis. Mat. Nat. Ser.*, I **7**(8), 143–69. *MR* 32 #4229.
Pettis, B. J. (1938). On integration in vector spaces. *Trans. Am. Math. Soc.* **44**, 277–304. *Zbl.*19.416.
Pfeffer, W. F. (1963). The Perron integral in topological spaces (Russian). *Casopis Pest. Mat.* **8**, 322–48. *MR* 32 #2546.
Pfeffer, W. F. (1964a). A definition of the integral in topological spaces (Russian). Casopis Pest. Mat. **89**, 129–47. *MR* 32 #1318.
Pfeffer, W. F. (1964b). A definition of the integral in topological spaces II (Russian). *Casopis Pest. Mat.* **89**, 257–77. *MR* 32 #1318.
Pfeffer, W. F. (1967). A note on the lower derivate of a set function and semihereditary systems of sets. *Proc. Am. Math. Soc.* **18**, 1020–5. *MR* 36 #1589.
Pfeffer, W. F. (1968). On the lower derivate of a set function. *Can. J. Math.* **20**, 1489–98. *MR* 38 #1229.
Pfeffer, W. F. (1968/1969). An integral in topological spaces. I. *J. Math. Mech.* **18**, 953–72. *MR* 39 #2938.

Pfeffer, W. F. (1969). An integral in topological spaces. *Bull. Am. Math. Soc.* **75**, 433–9 *MR* 39 #400.

Pfeffer, W. F. (1970a). An integral in topological spaces II. *Math. Scand.* **27**, 77–104. *MR* 43 #461.

Pfeffer, W. F. (1970b). Singular integrals are Perron integrals of a certain type. *Can. J. Math.* **22**, 260–4. *MR* 41 #1963.

Pfeffer, W. F. (1980). *The Riemann–Stieltjes Approach to Integration.* Nat. Res. Inst. Math. Sci. Pretoria, Tech. Rep. 187.

Pfeffer, W. F. (1982). The existence of locally fine simplicial subdivisions. *J. Australian Math. Soc.*, Ser. A, **33**, 114–24. *MR* 83j: 26008.

Pfeffer, W. F. (1983a). Integration by parts for the generalized Riemann–Stieltjes integral. *J. Australian Math. Soc.*, Ser. A. **34**, 229–33. *MR* 84 e: 26010.

Pfeffer, W. F. (1983b). The generalized Riemann integral in higher dimensions. *Lecture Notes Math.* **1033**, 269–75. *MR* 85g: 26006.

Pfeffer, W. F. (1984a). *The Divergence Theorem.* Tech. Rep. #64, Dept. Math. Sci. Univ. Petroleum Minerals, Dharhan, Saudi Arabia.

Pfeffer, W. F. (1984b). Une intégrale Riemannienne et le théorème de divergence. *C.R. Acad. Sci. Paris, Sér. I. Math.* **299**, 299–301. *MR* 85k: 26021.

Pfeffer, W. F. (1986). The divergence theorem. *Trans. Am. Math. Soc.* **295**, 665–85. *MR* 87f: 26015.

Pfeffer, W. F. (1987a). The multidimensional fundamental theorem of calculus. *J. Australian Math. Soc.* Ser. A. **43**, 143–70. *MR* 89b: 26013

Pfeffer, W. F. (1987b). A Riemann-type integration and the fundamental theorem of calculus. *Rend. Circ. Math. Palermo*, **36**, 482–506. *MR*

Pfeffer, W. F. (1988a). Stokes theorem for forms with singularities. *C.R. Acad. Sci. Paris Sér. I. Math.* **306**, 589–92. *MR* 89k: 26008

Pfeffer, W. F. (1988b). A note on the generalized Riemann integral. *Proc. Am. Math. Soc.* **103**, 1161–6. *MR* 89i: 26008.

Pfeffer, W. F. (1988/1989a). On the generalized Riemann integral defined by means of special partitions. *Real Analysis Exchange* **14**, 506–11. *MR*

Pfeffer, W. F. (1988/1989b). The Gauss–Green theorem. *Real Analysis Exchange* **14**, 523–7. *MR*

Pfeffer, W. F. and Wilbur, W. J. (1970). On the measurability of Perron integrable functions. *Pacific J. Math.* **34**, 131–44. *MR* 43 #458.

Pfeffer, W. F. and Yang, W.-C. (1988/1989). The multidimensional variational integral and its extensions. *Real Analysis Exchange* **14**, 50. *MR*

Pfeffer, W. F. and Yang, W.-C. (1989/1990). A multidimensional variational integral and its extensions. *Real analysis Exchange* **15**, 111–69.

Phillips, R. S. (1940). Integration in a convex linear topological space. *Trans. Am. Math. Soc.* **47**, 114–45. *MR* 2–103.

Pierson-Gorez, C. (1988/1989). Integrals in unbounded intervals and connection with convergence of double series. *Real Analysis Exchange* **14**, 44–7. *MR*

Plant, A. T. (1974). Hölder continuous nonlinear product integrals. *Lecture Notes in Math.* **415**, 417–21. *MR* 58 #2514.

Plessner, A. (1923). Zur Theorie der conjugieren trigonometrische Reihen. *Mitt. Math. Seminar Univ. Giessen* **10**, 1–36.

Pollard, S. (1923). The Stieltjes integral and its generalizations. *Quart. J. Math. Oxford* **49**, 73–138.

Pollard, S. (1928). The summation of Denjoy–Fourier series. *Proc. Lond. Math. Soc.* **27**(2), 209–22.

Ponomarev, S. P. (1986). *Functions with a D_*-integrable symmetric derivative* (Russian). *Mat. Zametki* **39**, 221–7, 302. English translation *Math. Notes* **39**, 221–7. MR 87f:26005.

Postolică, V. (1981). On the primitives of functions with bounded-p-variation (Romanian). *Stud. Cerc. Mat* **33**, 141–6. MR 83g:26016.

Pratt, J. W. (1960). On interchanging limits and integrals. *Ann. Math. Statistics* **31**, 74–7. MR 23 #A997.

Preiss, D. and Thomson, B. S. (1988/1989). A symmetric covering theorem. *Real Analysis Exchange* **14**, 253–4. MR

Price, G. B. (1940). The theory of integration. *Trans. Am. Math. Soc.* **47**, 1–50. MR 1–239.

Privalov, I. I. (Privaloff) (1916). Sur la dérivation des series de Fourier. *Rend. Circ. Mat. Palermo* **41**, 202–6.

Przeworska-Rolewicz, D. and Rolewicz, S. (1966). On integrals of functions with values in a complete linear metric space. *Studia Math.* **26**, 121–31. MR 33 #564.

Pu, H. W. (1971). Concerning Riemann-complete integral. *Bull. Soc. Roy. Sci. Liège* **40**, 419–23. MR 46 #308.

Pu, H. W. (1972a). A Cauchy criterion and a convergence theorem for Riemann-complete integral. *J. Australian Math. Soc. Ser. A* **13**, 21–4. MR 45 #5287.

Pu, H. W. (1972b). Another proof that Riemann-complete integral includes the Lebesgue integral. *Bull. Soc. Roy. Sci. Liège*, **41**, 250–1. MR 46 # 9276.

Pu, H. W. (1973). On the derivative of indefinite RC-integral. *Colloq. Math.* **28**, 105–10. MR 49 #486.

Pych-Taberska, P. (1972). Theorems of the Romanovski type for the Denjoy–Perron integrals. *Comment. Math. Prace. Mat.* **16**, 125–31. MR 48 #476.

Pych-Taberska, P. (1973a). The Denjoy integral in some approximation problems II. *Fasc. Math.* **7**, 53–69. MR 53 #3572.

Pych-Taberska, P. (1973b). The Denjoy integral in some approximation problems III. *Proc. Conf. Constructive Theory of Functions*, Cluj.

Pych-Taberska, P. (1974). The Denjoy integral in some approximation problems I. *Functiones et Approximatio Comment. Math.* **1**, 91–105. MR 53 #3571.

Pych-Taberska, P. (1976a). The Denjoy integral in some approximation problems IV. *Comm. Math. Prace Mat.* **19**, 117–25. MR 56 #950.

Pych-Taberska, P. (1976b). Some properties of double Denjoy integrals. *Functiones et Approximatio Comment. Math.* **2**, 207–17. MR 56 #5812.

Pych-Taberska, P. (1976c). Theorems for the Denjoy-integrable functions possessing derivatives of positive order. *Functiones et Approximatio Comment. Math.* **4**, 61–9. MR 56 #6248.

Pych-Taberska, P. (1976d). The Denjoy integral in some approximation problems. *Rev. Anal. Numérique Théorie Approximation* **5**, 79–86. MR 57 # 10336.

Radon, J. (1913). Theorie und Anwendungen der absolut additiven Mengen-funktionen. *Sitzungsberichte der Kais. Ak. der Wiss. in Wien.* **122**, Abt IIa, 28–33.

Rădulescu, M. (1969). Une définition de l'intégrale. *Studia Univ. Babeş-Bolyai Sér. Math.-Phys.* **14**(1), 23–34. MR 40 #5801.

Rădulescu, M. (1970). L'intégrale M_*. *Studia Univ. Babeş-Bolyai Ser. Math.-Phys.* **15**(1), 23–34. MR 42 #437.

Rădulescu, M. (1971). L'intégrabilité M_* des séries trigonométriques presque partout convergentes. *An. Univ. Timişoara Ser. Sti. Mat.* **9**, 99–108. MR 49 #5250.

Reichelderfer, P. V. and Ringenberg, L. A. (1941). The extension of rectangle functions. *Duke Math. J.* **8**, 231–42. MR 3–74.

Rennie, B. C. (1974). Repeated Riemann integrals. *Proc. Camb. Phil. Soc.* **76**, 187–9. *MR* 50 #4857.

Rickart, C. E. (1942). Integration in a convex linear topological space. *Trans. Am. Math. Soc.* **52**, 498–521. *MR* 4–162.

Rickart, C. E. (1944). An abstract Radon–Nikodym theorem. *Trans. Am. Math. Soc.* **56**, 50–66. *MR* 6–70.

Ridder, J. (1929). Sur un théorème dans la théorie de la totalisation. *Nieuw Arch. Wisk.* **16**, 76–9. *Zbl*.2.20.

Ridder, J. (1930). Über approximativ Ableitungen bei Punkt- und Intervallf unkyionen. *Fundamenta Math.* **15**, 324.

Ridder, J. (1931a) Quelques théorèmes sur les fonctions primitives. *Nieuw Arch. Wisk.* **17**, 169–72. *Zbl*.4.206.

Ridder, J. (1931b). Über den Perronschen Integralbegriff und seine Peziehung zu den R-, L-, und D-Integralen. *Math. Zeit.* **34**, 234–69. *Zbl*.2.386.

Ridder, J. (1932). Über die Bedingung (N) von Lusin und das allgemeine Denjoysche Integral. *Math. Zeit.* **35**, 51–7. *Zbl*.4.206.

Ridder, J. (1933a). Über approximativ stetige Denjoy-Integrale. *Fundamenta Math.* **21**, 1–10. *Zbl*.8.109.

Ridder, J. (1933b). Über das allgemeine Denjoysche Integral. *Fundamenta Math.* **21**, 11–19. *Zbl*.8.109.

Ridder, J. (1933c). Das Riemann–Stieltjessche Integral. *Prace Mat. Fiz.* **41**, 65–95. *Zbl*.9.206.

Ridder, J. (1933d). Der Perronsche Integralbegriff. *Math. Zeit.* **37**, 161–9. *Zbl*.7.058.

Ridder, J. (1934a). Über die gegenseitigen Beziehungen verschiediener approximativ stetiger Denjoy-Perron-Integrale. *Fundamenta Math.* **22**, 136–62. *Zbl*.9.057.

Ridder, J. (1934b). Über die T- und N- Bedingungen und die approximativ stetigen Denjoy–Perron-Integrale. *Fundamenta Math.* **22**, 163–79. *Zbl*.9.058.

Ridder, J. (1935a). Über Denjoy–Perron Integration von Funktionen zweier Variablen. *C.R. Soc. Sci Varsovie* **28**, 5–16. *Zbl.* 13.007.

Ridder, J. (1935b). Über Perron–Stieltjessche und Denjoy–Stieltjessche Integration. *Math. Zeit.* **40**, 127–60. *Zbl*.11.108.

Ridder, J. (1936). Denjoysche und Perronsche Integration. *Math. Zeit.* **41**, 184–99. *Zbl*.14.055.

Ridder, J. (1937a). Çesàro–Perron Integration. *C.R. Soc. Sci. Varsovie* **29**, 126–52. *Zbl*.17.060.

Ridder, J. (1937b). Über die gegenseitigen Beziehungen einiger "trigonometrischer" Integrationen. *Math. Zeit.* **42**, 322–36. *Zbl*.15.400.

Ridder, J. (1938a). Das spezielle Perron–Stieltjessche Integral. *Math. Zeit.* **43**, 637–81. *Zbl*.18.249.

Ridder, J. (1938b). Das allgemeine Perron–Stieltjessche Integral (PS-Integration II). *Math. Ann.* **116**, 76–103. *Zbl*.19.202.

Ridder, J. (1939a). Sur la totalisation par rapport à une fonction à variation bornée généralisée. *C.R. Acad. Sci. Paris* **209**, 623–5. *MR* 1–110.

Ridder, J. (1939b). Nouvelles propriétés de la totalisation par rapport à une fonction à variation bornée généralisée. *C.R. Acad. Sci. Paris* **209**, 670–2. *MR* 1-110.

Ridder, J. (1941). Mass- und Integrations theorie in strukturen I, *Acta Math.* **73**, 131–73. *MR* 3–206.

Ridder, J. (1943). Denjoy–Stieltjessche und Perron–Stieltjessche Integration in k-dim. *Euklidischen Raum. Nieuw Arch. Wisk.* **21**(2), 212–41. *Indag. Math. MR* 7–421.

Ridder, J. (1946). Zur mass- und Integrations theorie in strukturen, *Indag. Math.* **8**, I, 64–71, II 72–81, *MR* 7-513.
Ridder, J. (1947a). Ueber Definitionen von Perron-Integralen I. *Indag. Math.* **9**, 227–35. *MR* 8–506.
Ridder, J. (1947b). Ueber Definitionen von Perron-Integralen II. *Indag. Math.* **9**, 280–9. *MR* 9–19.
Ridder, J. (1957). Integration von Diffierentialkoeffizienten höherer Ordnung. *Indag. Math.* **19**, 364–8. *MR* 20 #950.
Ridder, J. (1965a). Ein einheitliches Verfahren zur Definition von absolut- und bedingtkonvergenten Integralen I. *Indag. Math.* **27**, 1–13. *MR* 31 #308.
Ridder, J. (1965b). Ein einheitliches Verfahren zur Definition von absolut- und bedingtkonvergenten Integralen II. *Indag. Math.* **27**, 14–30, *MR* 31 # 308.
Ridder, J. (1965c). Ein einheitliches Verfahren zur Definition von absolut- und bedingtkonvergente Integralen II bis. *Indag. Math.* **27**, 31–9. *MR* 31 #308.
Ridder, J. (1965d). Ein einheitliches Verfahren zur Definition von absolut- und bedingtkonvergenten Integralen III. *Indag. Math.* **27**, 165–77. *MR* 35 # 4354.
Ridder, J. (1965e). Ein einheitliches Verfahren zur Definition von absolut- und bedingtkonvergenten Integralen IV. *Indag. Math.* **27**, 365–75. *MR* 32 # 187a.
Ridder, J. (1965f). Ein einheitliches Verfahren zur Definition von absolut- und bedingtkonvergenten Integralen IV bis. *Indag. Math.* **27**, 376–87. *MR* 32 #187b.
Ridder, J. (1965g). Ein einheitliches Verfahren zur Definition von absolut- und bedingtkonvergenten Integralen Va. *Indag. Math.* **27**, 705–21. *MR* 32 # 5836.
Ridder, J. (1965h). Ein einheitliches Verfahren zur Definition von absolut- und bedingtkonvergenten Integralen Vb. *Indag. Math.* **27**, 722–35. *MR* 32 #5836.
Ridder, J. (1965i). Ein einheitliches Verfahren zur Definition von absolut- und bedingtkonvergenten Integralen Vc. *Indag. Math.* **27**, 736–45. *MR* 32 # 5836.
Ridder, J. (1966). Ein einheitliches Verfahren zur Definition von absolut- und bedingtkonvergenten Integralen VI. *Indag. Math.* **28**, 248–57. *MR* 33 # 5833.
Ridder, J. (1967a). Ein einheitliches Verfahren zur Definition von absolut- und bedingtkonvergenten Integralen VII. *Indag. Math.* **29**, 1–7. *MR* 35 #4355a.
Ridder, J. (1967b). Ein einheitliches Verfahren zur Definition von absolut- und bedingtkonvergenten Integralen VII bis. *Indag. Math.* **29**, 8–17. *MR* 35 #4355b.
Ridder, J. (1967c). Ein einheitliches Verfahren zur Definition von absolut- und bedingtkonvergenten Integralen VIII. *Indag. Math.* **29**, 305–16. *MR* 36 #323.
Ridder, J. (1968a). Die allgemeine Riemann-integration in topologischen Räumen A. *Indag. Maths.* **30**, 12–23. 37 #1553.
Ridder, J. (1968b). Die allgemeine Riemann-Integration in topologischen Räumen B. *Indag. Math.* **30**, 137–48. *MR* 37 #1553.
Ridder, J. (1968c). Die allgemeine Riemann-Integration in topologischen Räumen C. *Indag. Math.* **30**, 239–52. *MR* 37 #4229.
Ridder, J. (1968d). Die allgemeine Riemann-Integration in topologischen Räumen D. *Indag. Math.* **30**, 363–76. *MR* 39 #402.
Ridder, J. (1969a). Ein einheitliches Verfahren zur Definition von absolut- und bedingtkonvergenten Integralen IX. *Indag. Math.* **31**, 10–17. *MR* 39 # 1619.
Ridder, J. (1969b). Äquivalenz von Integraldefinitionen im Sinne von Denjoy, von Perron und von Riemann. *Indag. Math.* **31**, 201–12. *MR* 40 #290.
Ridder, J. (1969c). Anwendung von Riemann–Summen in Definitionen von Integralen. *Indag. Math.* **31**, 309–26. *MR* 41 #1945.
Riečan, B. (1986). On the Kurzweil integral in compact topological spaces. *Rad. Mat.* **2**, 15–163. *MR* 88d:26014.

REFERENCES

Riemann, G. F. B. (1868). Über die Darstellbarkeit einer Funktion durch eine trigonometrische. *Reihe. Abh. Gesell. Wiss. Göttingen* . **13**. Math. Kl. 87–132. *Oeuvres mathématiques de Riemann*, 1898, Paris, reprinted 1968, Paris and Cleveland. *MR* 36 #4952.

Riesz, F. (1911). *Ann. Éc. Norm. Sup.* **28**(3), 38–68.

Riesz, F. (1914). Les opérations fonctionnelles linéaires. *Ann. Ec. Norm. Sup.* **31**(3), 9–14.

Riesz, F. (1919). Sur l'intégrale de Lebesgue. *Acta Math.* **42**, 191–205.

Riesz, F. (1932). Sur l'existence de la dérivée des fonctions d'une variable reelle et des fonctions d'intervalle. *Verhandlungen des internationalen Mathematiker-Kongress Zurich*, **I**, 258–69. Zbl.6.341.

Riesz, F. (1936). Sur l'intégrale de Lebesgue comme l'opération inverse de la dérivation. *Scuol. Norm. Sup. Pisa. Ann. Sci. Fis. Mat.* Ser. 2, **5**, 191–212. Zbl.14.206.

Riesz, F. (1940). Sur quelques notions fondamentales dans la théorie générale des opérations linéaires. *Ann. Math.* **41**, 174–206. *MR* 1–147.

Ringenberg, L. A. (1947). On the extension of interval functions *Trans. Am. Math. Soc.* **61**, 134–46. *MR* 8–320.

Ringenberg, L. A. (1948). The theory of the Burkill integral. *Duke Math. J.* **15**, 239–70. *MR* 9–575.

Robbins, H. E. (1948). Mixture of distributions. *Ann. Math. Statist.* **19**, 360–9. *MR* 10–108.

Robertson, A. P. and Robertson, W. (1966). *Topological Vector Spaces*, (Cambridge University Press). *MR* 50 #2854.

Roger, F. (1939). Sur l'extension à l'ordre n des théorèmes de M. Denjoy sur les nombres dérivés du premier ordre. *C.R. Acad. Sci. Paris* **209**, 11–14. *MR* 1–47.

Romanovskiĭ, P. (1932). Essai d'une exposition de l'intégrale de Denjoy sans nombres transfinis. *Fundamenta Math.* **19**, 38–44. Zbl.5.392. *MR* 2–354.

Romanovskii, P. (1941a). Intégrale de Denjoy dans les espaces abstraits . *Mat. Sb.* **9**(51), 67–120. *MR* 2–354.

Romanovskii, P. (1941b). Intégrale de Denjoy dans l'éspace à n dimensions. *Mat. Sb.* **9**(51), 281–307. *MR* 2–354.

Romanovskii, P. (1941c). Intégrale relative à une réseau. *Mat. Sb.* **9**(51), 309–16. *MR* 2–354.

Romanovskii, P. (1941d). Integral Denjoy–Stieltjes'a na proizvol'nom uporyadočennom, množestve. *Byull. Moskov. Gosp. Univ. Mat.* **11**, 1–11.

Rudin, W. (1974). *Real and Complex Analysis* (2nd edn), McGraw-Hill, New York. *MR* 49 #8783.

Russell, A. M. (1970). Functions of bounded second variation and Stieltjes-type integrals. *J. Lond. Math. Soc.* **2**(2), 193–208. *MR* 43 #440.

Russell, A. M. (1971). On functions of bounded kth variation. *J. Lond. Math. Soc.* **3**(2), 742, 746. *MR* 44 #4163.

Russell, A. M. (1973). Functions of bounded kth variation. *Proc. Lond. Math. Soc.* **26**(3), 547–63. *MR* 47 #3610.

Russell, A. M. (1974a). An integral representation for a generalized variation of a function. *Bull. Australian Math. Soc.* **11**, 225–9. *MR* 51 #3372.

Russell, A. M. (1974b). Functions of bounded kth variation and Stieltjes type integrals. *Bull. Australian Math. Soc.* **11**, 475–6.

Russell, A. M. (1975). Stieltjes-type integrals. *J. Australian Math. Soc.*, Ser. A, **20**, 431–48. *MR* 52 #14189.

Russell, A. M. (1976). A Banach space of functions of generalized variation. *Bull. Australian Math. Soc.* **15**, 431–8. *MR* 55 #3187.

Russell, A. M. (1977). Applications of Riemann–Stieltjes integrals of order k in functional analysis. *SIAM J. Math. Anal.* **8**, 879–90. *MR* 56 #9245.

Russell, A. M. (1978a). Necessary and sufficient conditions for the existence of a generalized Stieltjes integral. *J. Australian Math. Soc.,* Ser. A **26**, 501–10. *MR* 80f:26007.

Russell, A. M. (1978b). Further results on integral representation of functions of generalized variation. *Bull. Australian Math. Soc.* **18**, 407–20. *MR* 80a:26002.

Russell, A. M. (1979). A commutative Banach algebra of functions of generalized variation. *Pacific J. Math.* **84**, 455–63. *MR* 81h:46055.

Rutledge, D. and Worth, R. (1987/1988). A comparison of two generalizations of the Riemann integral. *Real Analysis Exchange* **13**, 432–5. *MR* 89f:26009.

Sain, D. N. and Mukhopadhyay, S. N. (1986). Linear functionals on the space of Çesàro–Perron integrable functions. *Bull. Inst. Math. Acad. Sinica* **14**, 417–25. *MR* 88g:28011.

Saks, S. (1923). La condition (N) et l'intégrale de M. M. Denjoy–Perron. *Fundamenta Math.* **13**, 218–27.

Saks, S. (1924). Sur les nombres dérivées des fonctions. *Fundamenta Math.* **5**, 98–104.

Saks, S. (1927). Sur les fonctions d'intervalle. *Fundamenta Math.* **10**, 211–24 (p. 124), *Jbuch* **53**, 233.

Saks, S. (1930). Sur l'intégrale de M. Denjoy. *Fundamenta Math.* **15**, 242–62.

Saks, S. (1931). Sur certaines classes de fonctions continues. *Fundamenta Math.* **17**, 124–51. *Zbl.*3.109.

Saks, S. (1932). On the generalized derivatives. *J. Lond. Math. Soc.* **7**, 247–51. *Zbl.*5.352.

Saks, S. (1935). On the strong derivatives of functions of intervals. *Fundamenta Math.* **25**, 235–52. *Zbl.*12.059.

Saks, S. (1936). On derivates of functions of rectangles. *Fundamenta Math.* **27**, 72–6. *Zbl.*15.105.

Saks, S. (1937). *Theory of the Integral* (2nd English edn). Warsaw. *Zbl.*17.300. *MR* 29:4850.

Saks, S. (1938). Integration in abstract metric spaces. *Duke Math. J.* **4**, 408–11. *Zbl.*19.170.

Saks, S. and Zygmund, A. (1934). On functions of rectangles and their application to the analytic functions. *Ann. Scuola norm. super. Pisa* **3**, 1–6.

Sargent, W. L. C. (1929). Change of variable in the Denjoy integral. *J. Lond. Math. Soc.* **4**, 27–32.

Sargent, W. L. C. (1935). On the Çesàro derivates of a function. *Proc. Lond. Math. Soc.* **40**(2), 235–54. *Zbl.*12,345.

Sargent, W. L. C. (1941). On sufficient conditions for a function integrable in the Çesàro–Perron sense to be monotonic. *Quart. J. Math., Oxford*, **12**, 148–53. *MR* 3–228.

Sargent, W. L. C. (1942). A descriptive definition of Çesàro–Perron integrals. *Proc. Lond. Math. Soc.* **47**(2), 212–47. *MR* 3–228.

Sargent, W. L. C. (1946a). A mean value theorem involving Çesàro means. *Proc. Lond. Math. Soc.* **49**(2), 227–40. *MR* 8–260.

Sargent, W. L. C. (1946b). On the order of magnitude of the Fourier coefficients of a function integrable in the $C_\lambda L$ sense. *J. Lond. Math. Soc.* **21**, 198–203. *MR* 8–576.

Sargent, W. L. C. (1948a). On the integrability of a product. *J. Lond. Math. Soc.* **23**, 28–34. *MR* 10–108.

Sargent, W. L. C. (1948b). On the summability (C) of allied series and the existence of

$$(CP)\int_0^\pi \frac{f(x+t)-f(x-t)}{t}\,dt.$$

Proc. Lond. Math. Soc. **50**(2), 330–48. *MR* 10–187.

Sargent, W. L. C. (1949). On fractional integrals of a function integrable in the Cesàro–Perron sense. *Proc. Lond. Math. Soc.* **51**(2), 46–80. *MR* 10–516.

Sargent, W. L. C. (1950). On linear functionals in spaces of conditionally integrable functions. *Quart. J. Math. Oxford* **1**(2), 288–98. *MR* 12–616.

Sargent, W. L. C. (1951a). On the continuity (C) and integrability (CP) of fractional integrals. *Proc. Lond. Math. Soc.* **52**(2), 253–70. *MR* 12–599.

Sargent, W. L. C. (1951b). On generalized derivatives and Cesàro–Denjoy integrals. *Proc. Lond. Math. Soc.* **52**(2), 365–76. *MR* 12–811.

Sargent, W. L. C. (1951c). Some properties of C_λ-continuous functions. *J. Lond. Math. Soc.* **26**, 116–21. *MR* 12–810.

Sargent, W. L. C. (1951d). On the integrability of a product (II). *J. Lond. Math. Soc.* **26**, 278–85. *MR* 13–449.

Sargent, W. L. C. (1953). On some theorems of Hahn, Banach and Steinhaus. *J. Lond. Math. Soc.* **28**, 438–51. *MR* 15–134.

Sarkhel, D. N. (1973). On ω-approximately continuous Denjoy–Stieltjes integral. *Colloq. Math.* **28**, 111–131. *MR* 49 #487.

Sarkhel, D. N. (1974). On ω-approximately continuous Perron–Stieltjes and Denjoy–Stieltjes integral. *J. Australian Math. Soc.* Ser. A. **18**, 129–52. *MR* 58 #1055.

Sarkhel, D. N. (1978). A criterion for Perron integrability. *Proc. Am. Math. Soc.* **71**, 109–12. *MR* 58 #17006.

Sarkhel, D. N. (1981). Topological aspect of Cesàro-continuity. *Proc. Indian Sci. Congress*, Abstract 14.

Sarkhel, D. N. (1986a). A wide Perron integral. *Bull. Australian Math. Soc.* **34**, 233–51. *MR* 88b:26013.

Sarkhel, D. N. (1986b). A note on the restricted Denjoy integral. *Ranchi Univ. Math. J.* **17**, 31–2. *MR* 88k:26009.

Sarkhel, D. N. (1987). A wide constructive integral. *Math. Japon.* **32**, 295–309. *MR* 88g:26011.

Sarkhel, D. N. and De, A. K. (1981). The proximally continuous integrals. *J. Australian Math. Soc.* Ser. A **31**, 26–45. *MR* 82h:26014.

Sarkhel, D. N. and Kar, A. B. (1984). (PVB) functions and integration. *J. Australian Math. Soc.* Ser. A, **36**, 335–53. *MR* 85b:26013.

Scanlon, C. H. (1973). Additivity and indefinite integration for McShane's *P*-integral. *Proc. Am. Math. Soc.* **39**, 129–134. *MR* 47 #3609.

Schurle, A. W. (1984/1985). Perron–Stieltjes integrability with respect to gap functions. *Real Analysis Exchange* **10**, 279–93. *MR* 86h:26009.

Schurle, A. W. (1985). Perron integrability versus Lebesgue integrability. *Can. Math. Bull.* **28**, 463–8. *MR* 87b:26011.

Schurle, A. W. (1986a). A function is Perron integrable if it has locally small Riemann sums. *J. Australian Math. Soc.*, Ser. A. **41**, 224–32. *MR* 87f:26009.

Schurle, A. W. (1986b). A new property equivalent to Lebesgue integrability. *Proc. Am. Math. Soc.* **96**, 103–6. *MR* 86m:26007.

REFERENCES

Schuss, Z. (1969). A new proof of a theorem concerning the relationship between the Lebesgue and RC-integral. *J. Lond. Math. Soc.* **44**, 365–8. *MR* 38 #6020.
Schwabik, Š. (1973a). On the relation between Young's and Kurzweil's concept of Stieltjes integral. *Casopis Pěst. Mat.* **98**, 237–51, 315. *MR* 48 #477.
Schwabik, Š. (1973b). On a modified sum integral of Stieltjes type. *Časopis Pěst. Mat.* **98**, 274–7 *MR* 48 #478.
Schwartz, L. (1950). Théorie des distributions I, (Hermann et Cie, Paris) *MR* 12–31.
Shapiro, V. L. (1958). The divergence theorem for discontinuous vector fields *Ann. Math*, **68**, 604–24.
Sharma, P. L., Jaiswal, A. and Singh, B. (1975). An integral of Çesàro–Perron type. *Math. Japon.* **20**, 107–11. *MR* 52 #8352.
Shohat, J. A. (1930). Stieltjes integrals in mathematical Statistics. *Ann. Math. Stat.* **1**, 73–95.
Sierpiński, W. (1920). Sur un problème concernant les ensembles mesurables superficiellement. *Fundamenta Math.* **1**, 112–15.
Sierpinski, W. (1923). Un lemme métrique. *Fundamenta Math.* **4**, 201–3.
Skljarenko, V. A. (1971). Relation between (H^k)- and (P)- integrals (Russian). *Vestnik Moskov. Univ. Ser. I Mat. Mel.* **26**, 69–77. *MR* 43 #7567.
Skljarenko, V. A. (1972). Certain properties of the P^2-primitive (Russian). *Mat. Zametki.* **12**, 693–700. English translation *Math. Notes* **12**, 856–60. *MR* 47 #8785.
Skljarenko, V. A. (1973). On Denjoy integrable sums of everywhere convergent trigonometric series (Russian). *Dokl. Akad. Nauk SSSR* **210**, 533–6. Eng. trans. *Soviet Math. Dokl.* **14**, 771–5. *MR* 49 #578.
Skljarenko, V. A. (1980). Integration by pàrts in the SCP Burkill integral (Russian). *Mat. Sb.* **112**(154), 630–46. Eng. trans. *Math. USSR Sbornik* **40**(1981),567–83. *MR* 81k:26009.
Skvorcov, V. A. (Skvortsov) (1959). Interrelation between Denjoy's general integral and totalisation ($T_{2s})_0$ (Russian). *Dokl. Akad. Nauk SSSR* **127**, 975–6. *MR* 22 #755.
Skvorcov, V. A. (1960). Interrelation between general Denjoy integrals and totalisation ($T_{2s})_0$ (Russian). *Mat. Sb.* **52**, 551–78. *MR* 22 #12194.
Skvorcov, V. A. (1962). On the relation between the D-integral and the totalisation (T_{2s}) (Russian). *Vestnik Moskov. Univ. Ser. 1 Mat. Meh.* **6**, 20–5. *MR* 26 #3870.
Skvorcov, V. A. (1963). Some properties of the CP-integral (Russian). *Mat. Sb.* **60**(102), 304–24. Eng. trans. *Am. Math. Soc. Trans.* **54**(2), (1966). 233–54. *MR* 27 #264.
Skvorcov, V. A. (1964). On integrating the exact Schwarzian derivative (Russian). *Mat. Sb.* **63**(105), 329–40. *MR* 28 #4073.
Skvorcov, V. A. (1966a). Interconnection between Taylor's AP-integral and James' P^2-integral (Russian). *Mat. Sb.* **70**(112), 380–93. *MR* 34 #305.
Skvorcov, V. A. (1966b). Concerning definitions of P^2- and SCP-integrals (Russian). *Vestnik Moskov. Univ. Ser. 1*, **2**, 12–19. *MR* 34 #7765.
Skvorcov, V. A. (1967). A relation between certain integrals (Russian). *Vestnik Moskov. Univ. Ser. 1*, **22**, 68–72. *MR* 36 #324.
Skvorcov, V. A. (1968a). The generality of Cross' H^k-integral (Russian). *Vestnik Moskov. Univ. Ser. 1*, **23**, 19-22. *MR* 40 #7714.
Skvorcov, V. A. (1968b). Calculation of the coefficients of an everywhere converging Haar series (Russian). *Mat. Sb.* **75**(117), 349–60. *MR* 37 #1884.
Skvorcov, V. A. (1969). A certain generalization of the Perron integral (Russian). *Vestnik Moskov. Univ. Ser. 1*, **24**, 48–51. *MR* 40 #5802.
Skvorcov, V. A. (1972). The Marcinkiewicz–Zygmund integral and its connection with the SCP-integral of Burkill (Russian). *Vestnik Moskov. Univ. Ser. 1*, **27**, 78–82. English translation *Mosc. Univ. Math Bull.* **27**, 66–70. *MR* 49 #5251.

REFERENCES

Skvorcov, V. A. (1982). Constructive variant of the definition of an HD-integral (Russian). *Vestnik Moskov. Univ.* Ser. 1, **127**, 41–5. *MR* 84f: 26012.

Skvorcov, V. A. (1986/1987). Generalized integrals in theory of trigonometric, Haar and Walsh series. *Real Analysis Exchange* **12**, 59–62.

Ślęzàk, W. (1984). On the module of strong derivatives over the ring of continuous functions of two variables. *Demonstratio Math.* **17**, 853–67. *MR* 87b:26020.

Šmidov, F. I. (1955). On the theory of the integral (Russian). *Dokl. Akad. Nauk SSSR* **101**, 31–4. *MR* 16–805.

Šmidov, F. I. (1966). On the question of inclusion of a given function in the class of VBG_* functions (Russian). *Izv. Vysš. Učebn. Zaved. Mat.* **50**, 170–73. *MR* 33 #5799.

Smith, E. S. (1979). A comparison of the Riemann-complete and Lebesgue integrals. *Eleutheria*, 493–9. *MR* 82h:26016.

Smith, H. J. S. (1875). On the integration of discontinuous functions. *Proc. Lond. Math Soc.* **6**, 140–53.

Smith, H. L. (1925). On the existence of the Stieltjes integral. *Trans. Am. Math. Soc.* **27**, 491–515.

Smoluchowski, M. (1906) *Ann. d. Physik* **21**, 756.

Soeparno, D. (1988). An Orlicz scale of integrals. *Southeast Asian Bull. Math.* **12**, 123–33.

Solomon, D. W. (1967). On a constructive definition of the restricted Denjoy integral. *Trans. Am. Math. Soc.* **128**, 248–56. *MR* 35 #1728.

Solomon, D. W. (1969a). Denjoy integration in abstract spaces. *Mem. Am. Math. Soc.* **85**. *MR* 39 #404.

Solomon, D. W. (1969b). On separation in measure and metric density in Romanovski spaces. *Duke Math. J.* **36**, 81–90. *MR* 38 #6019.

Solomon, D. W. (1969c). On non-measurable sets. *Duke Math. J.* **36**, 183–91. *MR* 39 #7050.

Steffensen, J. F. (1932). On Stieltjes' integral and its applications to actuarial questions. *J. Inst. Actuaries* **63**, 443–83. *Zbl.*6.021.

Steib, M. L. (1981). A Cauchy type condition and integrability. *Czechoslovak Math. J.* **31**(106), 241–74. *MR* 82h:26017.

Stieltjes, T. J. (1894). Recherches sur les fractions continues. *Ann. Fac. Sci. de Toulouse* **8**, J1–22

Stone, M. H. (1948). Notes on integration. *Proc. Nat. Acad. Sci. USA*, **34**, I, 336–42. *MR* 10–24; II, 447–55, *MR* 10–107; III, 483–90, *MR* 10–239.

Stone, M. H. (1949). Notes on integration. IV. *Proc. Nat. Acad. Sci. USA*, **35**, 50–8. *MR* 10–360.

Šul'man, V. S. (1971). An example of a function that is integrable in the sense of Denjoy and nonsummable in the sense of Hinčin (Russian). *Mat. Zametki* **10**, 295–300. *MR* 44 #4186.

Sunouchi, G. and Utagawa, M. (1949). The generalized Perron integrals. *Tôhoku Math. J.* **1**(2), 95-9. *MR* 11-90.

Taylor, S. J. (1955). An integral of Perron's type defined with the help of trigonometric series. *Quart. J. Math., Oxford* **6**(2), 255–74. *MR* 19–255.

Temple, G. (1971). *The Structure of Lebesgue Integration Theory*. Clarendon Press, Oxford.

Thomae, J. (1875). Die partielle Integration. *Z.f. Math.* **20**, 475–8.

Thomas, E. G. F. (1976). Totally summable functions with values in locally convex spaces. *Lecture Notes in Math.* **541**, 117–31. *MR* 56 #8799.

Thomson, B. S. (1970a). Constructive definitions for non-absolutely convergent integrals. *Proc. Lond. Math. Soc.* **20**(3), 699–716. *MR* 42 #3248.

Thomson, B. S. (1970b). Spaces of conditionally integrable functions. *J. Lond. Math. Soc.* **2**(2), 358–60. *MR* 41 #4223.

Thomson, B. S. (1971a). Construction of measures and integrals. *Trans. Am. Math. Soc.* **160**, 287–96. *MR* 43 #6385.

Thomson, B. S.(1971b). Covering systems and derivates in Henstock division spaces. *J. Lond. Math. Soc.* **4**(2), 103–8. *MR* 45 #3664.

Thomson, B. S. (1971c). On the Henstock strong variational integral. *Can. Math. Bull.* **14**, 87–99. *MR* 45 #8799.

Thomson, B. S. (1972a). A theory of integration. *Duke Math. J.* **39**, 503–9. *MR* 46 #3736.

Thomson, B. S. (1972b). On McShane's vector-valued integral. *Duke Math. J.* **39**, 511–19. *MR* 46 #3742.

Thomson, B. S. (1975). Covering systems and derivates in Henstock division spaces II. *J. Lond. Math. Soc.* **10**(2), 125–8. *MR* 51 #840.

Thomson, B. S. (1976). Construction of measures in metric spaces. *J. Lond. Math. Soc.* **14**(2), 21–4. *MR* 54 #10537.

Thomson, B. S. (1977a). A characterisation of the Lebesgue integral. *Can. Math. Bull.* **20**, 353–7. *MR* 57 #3332.

Thomson, B. S.(1977b). Measures generated by a differentiation basis. *Bull. Lond. Math. Soc.* **9**, 279–82. *MR* 57 #16524.

Thomson, B. S. (1980). On the derived numbers of VBG_* functions. *J. Lond. Math. Soc.* **22**(2), 473–85. *MR* 81m:26004.

Thomson, B. S. (1980/1981). On full covering properties. *Real Analysis Exchange* **6**, 77–93. *MR* 82c:26008.

Thomson, B. S. (1981a). On the total variation of a function. *Can. Math. Bull.* **24**, 331–40. *MR* 83a:26016.

Thomson, B. S. (1981b). Outer measures and total variation *Can. Math. Bull.* **24**, 341–5. *MR* 83a:26017.

Thomson, B. S. (1982/1983). Derivation bases on the real line. *Real Analysis Exchange* **8**, I, 67–207. II, 278–442. *MR* 84c:26008a, b.

Thomson, B. S. (1985). *Real Functions, Lecture Notes in Mathematics* 1170, Springer-Verlag, Berlin, Heidelberg, New York. *MR* 87f:26001.

Thomson, B. S. (1986/1987). Some remarks on differential equivalence. *Real Analysis Exchange* **12**, 294–312. *MR* 88c:26007.

Thomson, H. B. (1985). Taylor's theorem with the integral remainder under very weak differentiability assumptions. *Australian Math. Gaz.* **12**, 1–6. *MR* 86f:26002.

Titchmarsh, E. C. (1924). The order of magnitude of the coefficients in a generalized Fourier series. *Proc. Lond. Math. Soc.* **22**, xxv–xxvi.

Titchmarsh, E. C. (1929). On conjugate functions. *Proc. Lond. Math. Soc.* **29**(2), 49–80.

Titchmarsh, E. C. (1948). *Introduction to the Theory of Fourier Integrals*, Oxford University Press.

Tolstov, G. P. (Tolstoff) (1939a). Sur quelques propriétés des fonctions approximativement continues (French). *Math. Sb.* **5**(47), 637–45. *MR* 1–206.

Tolstov, G. P. (1939b). Sur l'intégrale de Perron (French). *Mat. Sb.* **5**(47), 647–60. *MR* 1–208.

Tolstov, G. P. (1939c). La méthode de M. Perron pour l'intégrale de M. Denjoy. *Dokl. Akad. Nauk SSSR* **25**, 470–2. *MR* 1–305.

Tolstov, G. P. (1940). La méthode de Perron pour l'intégrale de Denjoy. *Mat. Sb.* **8**(50), 149–68. *MR* 2–132.

Tolstov, G. P. (1948). In the interchange of integrations (Russian). *Dokl. Akad. Nauk SSSR* **63**, 3–6. *MR* 10–360.

Tolstov, G. P. (1949). The incorrectness of Fubini's theorem for the multidimensional regular Denjoy integral. *Mat. Sb.* **24**(66), 263–78. *MR* 10–690.

Tolstov, G. P. (1950a). Integral as a primitive. *Dokl. Akad. Nauk SSSR* **73**, 659–62. *MR* 12–167.

Tolstov, G. P. (1950b). On the curvilinear and iterated integral. *Trudy Mat. Inst. Steklov.* **35**, 1–103. *MR* 13–448.

Tolstov, G. P. (1961a). Majorants and minorants in various problems (Russian). *Studies of Modern Problems of Constructive Theory of Functions*, pp. 308–11. Fizmatgiz, Moscow. *MR* 32 #2535.

Tolstov, G. P. (1961b). The definition of the restricted Denjoy integral by classical means (Russian). *Studies of Modern Problems of Constructive Theory of Functions*, pp. 312–313. Fizmatgiz, Moscow. *MR* 32 #2548.

Tolstov, G. P. (1961c). Parametric differentiation and the narrow Denjoy integral (Russian). *Mat. Sb.* **53**(95), 387–92. *MR* 25 #3144.

Tolstov, G. P. (1962). The derivative and the integral from a general viewpoint (Russian). *Dokl. Akad. Nauk SSSR* **142**, 1040–2. English translation *Soviet Math. Dokl.* **3** (1962), 253–5. *MR* 25 #154.

Tolstov, G. P. (1964a). An elementary integration which can be reduced to a non-measurable indefinite integral (Russian). *Mat. Sb* **65**(107), 454–7. *MR* 30 #1228.

Tolstov, G. P. (1964b). Three types of abstract integrals (Russian). *Dokl. Akad. Nauk SSSR* **158**, 536–9. English translation *Soviet Math. Dokl.* **5** (1965), 1275–8. *MR* 30 #2118.

Tolstov, G. P. (1965). Derivative and integral (the axiomatic approach) (Russian). *Mat. Sb.* **66**(108), 608–30. *MR* 30 #4895.

Tolstov, G. P. (1966). Diferentiation and integration in abstract spaces (Russian). *Mat. Sb.* **71**(113), 420–2. *MR* 34 #302.

Tolstov, G. P. (1969). On the Radon–Nikodym theorem (Russian). *Mat. Sb.* **80**(122), 334–8. *MR* 40 #2804.

Tonelli, L. (1923). Sulla Nozione di Integrale. *Annali di Mat.* (IV) **1**, 105–45.

Tonelli, L. (1926). Sur la quadrature des surfaces. *C.R. Acad. Sci. Paris* **182**, 1198–200.

Topsøe, F. (1970). *Topology and Measure*, Lecture Notes in Mathematics 133, Springer-Verlag, Berlin. *MR* 54 #10546.

Trjitzinsky, W. J. (1954). *Les problèmes de totalisation se rattachant aux Laplaciens non-sommables.* Mémor. Sci. Math. 125. *MR* 16–471.

Trjitzinsky, W. J. (1955). Les Laplaciens généralisés nonsommables. *J. Math. Pures Appl.* **34**, 1–136. *MR* 20 #3389.

Trjitzinsky, W. J. (1963). *Totalisations dans les Espaces Abstraits.* Mémor. Sci. Math. 155. *MR* 28 #4078.

Trjitzinsky, W. J. (1965). *Totalisation des séries dans les espaces abstraits.* Mémor. Sci. Math. 161, Gauthier-Villars, Paris. *MR* 34 #2826.

Trjitzinsky, W. J. (1968). *La totale-D de Denjoy et la totale-S symétrique.* Mémor. Sci. Math. 166, Gauthier-Villars, Paris. *MR* 39 #1604.

Trjitzinsky, W. J. (1969). Totalisations abstraites dans les espaces vectoriels. *Ann. Mat. Pure Appl.* **82**, 275–379. *MR* 55 #8274.

Tychonoff, A (1929) 'Über die topologische Erweiterung von Räumen', *Math. Ann.* **102**, 544–61.

Ugrin-Šparac, D. (1970). A generalization of Taylor's formula and its application in elementary theory of distributions. *Glasnik Mat.* **5**(25), 71–80. MR 43 #2170.
Uhl, Jr., J. J. (1972). A characterization of strongly measurable Pettis integrable functions. *Proc. Am. Math. Soc.* **34**, 425–7. MR 47 #5222.
Vallée Poussin, Ch. de la (1892a) Étude des intégrales a limites infinies pour lesquelles la fonction sous le signe est continue. *Ann. Soc. Sci. Bruxelles* **16** (2nd. part), 150–180.
Vallée Poussin, Ch. de la (1892b). Recherches sur la convergence des intégrales définies. *J. de Math. pures et appl.* **8**(4), Fasc.4, 421–67 (particularly on and after p. 453).
Vallée Poussin, Ch. de la (1912). Sur l'unicité du développement trigonométrique. *Bull. Acad. Roy. Belg.*, 702–718.
Vallée Poussin, Ch. de la (1930). *Cours d'Analyse*, Vols. I, II., Paris.
Vallée Poussin, Ch. de la (1934). *Intégrales de Lebesgue, fonctions d'ensemble, classes de Baire*: (2nd edn.), Paris. Zbl.9.206.
Vanderlijn, G. (1941). Une généralisation de la intégrale de Radon. *Bull. Soc. Roy. Sci. Liège* **10**, 168–75. MR 7–251.
Verblunsky, S. (1930a). *Proc. Lond. Math. Soc.* **31**(2), 370–86.
Verblunsky, S. (1930b). The generalized third derivative and its application to the theory of trigonometric series. *Proc. Lond. Math. Soc.* **31**(2), 387–406.
Verblunsky, S. (1931). The generalized fourth derivative. *J. Lond. Math. Soc.* **6**, 82–4. Zbl.1.330
Verblunsky, S. (1934). On the theory of trigonometric series VII. *Fundamenta Math.* **23**, 193–236. Zbl.10.019.
Verblunsky, S. (1949). On Green's formula. *J. Lond. Math. Soc.* **24**, 146–8.
Verblunsky, S. (1971a). On a descriptive definition of Çesàro–Perron integrals. *J. Lond. Math. Soc.* **3**(2), 326–33. MR 44 #4161.
Verblunsky, S. (1971b). On the Peano derivatives. *Proc. Lond. Math. Soc.* **22**(3), 313–24. MR 44 #2896.
Vicente, G. J. (1941). Sur la primitive des différentielles totales. *Rev Fac. Ci. Univ. Coimbra* **9**, 65–8. MR 8–141.
Vinogradova, I. A. and Sklyarenko, V. A. (1984). Certain properties of LG^*-integrable functions. *Anal. Math.* **10**, 183–91. MR 86d:26014.
Vinogradova, I. A. and Skvorcov, V. A. (1973). Generalized integrals and Fourier series. (Russian). *J. Soviet Math.* **1**, 677–703. MR 51 #8715.
Vitali, G. (1907). Sull' integrazione per serie. *Rend. del Circ. mat. di Palermo* **23**, 137–55 (151).
Vitali, G. (1908). Sui gruppi di punti e sulle funzioni di variabili reali. *Atti Accad. Sci. Torino* **43**, 75–92.
Vulih, B. Z. (1941). On the Stieltjes integral of functions whose values belong to a semiordered space (Russian). *Leningrad State University Ann.* (*Math. ser.* 12) **83**, 3–29. MR 8–30.
Vybnornỳ, R. (1977). *Notes on Integration*, Lecture Notes 3, Department of Mathematics, University of Queensland. MR
Vybnorny, R. (1981a). *Nordisk. Mat. Tidskr. Normat*, **29** 2, 72–4. MR
Vybnorny, R. (1981b). Kurzweil's integral and arc length. *Australian Math, Soc. Gaz.* **8**, 19–22. MR 83b:26011.
Wang, F. T. (1934). On the convergence factor of the Fourier–Denjoy series. *Proc. Imp. Acad. Japan* **10**, 53–6. Zbl.8-350
Ward, A. J. (1936a). The Perron–Stieltjes integral. *Math. Zeit.* **41**, 578–604. Zbl.14.397.

Ward, A. J. (1936b). On the differentiation of additive functions of rectangles. *Fundamenta Math.* **26**, 167–82. Zbl.13.251.

Ward, A. J. (1936c). On the derivation of additive functions of intervals in m-dimensional space. *Fundamenta Math.* **28**, 265–79. Zbl.16.158.

Ward, A. J. (1937). A sufficient condition for a function of intervals to be monotone. *Fundamenta Math.* **29**, 22–5. Zbl.17.008.

Ward, A. J. (1938). Remark on the symmetrical derivates of additive functions of intervals. *Fundamenta Math.* **30**, 100–3. Zbl.18.114.

Webb, J. R. (1967). A Hellinger integral representation for bounded linear functionals. *Pacific J. Math.* **20**, 327–37. MR 34 #8169.

Wecken, F. (1939). Abstrakte Integrale und fastperiodische Funktionen. *Math. Zeit.* **45**, 377–404. MR 1–12.

Whyburn, W. M. (1932). On the integration of unbounded functions. *Bull. Am. Math. Soc.* **38**, 123–31. Zbl.4.055.

Wiener, N. (1923). Differential space. *J. Math. Phys.* **2**, 132–74.

Wiener, N. (1930). Generalized harmonic analysis. *Acta Math.* **55**, 117–258.

Wolf, F. (1939). On summable trigonometric series: an extension of uniqueness theorems. *Proc. Lond. Math. Soc.* **45**, 328–56. MR 1–225

Wolf, F. (1947). Contribution to a theory of summability of trigonometric series. *Univ. Cal. Publ. Math.* **1**, 159–227. MR 9-140.

Wright, F. M. and Baker, J. D. (1969) On integration-by-parts for weighted integrals. *Proc. Am. Math. Soc.* **22**, 42–52. MR 39 #7056.

Xu Dong Fu and Lu Shipan (1987/1988). Henstock integrals and Lusin's condition (N). *Real Analysis Exchange* **13**, 451–3. MR 89e:26021.

Yoneda, K. (1973). On generalized (A)-integrals. *Math. Japon* **18**, 149–67. MR 50 #14047.

Young, G. C. (1916a). A note on derivates and differential coefficients. *Acta Math.* **37**, 141–54.

Young, G. C. (1916b). On the derivates of a function. *Proc. Lond. Math. Soc.* **15**(2), 360–84.

Young, G. C. and Young, W. H. (1911) On the existence of a differential coefficient. *Proc. Lond. Math. Soc.* **9**(2), 325–35.

Young, L. C. (1927). The Theory of Integration, Cambridge.

Young, L. C. (1930). Note on the theory of measure. *Proc. Camb. Phil. Soc.* **26**, 88–93.

Young, L. C. (1936). An inequality of Hölder type connected with Stieltjes integration. *Acta Math.* **67**, 251–82. Zbl.16.104.

Young, R. C. (Mrs Tanner) (1927) Les fonctions additives d'ensemble les fonctions de point á variation bornée et la généralisation de la notion d'espace à n dimensions. *Enseign. Math.* nos. 1–2–3.

Young, R. C. (1928a). Functions of Σ defined by addition, or functions of intervals in n-dimensional formulation. *Math. Zeit.* **29**, 171–216.

Young, R. C. (1928b). On Riemann integration with respect to a continuous increment. *Math. Ann.* **29**, 217–33.

Young, R. C. (1929). On 'Riemann' integration with respect to an additive function of sets. *Proc. Lond. Math. Soc.* **29**(2), 479–89.

Young, W. H. (1903). Zur Lehre der nicht abgeschlossenen Punktmengen. *Ber. Verh. Sächs. Akad. Leipzig* **55**, 287–93.

Young, W. H. (1904a). On non-uniform convergence and term-by-term integration of series. *Proc. Lond. Math. Soc.* **1**(2), 89–102.

Young, W. H. (1904b). On upper and lower integration. *Proc. Lond. Math. Soc.* **2**(2), 52–66. *Jbuch* **35**, 310.

Young, W. H. (1905). On the general theory of integration. *Phil. Trans. Roy. Soc. London* **204**, 221–52. *Jbuch* **36**, 358.

Young, W. H. (1908). On the inequalities connecting the double and repeated upper and lower integrals of a function of two variables. *Proc. Lond. Math. Soc.* **6**(2), 247–9.

Young, W. H. (1909). Term-by-term integration of oscillating series. *Proc. Lond. Math. Soc.* **8**(2), 99–116.

Young, W. H. (1910a). On semi-integrals and oscillating successions of functions. *Proc. Lond. Math. Soc.* **9**(2), 286–324.

Young, W. H. (1910b). On a change of order of integration in an improper repeated integral. *Trans. Camb. Phil. Soc.* **21**(13), 361–76.

Young, W. H. (1910c). On parametric integration. *Monatsch. für Math. und Phys.* **21**, 125–49.

Young, W. H. (1911). On the conditions that a trigonometrical series should have the Fourier form. *Proc. Lond. Math. Soc.* **9**(2), 421–33.

Young, W. H. (1912). On classes of summable functions and their Fourier series. *Proc. Roy. Soc. Ser. A,* **87**, 225–9.

Young, W. H. (1914). On the integration with respect to a function of bounded variation. *Proc. Lond. Math. Soc.* **13**(2), 109–50.

Young, W. H. (1917a). On integrals and derivates with respect to a function. *Proc. Lond. Math. Soc.* **15**(2), 35–63.

Young, W. H. (1917b). On non-absolutely convergent, not necessarily continuous, integrals. *Proc. Lond. Math. Soc.* **16**(2), 175–218.

Young, W. H. (1918). On multiple integration by parts and the second theorem of the mean. *Proc. Lond. Math. Soc.* **16**(2), 273–93.

Young, W. H. (1926). The progress of mathematical analysis in the 20th century. *Proc. Lond. Math. Soc.* **24**(2), 421–34.

Young, W. H. and Young, G. C. (1915). On the reduction of sets of intervals. *Proc. Lond. Math. Soc.* **14**(2), 111–30.

Zaanen, A. C. (1953) *Linear Analysis* (Amsterdam).

Zahorski, Z. (1941). Über die Menge der Punkte in welchen die Ableitung unendlich ist. *Tôhoku Math. J.* **48**, 321–30. *MR* 10–359.

Zygmund, A. (1959). *Trigonometric Series* (2nd edn), I, II, Cambridge Unversity Press., *MR* 21 #6498.

NAME INDEX

Abel vi, 38
Aleksandroff 37
Alexiewicz 185, 199
Appling 34
Archimedes 32
Arley 34
Arzelà 121, Section 3.3, 195, 197, 207

Baire 16, 21, 22, 35, 53, 54, 205
Baker 40
Banach vi, 3, 5, 21, 22, 24, 25, 33, 35, 36, 40, 79, 86, 87, 90, 94, 104, 108, 121, 143, 149, 159
Bartle 36
Bauer 171
Benninghofen 38
Bernoulli, Daniel 32
Birkhoff 34, 35
Bliss 33
Bochner 3–5, 35, 36
Bogdanowicz (Bogdan) 36
Borchsenius 34
Borel 2, 17, 32, 36, 50, 57
Bosanquet 154, 173
Bose 173
Bourbaki 35, 201
Brooks 35, 36
Brown 158
Buczolich 53
Bullen 38, 53
Burkill, H. 173
Burkill, J. C. v, vi, 2, 34, 38, 40, 47, 48, Section 1.10, 100, 173, 184, 202, 203

Cameron 154
Carathéodory 34, 63, 69, 105, 108
Carmichael 33
Cartesian product 8, 9
Cauchy 6, 32, 36, 115, 116, 126, 184, 194
Çesari 34, 40
Çesaro vi, 38, 173, 184
Chandrasekhar 157
Chapman 157, 162
Chatfield 34
Chatterji 35
Copeland 34
Cousin 52
Cross vii

Daniell 34, 36
Darboux 6, 32, 33, 115
Davies vi, 6

Denjoy (general) v, 37, 38, Section 1.9, 115, 184
Denjoy (special) v, vi, 6, 36, 37, 50, 56, 115, 116, 119, 184, 201
Descartes 32
Dienes 34
Dinculeanu 36
Dobrakov 36
Dollard 34, 35
Dunford 33, 35, 74
Dushnik 34

Einstein 157
Ellis, H. W. 38
Ellis, R. 26
Enomoto (now Nakanishi) 36
Eudoxus 32
Euler 32, 203

Fan 36
Fatou 137, 138, 140, 195, 198
Feynman vii, 121, 157, 158, 166
Finetti de 33
Foglio 38, 183
Foran 53
Fréchet 34, 35
Freudenthal 36
Friedman 34, 35
Fubini property 44, 51
Fujita 36

Gage 173
Gelfand 35
Getchell 34
Gillespie 32
Glivenko 34
Goldstine 35
Gomes 36
Goursat 17
Gowurin 34–6
Graves 33

Hahn vi, 24, 35, 36
Hake 37, 171
Harnack 6, 37, 115–18
Hellinger 34, 40, 48
Helton, B. W. 34
Helton, J. C. 34
Henstock vi, ix, 1, 2, 6, 23, 30, 32, 34, 36–9, 46, 47, 49, 50, 52, 54–8, 69, 74, 80, 90, 92, 99, 100, 103, 110, 111, 115, 119, 121, 124–26, 132, 134, 138, 144, 146, 154, 156, 166, 171, 173, 174, 183, 202

NAME INDEX

Hildebrandt 34, 35
Hinčin (Khintchine) 37, 38, 56
Hobson 34, 115
Hölder 188–90, 192
Huygens, C. and L. 32
Hyslop 34

Izumi 35, 36, 38

Jacob 33
James vi, 173
Jarnik 56
Jeffery vi, 38, 174
Jessen 157, 163, 164
Johnson 166
Jordan 88, 120

Kaltenborn 34
Kappos 36
Kartàk 52
Kay 34
Kempisty 34, 38
Khintchine (Hinčin) 37, 38, 56
Kolmogorov 34, 157, 162
Krzyzański 38
Kunugi 36
Kurzweil vi, ix, 1, 2, 6, 32, 37, Section 1.4, Section 1.5, 56, 154

Lebesgue v, vi, ix, 1, 3–7, 32, 34–7, 43, 50, 54, 74, 95, 115, 119–21, 126, 128, Section 3.3, 147, 154, 156, 174, 184, 195, 197, 202, 203, 207
Lee, C. M. 38
Lee, P. Y. (controlled convergence) 134
Lee, T. W. 154
Leibnitz 1, 32, 120
Levi 6, 36, 126
Liu 53
Looman 37, 38, 171
Luxemburg 126
Luzin 37, 52

McCrudden 52
McGill 111
McGrotty 54, 69
McLeod 6
MacNeille 35, 37
MacNerney 34
McShane vi, 1, 6, 30, 35, 38, 47, 50, Section 1.6, 69, 171
Marcinkiewicz vi, 38, 173
Mařik 56

Martin 34, 154
Masani 34
Matsuyama 35
Matyska 56
Mawhin 38, 52, 56
Meinershagen 53
Mikusiński 35
Miller vi, 38, 174
Minkowski 21, 188
Moore 34, 36, 41, 49, 92
Morgan, de 9
Muldowney 100, 157, 166

Nakamura 35
Nakanishi (previously Enomoto) 36
Nakano 35
Natanson 37
Neuberger 34
Newton 1, 2, 5, 6, 32, 33, 37
Nikodym 108, 130

Okano 36
Olmsted 36
Orihara 35
Orlicz xi, Section 8.2

Perron vi, x, 6, 37, 38, 50, 56–8, 116, 119, Section 7.1, 173, 184, 201
Pettis 4, 35, 74
Pfeffer 53, 56, 184
Phillips 35, 36
Plant 34
Poisson 32, 38
Pollard 34, 49
Preiss 54
Price 35, 36
Przeworska-Rolewicz 125

Radon 35, 54, 108, 130
Rickart 35
Ridder 34, 36, 38
Riemann v, vi, 2, 6, 32–5, 43, 47–50, 53, 83, 115, 121, Section 3.3, 155, 158, 184, 185
Riesz 33, 35, 185, 199
Robbins 154
Robertson, A. P. and W. 25
Rolewicz 125
Romanovski 38

Saks 34, 37, 38, 69, 90, 171
Sargent vi, 23, 201
Sarkhel 171
Schuss vi, 6

NAME INDEX

Schwabik 56
Shapiro 56
Shohat 34
Sierpinski 114, 144, 148, 154
Skvorcov 38
Smith, H. J. S. 32
Smith, H. L. 33, 34, 36, 41, 49, 92
Smoluchowski 157, 162
Solomon 37
Stackwisch 34
Steffensen 33
Steinhaus vi, 24, 25
Stieltjes vi, 33, 35, 37, 38, 47, 48
Stone 34
Sunouchi 35

Taylor, A. E. 35
Taylor, S. J. vi, 38, 173
Temple 28
Thomson 23, 45, 46, 54, 69, 106, 108, 201
Tolstov 37, 56
Tonelli x, 35, Section 5.2
Tychonoff 111, 160

Uhl 35

Vallée Poussin, de la 36, 115, 120
Vanderlijn 33
Verblunsky 38, 55
Vitali 128, 144, 148
Vulih 33

Ward x, 37, 38, 50, Section 7.2, 174
Wecken 36
Weil 30
Wiener vii, 157, 158
Wright 40

Yang 56
Young, G. C. 52
Young, L. C. 34
Young, R. C. (Mrs. Tanner) 34
Young, W. H. xi, 34–6, 52, 114, 115, Section 8.2

Zaanen 198
Zygmund vi, 38, 173

SUBJECT INDEX

Abel–Poisson–Perron integral 38
absorbing 21
additive division space 1, 42, 52, Section 2.7
additive semigroup 12
algebra Section 0.3
approximate Perron v, 38, Section 1.10, 184
α-space 22, 201
associated point 40
associative 11

backwards martingale 163
balanced set 21
barrel 21
barrel space 21, 22, 25
base for topology 13
Boolean algebra 36
bounded N-variation 179
bounded variation 76
brick 47
brick-point 47
β-space 22, 25, 201

Cartesian product 8, 9
category
  first 13
  second 14
category theorem 16, 205
closed set 13
closed sphere 16
closure 13
cluster point 13
co-meagre 14
commutative 11, 12
compact 17
compatible division 1
compatible with topology 43
complement 8
complete metric space 16
complete normed linear space (LNC-space) 21
complete space 30
conditional expectation 164
contact point 13
continuous 17
continuous linear functional 23
controlled convergence 134
convergence-factor integrals, vi, vii, x, Section 7.3
convergent 16, 27
convex 20
co-partitional 41
coset 8

countable additivity 71, 73
cover 17

decomposable division system (space) 6, 43, 74, Section 2.4, Section 2.6
Denjoy (general) v, 37, 38, Section 1.9, 115, 184
Denjoy (special) v, vi, 6, 36, 37, 50, 56, 115, 116, 119, 184, 201
dense 13
dense-in-itself 14
density integration Section 8.4
derivates (upper and lower) 167
derivative 144
derived set 13
δ-fine 1
diagonal 41
directed for divisions 42
directed set 26
disjoint sets 9
distinct sets 8
distributive 12
divergence theorem 55, Section 1.8
divide 41, 42
division 1, 41
division space 1, 42, 52, Section 2.5
division system 42, Section 2.3

element 8
elementary set 41, 45, 61
empty set 8
equicontinuous 24
équilibré 21
equivalence class 8
equivalence relation 8

family 8
field 12
finite cover 17
finite intersection property 17
finitely additive class 63
finitely multiplicative 61
finitely submultiplicative (subadditive) 93
finitely supermultiplicative (superadditive) 93
first category 13
free almost everywhere 60, 65
free bounded variation 60, 65
free division system 41, Section 2.1
freely decomposable 41, Section 2.2
freely fully decomposable 41
freely measurably decomposable 41

SUBJECT INDEX

free measurable (Carathéodory) 63, 68
free norm variation 60
free variation 59
free variational integral 61
free variationally equivalent 60, 65
free variation set 65
free variation zero 60, 65
frontier star-set 43
F-space 21, 22, 25
Fubini division system (space) x, 44, 49, 51, 114, Section 5.1, 155, 161, 163, 164
Fubini property 44, 51
fully decomposable 6, 43, 74, 81
function 8
fundamental (A; E) 75
fundamental sequence 16, 30, 194

gauge 1
gauge integral ix, 2, 5–7, 32, Section 1.4, Section 1.5, 115, 116, 119, 132
generalized Riemann integral 41
generalized sequence 27
group 12
G-set 13

Hausdorff, Hausdorffian 18, 26–28, 66, 68, 75, 85

image 8
indicator 10
infinitely divisible 41
integration by parts 112
integration by substitution 82, 91
interior 13
intersection of sets 9
interval (generalized) 40, 59
interval-point pair 40
intrinsic topology 42, Section 2.9
invariant metric 21
inverse image 8
isolated point 14

join of sets 9

lattice 28
Lee, P. Y. (controlled convergence) 134
limit 16, 27
limiting free variation set 65
limiting variation set 76
limit point 13
linear topological space 20
linear (vector) space 12
linked 42

local base 13
locally convex (linear topological) space 21

mapping 8
meagre 13
measurable 3
measurably decomposable 43, 81
measure 3, 4
meet of sets 9
metric 16
metric space 16
metric topology 16
modulus 59
mutually disjoint sets 9
MZ-integral vi, 38, 173

N-continuous 179
neighbourhood 13
N-integral 38, 39, 176, 183
N-major (minor) function 175
non-meagre 14
non-overlapping 45
norm 20, 23
normal topology 64
normed (linear) space 20
norm of a division 48
norm topology 20
nowhere dense 13
nowhere dense-in-itself 14
null function 3
N-variation 69, 179, 183
N-variational integral vi, 38, 39, 115, 179, 183
N-variation zero 179

o-convergent 28
o-limit 28
open set 13
operator 8
order Section 0.5
order-Cauchy 30
order-complete 28
ordered pair 8
order-fundamental 29, 30
outer measure 72

pair 8
partial division 41
partial interval 41
partial set 41
partition 1, 41, 42
perfect 14
portion 14
precompact 24
product division system (space) 44, 45, 149

product of sets 9
product topology 19
proper partial set 41
proper subset 8
propositional function 8
pseudometric 15

rare 13
refine, refinement 41
refinement integral Section 1.3, 50
reflexive relation 8
regular measure 102
regular topological space 19
relation 8
relatively order-compact 28
residual 14
restricted division 1, 52
restriction, restriction property 42
Riemann (generalized) v, vi, xi, 6, 36–9, 50, 56, 57, 59, Section 2.3, 83, 87, 90, 94, 95, 116, 118, 166, 171, 178, Section 8.3
Riemann–Stieltjes integral 33, 35, 47, 48
ring 12

scalar 13
second category 14
semigroup 11
set 8
shrinks 41, 42, 59
singleton 8
S-integral 174, 175
sphere 16
stable 43
star-set 43, Section 2.9
step-function 111
strongly compatible 43
strong Perron major(minor) function 168, 169
strong Perron upper(lower) integral 168–70, 172, 173
strong topology 20
strong variational integral 87, 170
subbase 13
subset 8
substitution (integration by) 82, 91, 92
sum of sets 9
symmetric Çesaro–Perron integral 38
symmetric difference of sets 10
symmetric relation 8

tonneau 21
topological group 19, 26
topological space 13, 27
topological vector space 20
topology Section 0.4
transitive relation 8
triangle inequality 15

union of sets 9
universal set 8

variation 69
variational integral 39, Section 2.1, Section 2.3, 87, 171
variation set 69, 76
variation zero 76, 77
vector 13
vector space 12

weakly compatible 43
weakly integrable 74
weighted refinement integral 40, 42